PHIL NAUGHTON

REA's Test Prep Books A

(a sample of the hundreds of letters REA receives each year)

“ Your Fundamentals of Engineering Exam book was the absolute best preparation I could have had for the exam, and it is one of the major reasons I did so well and passed the FE on my first try. ”

Student, Sweetwater, TN

“ My students report your chapters of review as the most valuable single resource they used for review and preparation. ”

Teacher, American Fork, UT

“ Your book was such a better value and was so much more complete than anything your competition has produced (and I have them all!) ”

Teacher, Virginia Beach, VA

“ Compared to the other books that my fellow students had, your book was the most useful in helping me get a great score. ”

Student, North Hollywood, CA

“ Your book was responsible for my success on the exam, which helped me get into the college of my choice... I will look for REA the next time I need help. ”

Student, Chesterfield, MO

“ Just a short note to say thanks for the great support your book gave me in helping me pass the test... I'm on my way to a B.S. degree because of you! ”

Student, Orlando, FL

(more on next page)

(continued from front page)

" I just wanted to thank you for helping me get a great score on the AP U.S. History exam... Thank you for making great test preps! "

Student, Los Angeles, CA

" I did well because of your wonderful prep books... I just wanted to thank you for helping me prepare for these tests. "

Student, San Diego, CA

" I used your book to prepare for the test and found that the advice and the sample tests were highly relevant... Without using any other material, I earned very high scores and will be going to the graduate school of my choice. "

Student, New Orleans, LA

" What I found in your book was a wealth of information sufficient to shore up my basic skills in math and verbal... The section on analytical analysis was excellent. The practice tests were challenging and the answer explanations most helpful. It certainly is the Best Test Prep for the GRE! "

Student, Pullman, WA

" I really appreciate the help from your excellent book. Please keep up the great work. "

Student, Albuquerque, NM

" I am writing to thank you for your test preparation... your book helped me immeasurably and I have nothing but praise for your GRE preparation. "

Student, Benton Harbor, MI

(more on back page)

The Best Test Preparation & Review Course

FE/EIT

Fundamentals of Engineering / Engineer-in-Training

PM Exam in Mechanical Engineering

Richard V. Conte, P.E.
Associate Professor
Manhattan College
Riverdale, NY

Jeff Pieper, Ph.D.
Associate Professor
University of Calgary
Calgary, Canada

Orion P. Keifer, P.E.
Mechanical Engineer
Atlantic Beach, FL

And the Staff of REA
Dr. M. Fogiel, Director

Research & Education Association
61 Ethel Road West
Piscataway, New Jersey 08854

The Best Test Preparation and Review Course for the FE/EIT
(Fundamentals of Engineering/Engineer-in-Training)
PM Exam in Mechanical Engineering

Printed in the United States of America

Library of Congress Catalog Card Number 98-68367

International Standard Book Number 0-87891-262-2

Research & Education Association
61 Ethel Road West
Piscataway, New Jersey 08854

ABOUT RESEARCH & EDUCATION ASSOCIATION

Research & Education Association (REA) is an organization of educators, scientists, and engineers specializing in various academic fields. Founded in 1959 with the purpose of disseminating the most recently developed scientific information to groups in industry, government, high schools, and universities, REA has since become a successful and highly respected publisher of study aids, test preps, handbooks, and reference works.

REA's Test Preparation series includes study guides for all academic levels in almost all disciplines. Research & Education Association publishes test preps for students who have not yet completed high school, as well as high school students preparing to enter college. Students from countries around the world seeking to attend college in the United States will find the assistance they need in REA's publications. For college students seeking advanced degrees, REA publishes test preps for many major graduate school admission examinations in a wide variety of disciplines, including engineering, law, and medicine. Students at every level, in every field, with every ambition can find what they are looking for among REA's publications.

Unlike most test preparation books—which present only a few practice tests that bear little resemblance to the actual exams—REA's series presents tests that accurately depict the official exams in both degree of difficulty and types of questions. REA's practice tests are always based upon the most recently administered exams, and include every type of question that can be expected on the actual exams.

REA's publications and educational materials are highly regarded and continually receive an unprecedented amount of praise from professionals, instructors, librarians, parents, and students. Our authors are as diverse as the fields represented in the books we publish. They are well-known in their respective disciplines and serve on the faculties of prestigious universities throughout the United States.

ACKNOWLEDGMENTS

In addition to our authors, we would like to thank the following:

Dr. Max Fogiel, President, for his overall guidance which has brought this publication to its completion.

Nicole Mimnaugh, New Book Development Manager, for directing the editorial staff throughout each phase of the project.

Kelli A. Wilkins, Assistant Editorial Manager, for coordinating the development of the book.

Martin Perzan for typesetting the book. Gary DaGiau and George Wetzel for their editorial contributions.

CONTENTS

FE/EIT

FE: PM Mechanical Engineering Exam

CHAPTER 1

You Can Succeed on the FE: PM Mechanical Engineering Exam

CHAPTER 1

YOU CAN SUCCEED ON THE FE: PM MECHANICAL ENGINEERING EXAM

By reviewing and studying this book, you can succeed on the Fundamentals of Engineering Examination PM Portion in Mechanical Engineering. The FE is an eight-hour exam designed to test knowledge of a wide variety of engineering disciplines. The FE was formerly known as the EIT (Engineer-in-Training) exam. The FE Exam format and title have now replaced the EIT completely.

The purpose of REA's *Best Test Preparation and Review Course for the FE: PM Mechanical Engineering Exam* is to prepare you sufficiently for the afternoon portion of the Mechanical Engineering FE exam by providing 13 review chapters, including sample problems in each review, and two practice tests. The review chapters and practice tests reflect the scope and difficulty level of the actual FE: PM Exam. The reviews provide examples with thorough solutions throughout the text. The practice tests provide simulated FE exams with detailed explanations of answers. While using just the reviews or the practice tests is helpful, an effective study plan should incorporate both a review of concepts and repeated practice with simulated tests under exam conditions.

ABOUT THE TEST

The Fundamentals of Engineering Exam (FE) is one part in the four-step process toward becoming a professional engineer (PE). Graduating from an approved four-year engineering program and passing the FE qualifies you for your certification as an "Engineer-in-Training" or an "Engineer Intern." The final two steps towards licensing as a PE involve completion of four years of additional engineering experience and passing the Principles and Practices of Engineering Examination administered by the National Council of Examiners for Engineering and Surveying (NCEES). Registration as a professional engineer is deemed both highly rewarding and beneficial in the engineering community.

In order to register for the FE, contact your state's Board of Examiners for Professional Engineers and Land Surveyors. To determine the location for the Board in your state, contact the main NCEES office at the following address:

National Council of Examiners for Engineering and Surveying
PO Box 1686
Clemson, SC 29633-1686
(864) 654-6824
Website: http://www.ncees.org

TEST FORMAT

The FE consists of two distinct sections. One section is given in the morning (FE: AM) while the other is administered in the afternoon (FE: PM). This book will prepare you for the FE: PM exam in Mechanical Engineering.

The FE: PM is a *supplied reference exam,* and students are not permitted to bring reference material into the test center. Instead, you will be mailed a reference guide when you register for the exam. This guide will provide all the charts, graphs, tables, and formulae you will need. The same book will be given to you in the test center during the test administration.

You will have four hours to complete the exam. The FE: PM consists of 60 questions covering 13 different engineering subjects. The subjects and their corresponding percentages of questions are shown on the next page.

FE: PM MECHANICAL ENGINEERING SUBJECT DISTRIBUTION

Subject	Percentage of Problems
Automatic Controls	5
Computers	5
Dynamic Systems	10
Energy Conversion & Power Plants	5
Fans, Pumps, & Compressors	5
Fluid Mechanics	5
Heat Transfer	10
Material Behavior/Processing	5
Measurement & Instrumentation	10
Mechanical Design	10
Refrigeration & HVAC	5
Stress Analysis	10
Thermodynamics	10

Our review book covers all of these topics thoroughly. Each topic is explained in detail, with example problems, diagrams, charts, and formulae.

You may want to take a practice exam at various studying stages to measure your strengths and weaknesses. This will help you to determine which topics need more study. Take one test when you finish studying so that you may see how much you have improved. For studying suggestions that will help you to make the best use of your time, see the "Study Schedule" presented after this chapter.

SCORING THE EXAM

Your FE: PM score is based upon the number of correct answers you choose. No points are taken off for incorrect answers. A single score of 0 to 100 is given for the entire (both AM and PM sections) test. Both the AM and PM sections have an equal weight. The grade given is on a pass/

fail basis. The point between passing and failing varies from state to state, although 70 is a general reference point for passing. Thus this general reference point for the FE: PM section alone would be 35.

The pass/fail margin is not a percentage of correct answers, nor a percentage of students who scored lower than you. This number fluctuates from year to year and is reestablished with every test administration. It is based on previous exam administrations and relates your score to those of previous FE examinees.

Because this grading system is so variable, there is no real way for you to know exactly what you got on the test. For the purpose of grading the practice tests in this book, however, REA has provided the following formula to calculate your score on the FE: PM practice tests:

$$\left[\frac{\text{No. of questions answered correctly on the FE: PM}}{240}\right] \times 100 = \text{your score}$$

Remember that this formula is meant for the computation of your grade for the practice tests in this book. It does not compute your grade for the actual FE examination.

TEST-TAKING STRATEGIES

How to Beat the Clock

Every second counts, and you will want to use the available test time for each section in the most efficient manner. Here's how:

1. Bring a watch! This will allow you to monitor your time.
2. Become familiar with the test directions. You will save valuable time if you already understand the directions on the day of the test.
3. Pace yourself. Work steadily and quickly. Do not spend too much time on any one question. Remember, you can always return to the problems that gave you the most difficulty. Try to answer the easiest questions first, then return to the ones you missed.

Guessing Strategy

1. When all else fails, guess! The score you achieve depends on the number of correct answers. There is no penalty for wrong answers, so it is a good idea to choose an answer for all of the questions.

2. If you guess, try to eliminate choices you know to be wrong. This will allow you to make an educated guess. Here are some examples of what to look for when eliminating answer choices:

   Thermodynamics—check for signs of heat transfer and work

   Fluid Mechanics—check for signs of pressure reading

3. Begin with the subject areas you know best. This will give you more time and will also build your confidence. If you use this strategy, pay careful attention to your answer sheet; you do not want to mismatch the ovals and answers. It may be a good idea to check the problem number and oval number *each time* you mark down an answer.

4. Break each problem down into its simplest components. Approach each part one step at a time. Use diagrams and drawings whenever possible, and do not wait until you get a final answer to assign units. If you decide to move onto another problem, this method will allow you to resume your work without too much difficulty.

HOW TO STUDY FOR THE FE: PM EXAM

Two groups of people take the FE examination: college seniors in undergraduate programs and graduate engineers who decide that professional registration is necessary for future growth. Both groups begin their Professional Engineer career with a comprehensive exam covering the entirety of their engineering curriculum. How does one prepare for an exam of such magnitude and importance?

Time is the most important factor when preparing for the FE: PM. Time management is necessary to ensure that each section is reviewed prior to the exam. Once the decision to test has been made, determine how much time you have to study. Divide this time amongst your topics, and make up a schedule which outlines the beginning and ending dates for study of each exam topic and include time for a final practice test followed by a brief review. Set aside extra time for the more difficult subjects, and include a buffer for unexpected events such as college exams or business trips. There is never enough time to prepare, so make the most of the time that you have.

You can determine which subject areas require the most time in several ways. Look at your college grades: those courses with the lowest grades probably need the most study. Those subjects outside your major are generally the least used and most easily forgotten. These will require a good deal of review to bring you up to speed. Some of the subjects may not be familiar at all because you were not required to study them in college. These subjects may be impossible to learn before the examination, although some can be self-taught. One such subject is engineering economics; the mathematics may be not exceptionally difficult to you and most of the concepts are common sense.

Another way to determine your weakest areas is to take one of the practice tests provided in this book. The included simulated exams will help you assess your strengths and weaknesses. By determining which type of questions you answered incorrectly on the practice tests, you find the areas that need the most work. Be careful not to neglect the other subjects in your review; do not rule out any subject area until you have reviewed it to some degree.

You may also find that a negative attitude is your biggest stumbling block. Many students do not realize the volume of material they have covered in four years of college. Some begin to study and are immediately overwhelmed because they do not have a plan. It is important that you get a good start and that you are positive as you review and study the material.

You will need some way to measure your preparedness, either with problems from books or with a review book that has sample test questions similar to the ones on the FE: PM Mechanical Engineering Exam. This book contains sample problems in each section which can be used before, during, or after you review the material to measure your understanding of the subject matter. If you are a wizard in thermodynamics, for example, and are confident in your ability to solve problems, select a few and see what happens. You may want to perform at least a cursory review of the material before jumping into problem solving, since there is always something to learn. If you do well on these initial problems, then momentum has been established. If you do poorly, you might develop a negative attitude. Being positive is essential as you move through the subject areas.

The question that comes to mind at this point is: "How do I review the material?" Before we get into the material itself, let us establish rules which lead to **good study habits**. Time was previously mentioned as the most important issue. When you decide to study you will need blocks of uninterrupted time so that you can get something accomplished. Two hours should be the minimum time block allotted, while four hours should be the maximum. Schedule five-minute breaks into your study period and stay with your schedule. Cramming for the FE can give you poor results, including short-term memory and confusion when synthesis is required.

Next, you need to work in a quiet place, on a flat surface that is not cluttered with other papers or work that needs to be completed before the next day. **Eliminate distractions**—they will rob you of time while you pay attention to them. **Do not eat while you study**; few of us can do two things at once and do them both well. Eating does require a lot of attention and disrupts study. Eating a sensible meal before you study resolves the "eating while you work" problem. We encourage you to have a large glass of water available since water quenches your thirst and fills the void which makes you want to get up and find something to eat. In addition, **you should be well rested when you study**. Late nights and early mornings are good for some, especially if you have a family, but the best results are associated with adequate rest.

Lastly, **study on weekend mornings while most people are still asleep**. This allows for a quiet environment and gives you the remainder of the day to do other things. If you must study at night, we suggest two-hour blocks ending before 11 p.m.

Do not spend time memorizing charts, graphs, and formulae; the FE is a supplied reference exam, and you will be provided a booklet of equations and other essential information during the test. This reference material will be sent to you prior to your examination date. You can use the supplied reference book as a guide while studying, since it will give you an indication of the depth of study you will need to pursue. Furthermore, familiarity with the book will alleviate some test anxiety since you will be given the same book to use during the actual exam.

While you review for the test, use the review book supplied by the NCEES, paper, pencil, and a calculator. Texts can be used, but reliance on them should be avoided. The object of the review is to identify what you know, the positive, and that which requires work, the negative. As you review, move past those equations and concepts that you understand and annotate on the paper those concepts that require more work. Using this method you can review a large quantity of material in a short time and reduce the apparent workload to a manageable amount. Now go back to your time schedule and allocate the remaining time according to the needs of the subject under consideration. Return to the material that requires work and review it or study it until your are satisfied that you can solve problems covering this material. When you have finished the review, you are ready to solve problems.

Solving problems requires practice. To use the problems in this book effectively, you should cover the solution and try to solve the problem on your own. If this is not possible, map out a strategy to answer the problem and then check to see if you have the correct procedure. Remember, that most problems that are not solved correctly were never started correctly. Merely reviewing the solution will not help you to start the problem when you see it again at a later date. Read the problems carefully and in parts. Many people teach that reading the whole problem gives the best overview of what is to come; however, solutions are developed from small clues that are in parts of a sentence. **Read the problem and break it into manageable parts**. Next, **try to avoid numbers until the problem is well formulated.** Too often, numbers are substituted into equations early and become show stoppers. You will need numbers, just use them after the algebra has been completed. **Be mechanical**, list the knowns, the requirements of the problem, and check off those bits of knowledge you have as they appear. Checking off the intermediate answers and information you know is a positive attitude builder. Continue to solve problems until you are confident or you exceed the time allowed in you schedule for that subject area.

As soon as you complete one subject, move to the next. Retain all of you notes as you complete each section. You will need these for your final overall review right before the exam. After you have completed the entire review, you may want to take a practice test. Taking practice exams will test your understanding of all the engineering subject areas and will help you identify sections that need additional study. With the test and the notes that you retained from the subject reviews, you can determine weak areas requiring some additional work.

You should be ready for the exam if you follow these guidelines:

- Program your time wisely.
- Maintain a positive attitude.
- Develop good study habits.
- Review the material smartly and maximize the learning process.
- Do practice problems and practice tests.
- Review again to finalize your preparation.

GOOD LUCK!

STUDY SCHEDULE

The following is a suggested eight-week study schedule for the Fundamentals of Engineering: PM Exam in Mechanical Engineering. You may want to condense or expand the schedule depending on the amount of time remaining before the test. Set aside some time each week, and work straight through the activity without rushing. By following a structured schedule, you will be able to complete an adequate amount of studying, and be more confident and prepared on the day of the exam.

Week 1	Acquaint yourself with this FE: PM Mechanical Engineering Test Preparation Book by reading the first chapter: "You Can Succeed on the FE: PM Mechanical Engineering Exam." Take Practice Test 1. When you score the test, be sure to look for areas where you missed many questions. Pay special attention to these areas when you read the review chapters.
Week 2	Begin reviewing Chapters 2 and 3. As you read the chapters, try to solve the examples without aid of the solutions. Use the solutions to guide you through any questions you missed.
Week 3	Study and review Chapters 4 and 5. Take notes as you read the chapters; you may even want to write concepts on index cards and thumb through them during the day. As you read the chapters, try to solve the examples without the aid of the solutions.
Week 4	Review any notes you have taken over the last few weeks. Study Chapters 6 and 7. As you read the chapters, try to solve the examples without the aid of the solutions.
Week 5	Study Chapters 8 and 9 while continuing to review your notes. As you read the chapters, try to solve the examples without the aid of the solutions.
Week 6	Study Chapters 10 and 11. As you read the chapters, try to solve the examples without the aid of the solutions to guide you through any questions you missed.

Week 7	Study Chapters 12, 13, and 14. As you read, try to solve the examples without the aid of the solutions. Use the solutions to guide you through any questions you missed.
Week 8	Take Practice Test 2. When you score the text, be sure to look for any improvement in the areas that you missed in Practice Test 1. If you missed any questions in any particular area, go back and review those areas. Be patient and deliberate as you review; with careful study, you can only improve.

FE/EIT

FE: PM Mechanical Engineering Exam

CHAPTER 2

Automatic Controls

CHAPTER 2

AUTOMATIC CONTROLS

AUTOMATIC CONTROL SYSTEMS

One definition of a system is a group of different but interacting elements that are integrated together to accomplish a particular objective. The objective can be as different as fabricating a car, generating and distributing electricity, controlling the space shuttle, or running a manufacturing plant. In order to succeed, there are no restraints on the interacting elements. The elements can be electrical, mechanical, hydraulic, pneumatic, thermal, or human.

System engineers concern themselves with the system's performance, stability, efficiency, reliability, and economy. Since the system is composed of devices and subsystems which can be represented by a block marked with the function performed, the system engineer doesn't focus on the component design. The complete system can be represented by connecting and combining many such blocks. The blocks simplify the system by including only the functional operations of the system without too much detail.

There are two types of automatic control systems, open and closed loops. *Open loop systems* have output quantities that do not effect the input action performed by the system—an input signal or command is applied, possibly amplified, and a power output results.

An automatic control on a system is used to produce a desired output when the system inputs are changed. Commands are provided as inputs which the output is to follow and the automatic control is supposed to minimize the disturbances. An open loop system is shown in Figure 1.

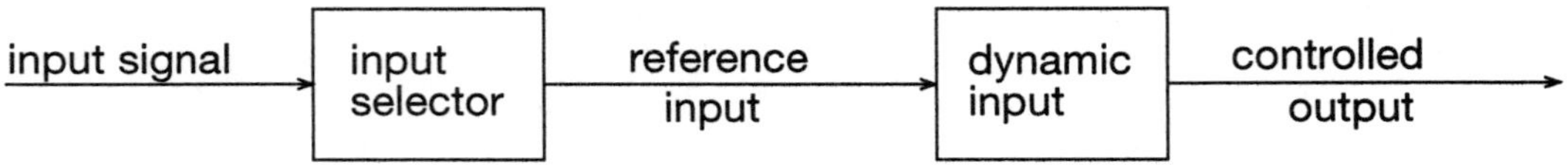

Figure 1. Open Loop Feedback System

The input signal actuates this system and specifies the desired output. The input signal has to be converted into a signal that is acceptable to the system, the reference input. The input converter converts the input signal into the reference input, which causes the dynamic unit to be activated and causes the controlled output.

PROBLEM 1:

All houses come with central heat. Your heating system comes with a control system to regulate the temperature. Draw a functional block diagram of this system. Indicate inputs, and outputs, where the desired temperature and the actual temperature are compared.

SOLUTION:

A possible answer is shown below.

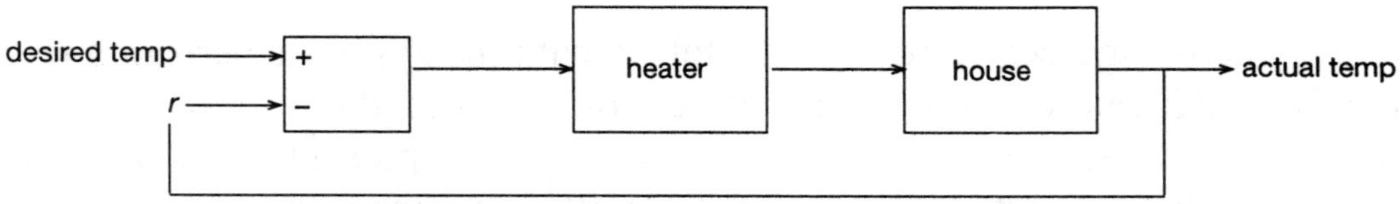

A meat broiler is an example of an open loop system. The setting for the broiler is the cooked level of the meat. A selector allows you to select how well you want to have the meat cooked, rare, medium, and well done. This sets the input selector. This selects the temperature and time the heating coils (dynamic input) will be on. The level of cooking of the meat is the controlled output. A broiler cannot set itself. A person must decide on the thickness and color for the meat by setting the input signal.

A *closed-loop control system,* shown in Figure 2, is composed of a process, a measurement of the controlled variable, and a controller which compares the actual measurement with the desired value and uses the difference between them (feedback) to automatically adjust one of the inputs to the process.

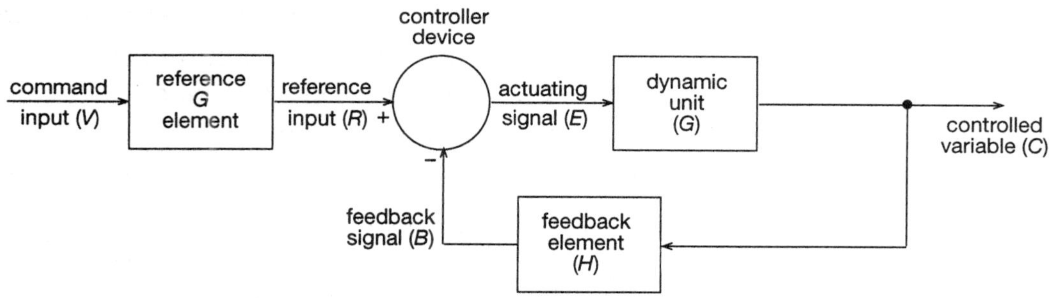

Figure 2. Closed–Loop Control System

FEEDBACK CONTROL SYSTEMS

The following symbols have been established by AIEE as the standard nomenclature for feedback control systems:

v = command and input determined externally; it is independent of the feedback control system

r = reference input derived from the command and actual system signal input

c = controlled variable, quantity directly measured and controlled, and system output

b = primary feedback signal, a function of the controlled variable, and compared with the reference input, resulting in the actuating signal (E)

E = actuating signal, obtained from comparison measuring device equal to the reference input minus the feedback

G_v = reference input element that generates a signal proportional to the command

G = dynamic unit comprised of control elements and systems

H = proportionality device comprised of feedback elements which produce feedback from, and can modify the characteristics of the controlled variable

A feedback control system uses a feedback signal, a function of the output signal, to compare with the reference input signal. The difference produces an actuating signal, an error signal. This signal is input to the dynamic device and produces the controlled output, a function of the reference and control signals.

It is common for feedback control systems to be classified according to the method used to control the power output from the feedback signal: on-off and step controllers and servomechanisms. On-off controllers activate or deactivate the power when the feedback signal reaches a pre-set value. Temperature control using thermostats or selectors, used in air conditioners and freezers, are examples of these types of systems. A step control system applies power in steps at pre-set times based on a feedback signal input, similar to furnace control using pulses to prevent overshooting the set point.

The *physical system* to be controlled can be electrical, thermal, hydraulic, pneumatic, gaseous, mechanical, or described by any other physical or chemical process. Generally, the control system will be designed to meet one of two objectives. A *servomechanism* is designed to follow changes in set point as closely as possible. Many electrical and mechanical control systems are servomechanisms. A *regulator* is designed to keep output constant despite changes in load or disturbances. Regulatory controls are widely used on chemical processes.

The *control components* can be actuated pneumatically, hydraulically, electronically, or digitally. Only in very few applications does actuation affect the ability to control. *Actuation* is chosen on the basis of economics.

The *purpose* of the control system must be defined. A large capacity or inertia will make the system sluggish for servo operation but will help to minimize the error for regulator operation. A *process* can be part of a control system either as a load on the servo or the process to be controlled. Thus, the process must be designed as part of the system. The process is *modeled* in terms of its static and dynamic gains in order that it be incorporated into the system diagram. Modeling uses Ohm's and Kirchoff's laws for electrical systems, Newton's laws for mechanical systems, mass balances for fluid flow systems, and energy balances for thermal systems.

Feedback control systems can be analyzed in two ways: first, solving for the roots of the differential equation representing the motion of the system and preparing and inspecting the curves representing the performance of the system; and second, evaluating the transfer functions and differential equations, representing the system.

The mathematical equations representing feedback system characteristics are similar in form, whether the components are mechanical, thermal, electrical, etc. For the simplest type of feedback control systems, consisting of inertia, friction, etc., the differential equations reduce to a second order equation and its roots are solvable.

$$A_n \frac{d^n \Delta_0}{dt^n} + ... + A_2 \frac{d^2 \Delta_0}{dt^2} + A_1 \frac{d\Delta_0}{dt} + A_0 \Delta_1 + A + \int \Delta_0 dt = F(e)$$

where:

Δ_0 = output quantity

Δ_1 = input quantity

e = error $(\Delta_1 - \Delta_0)$

A = functions of system parameters

The equations describing the closed loop feedback control system are as follows:

$$C(s) = G(s)E(s)$$

$$B(s) = H(s)C(s)$$

$$E(s) = R(s) - B(s)$$

The overall system transfer functions are:

$$\frac{\text{Output}}{\text{Input}} = \frac{C(s)}{R(s)} = \frac{G(s)}{1 + G(s)H(s)}$$

The characteristic equation of the system is equal to the denominator when set to zero and is used to calculate the stability and response of the closed loop system.

$$1 + G(s)H(s) = 0$$

The open loop transfer function is:

$$\frac{B(s)}{E(s)} = G(s)H(s)$$

The forward transfer function is:

$$\frac{C(s)}{E(s)} = G(s)$$

In order to solve for the system's transfer function, algebraic manipulations are performed at the different points of the block diagram (see Figure 3) components as follows:

Transfer function: $C = GE$

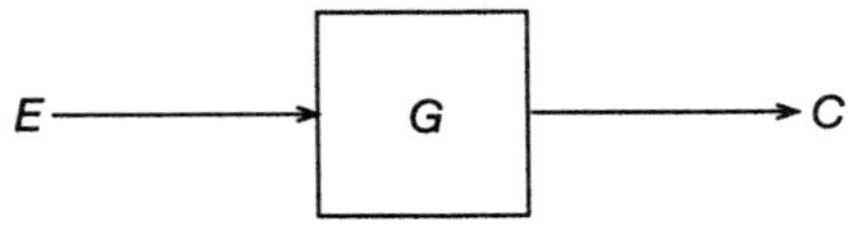

Summit point: $E = R - B$

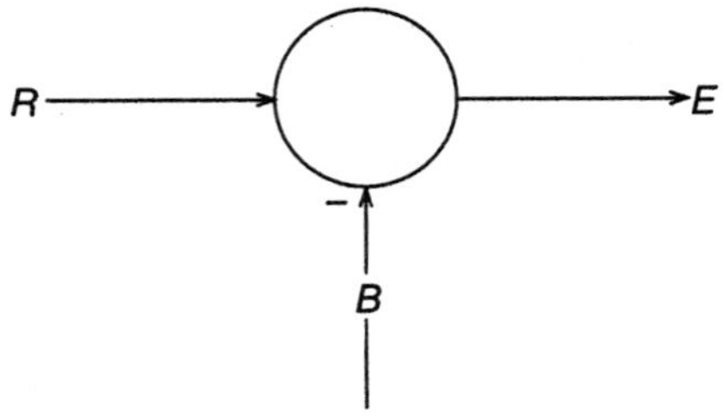

Pick-off point: $C = C = C$

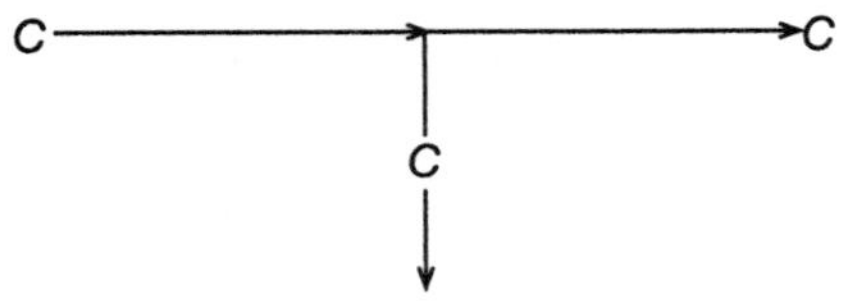

Figure 3. Block Diagrams

Applicable Rules

1. A closed–loop system can be substituted by an equivalent open-loop system.

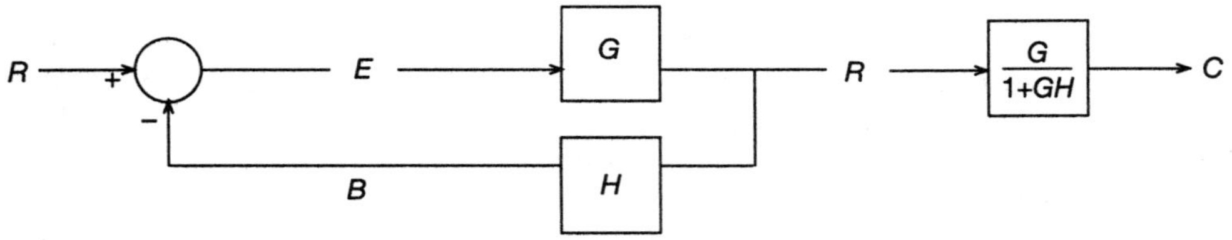

2. Gain of series blocks is the product of each gain.

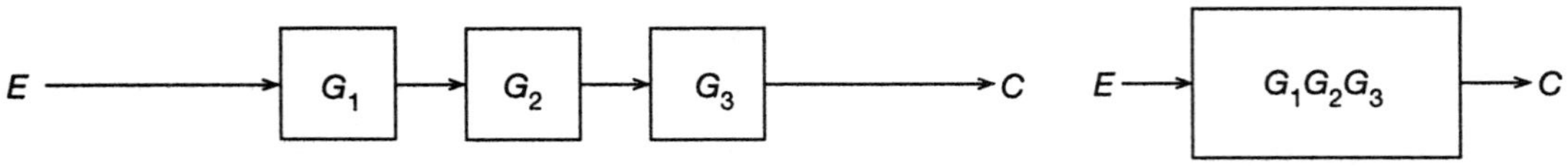

3. Order of addition does not affect results.

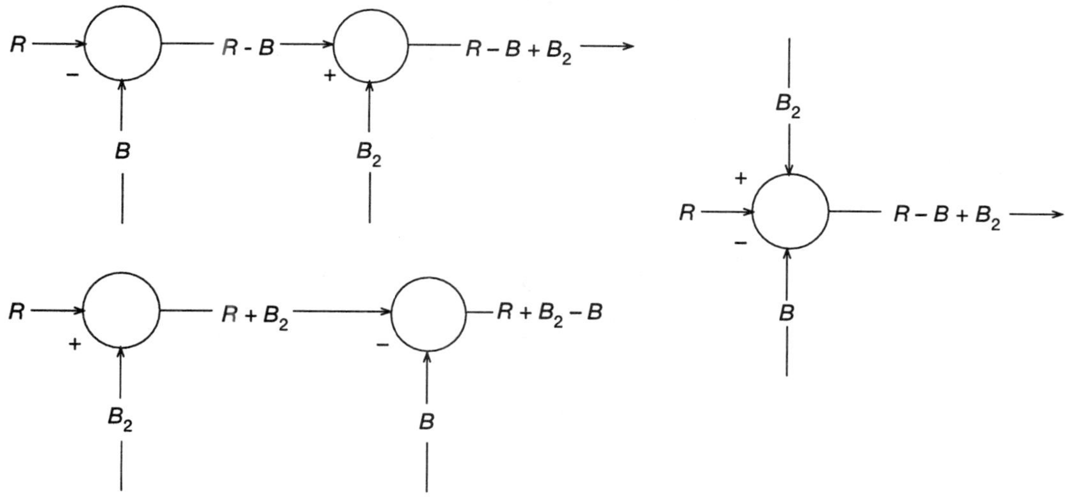

4. To shift a summing point beyond a gain block requires the insertion of gain in the variable added.

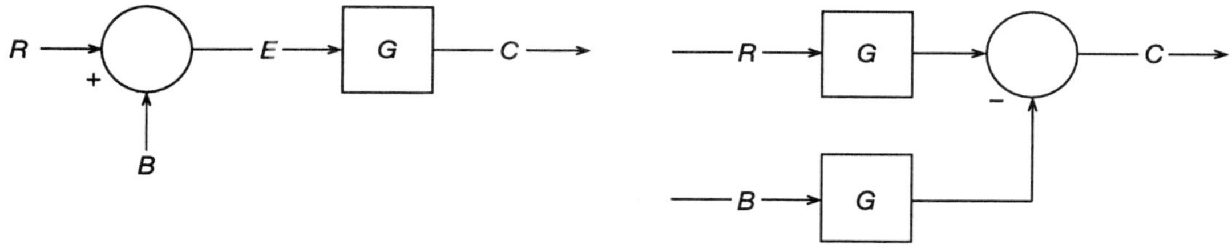

$C = GE = G\,(R - B)$

$C = RG - BG$

$C = G\,(R - B)$

5. To shift a pick-off beyond a gain block requires the insertion of $1/G$ in the variable picked off.

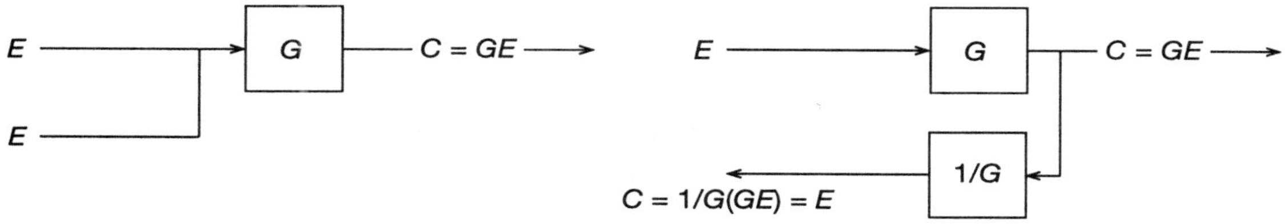

Figure 4. Equivalent Block Diagrams

PROBLEM 2:

Find the transfer function for each of the following block diagrams.

(a)

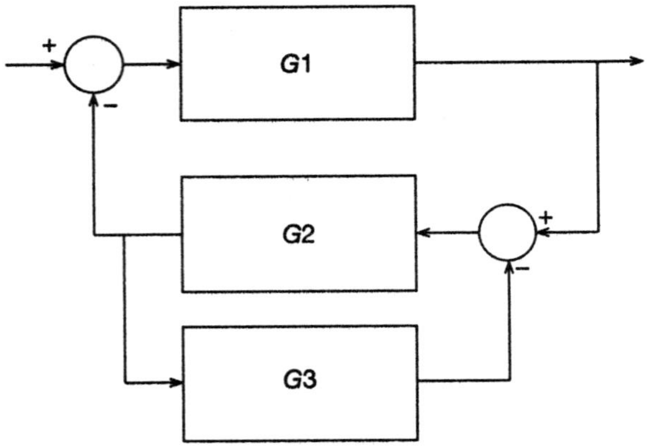

(b)

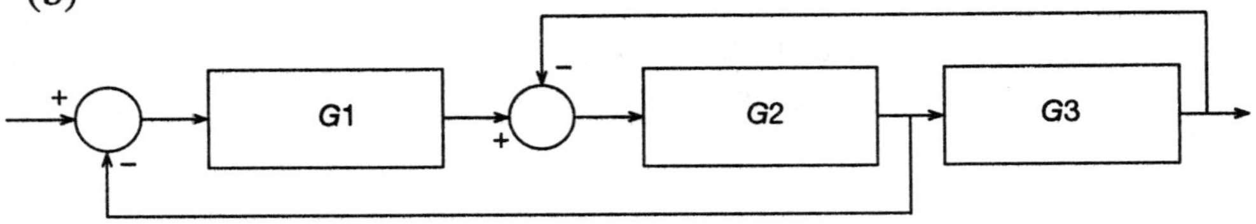

SOLUTION:

(a) Values are placed at representative positions. Equations are written for each summation point and block.

$$C(s) = G_1E(s) \tag{1}$$

$$I(s) = G_2F(s) \tag{2}$$

$$D(s) = G_3I(s) \tag{3}$$

$$F(s) = C(s) - D(s) \tag{4}$$

$$E(s) = R(s) - I(s) \tag{5}$$

The solution is found by solving the equations simultaneously for:

$$\frac{C(s)}{R(s)}$$

Solving Equations (2) and (3):

$$D(s) = G_2G_3F(s)$$

Solving Equations (1) and (5):

$$\frac{C(s)}{G_1} = R(s) - G_2F(s) \tag{6}$$

Equation 4 can be written as follows:

$$F(s) = C(s) - D(s) = C(s) - G_2G_3F(s)$$

from which:

$$F(s) = \frac{C(s)}{[1 + G_2G_3]} \tag{7}$$

Substituting Equation (7) into Equation (6):

$$\frac{C(s)}{G_1} = R(s) - G_2\left[\frac{C(s)}{1 + G_2G_3}\right]$$

$$C(s) = G_1R(s) - G_1G_2\left[\frac{C(s)}{1 + G_2G_3}\right]$$

$$C(s) + \frac{G_1G_2C(s)}{1 + G_2G_3} = G_1R(s)$$

$$C(s)\left[1 + \frac{G_1G_2}{1 + G_2G_3}\right] = G_1R(s)$$

$$\frac{C(s)}{R(s)} = \frac{G_1}{\left[1 + \dfrac{G_1G_2}{1 + G_2G_3}\right]} = \frac{G_1}{\dfrac{1 + G_2G_3 + G_1G_2}{1 + G_2G_3}} = \frac{G_1(1 + G_2G_3)}{1 + G_2G_3 + G_1G_2}$$

(b) The five simultaneous equations to be solved are as follows:

$$E(s) = R(s) - D(s) \tag{1}$$

$$I(s) = F(s) - C(s) \tag{2}$$

$$F(s) = G_1E(s) \tag{3}$$

$$D(s) = G_2I(s) \tag{4}$$

$$C(s) = G_3D(s) \tag{5}$$

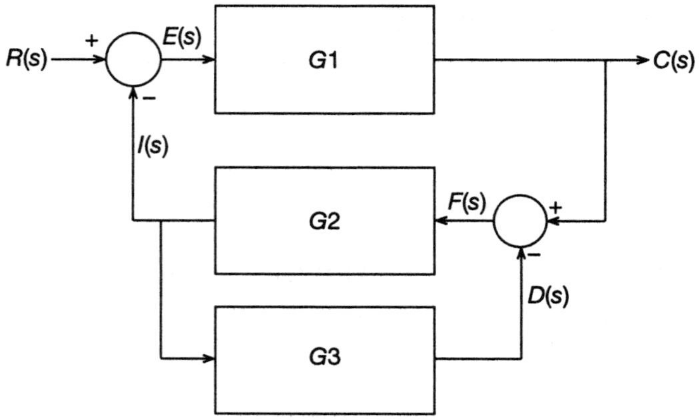

Solving these equations simultaneously yields:

$$\frac{C(s)}{R(s)} = \frac{G_1 \times G_2 \times G_3}{1 + G_1 \times G_2 + G_2 \times G_3}$$

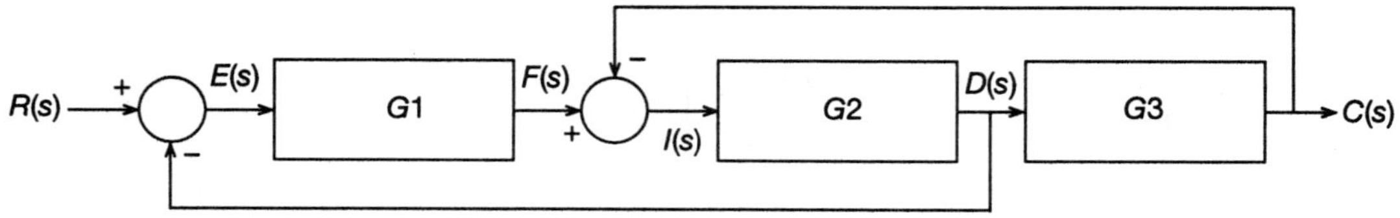

LAPLACE TRANSFORMS

For more complicated systems, the use of the method of transfer function analysis is preferred. If the system is initially at rest, the Laplace Transform of the differential equation can be used to obtain the transfer function and solve for the roots. The Laplace Transform is an algebraic operation used to convert functions of time into functions of s, the complex variable, of the form $\rho + j\omega$, and considers the initial (boundary) conditions and transient and steady-state components. Time is saved by using established tables of transforms.

The Laplace Transform is:

$$F(s) \equiv \ell[f(t)] = \int_0^\infty f(t) e^{-st} dt \text{ and } \ell \equiv \int_0^\infty e^{-st} dt$$

When solved through calculus, it results in:

$$\ell\left\{\frac{d}{dt}[f(t)]\right\} = SF(s) - f(0^+)$$

where $f(0^+)$ is the initial value of $f(t)$ as t goes to 0 from positive values.

After the transform of the differential is calculated, the response function is found using algebra. The complete solution is found solving for the inverse form of the Laplace Transforms, which are listed in tables.

$$f(t) = \ell^{-1}[F(S)]$$

If a transform of response is not listed in a table, it can be found by using partial fractions with constant coefficients. The inverse transform is the total of the inverse transforms of each fraction.

PROBLEM 3:

Solve the following differential equation using Laplace Transforms. Assume:

$$f = f(t) \quad \text{and} \quad F = F(s)$$

$$\frac{d^3 f}{dt^3} - 3\frac{d^2 f}{dt^2} + 3\frac{df}{dt} - f = t^2 e^t$$

$$f(0) = 1; \quad f'(0) = 0; \quad f''(0) = -2$$

SOLUTION:

Take the Laplace Transform of both sides:

$$\ell\left\{\frac{d^3 f}{dt^3}\right\} - 3\ell\left\{\frac{d^2 f}{dt^2}\right\} + 3\ell\left\{\frac{df}{dt}\right\} - \ell(f) = \ell\{t^2 e^t\}$$

Using the Laplace Transform pairs from Table 1:

Using item 6,

$$\ell\left\{\frac{d^3 f}{dt^2}\right\} = s^3F - s^2 f(0) - sf'(0) - f''(0) = s^3F - s^2 + 2$$

Using item 5,

$$\ell\left\{\frac{d^2 f}{dt^2}\right\} = s^2F - sf(0) - f'(0) = s^2F$$

Using item 4,

$$\ell\left\{\frac{df}{dt}\right\} = sF - f(0) = sF - 1$$

$$\ell\ \{f(t)\} = F(s)$$

Using item 13,

$$\ell\ \{t^2e^t\} = \frac{2!}{(s-1)^3} = \frac{2}{(s-1)^3}$$

$$s^3F - s^2 + 2 - 3(s^2F) + 3(sF - 1) - F = \frac{2}{(s-1)^3}$$

$$(s^3 - 3s^2 + 3s - 1)F - s^2 + 3s - 1 = \frac{2}{(s-1)^3}$$

$$(s^3 - 3s^2 + 3s - 1)F = \frac{2}{(s-1)^3} + s^2 - 3s + 1$$

$$(s - 1)^3\mathrm{F} = \frac{2}{(s-1)^3} + \mathrm{s}^2 - 3\mathrm{s} + 1$$

$$F = \frac{2}{(s-1)^6} + \frac{s^2}{(s-1)^3} - \frac{3s}{(s-1)^3} + \frac{1}{(s-1)^3}$$

$$= \frac{2}{(s-1)^6} + \frac{s^2 - 3s + 1}{(s-1)^3}$$

$$= \frac{2}{(s-1)^6} + \frac{(s-1)^2 - (s-1) - 1}{(s-1)^3}$$

$$F = \frac{1}{s-1} - \frac{1}{(s-1)^2} - \frac{1}{(s-1)^3} + \frac{2}{(s-1)^6}$$

Using Table 1 and applying the inverse transform:

$$f(t) = e^t - te^t - \frac{t^2e^t}{2} + \frac{t^5e^t}{60}$$

Table 1: Laplace Transform Pairs

	Functions of Time: $f(t)$ for $0 \le t$	Laplace Transforms: $\ell\ |f(t)|$
1.	$f(t)$	$\int_0^x f(t)e^{-st}dt = F(s)$
2.	$x(t) + y(t)$	$X(s) + Y(s)$
3.	$Kf(t)$	$KF(s)$
4.	$\frac{df(t)}{dt}$	$sF(s) - f(0)$
5.	$\frac{d^2f(t)}{dt^2}$	$s^2F(s) - sf(0) - \frac{df(0)}{dt}$
6.	$\frac{d^nf(t)}{dt^n}$	$s^nF(s) - \sum_{i=1}^{n} s^{n-i}\frac{d^{i-1}f(0)}{dt^{i-1}}$
7.	$\int_0^t f(t)dt$	$\frac{1}{s}F(s)$
8.	1 or $u(t)$	$\frac{1}{s}$
9.	t	$\frac{1}{s^2}$
10.	t^n for $n > -1$	$\frac{n!}{s^{n+1}}$
11.	e^{-at}	$\frac{1}{s+a}$
12.	te^{-at}	$\frac{1}{(s+a)^2}$
13.	t^ne^{-at}	$\frac{n!}{(s+a)^{n+1}}$
14.	$1 - e^{-at}$	$\frac{a}{s(s+a)}$
15.	$e^{-at} - e^{-bt}$	$\frac{b-a}{(s+a)(s+b)}$

	Functions of Time: $f(t)$ for $0 \le t$	Laplace Transforms: $\ell\ \lvert f(t) \rvert$
16.	$ae^{-at} - be^{-bt}$	$\dfrac{(a-b)s}{(s+a)(s+b)}$
17.	$\sin \alpha t$	$\dfrac{\alpha}{s^2 + \alpha^2}$
18.	$\cos \alpha t$	$\dfrac{s}{s^2 + \alpha^2}$
19.	$t \sin \alpha t$	$\dfrac{2\alpha s}{\left(s^2 + \alpha^2\right)^2}$
20.	$t \cos \alpha t$	$\dfrac{s^2 - \alpha^2}{\left(s^2 + \alpha^2\right)^2}$
21.	$e^{-at} \sin \alpha t$	$\dfrac{\alpha}{(s+a)^2 + \alpha^2}$
22.	$e^{-at} \cos \alpha t$	$\dfrac{s+a}{(s+a)^2 + \alpha^2}$
23.	$e^{-\zeta\omega_n t} \sin \omega_n \left(1-\zeta^2\right)^{1/2} t$ for $\zeta < 1$	$\dfrac{\omega_n \left(1-\zeta^2\right)^{1/2}}{s^2 + 2\zeta\omega_n s + \omega_n^2}$
24.	$e^{-\zeta\omega_n t} \sinh \omega_n \left(\zeta^2 - 1\right)^{1/2} t$ for $\zeta > 1$	$\dfrac{\omega_n \left(\zeta^2 - 1\right)^{1/2}}{s^2 + 2\zeta\omega_n s + \omega_n^2}$
25.	$1 - \dfrac{e^{-\zeta w_n t}}{\sqrt{1-\zeta^2}} \sin\left(\omega_n \sqrt{1-\zeta^2}\, t + \phi\right)$ $\phi = \tan^{-1} \dfrac{\sqrt{1-\zeta^2}}{\zeta}$ for $\zeta < 1$	$\dfrac{{\omega_n}^2}{s\left(s^2 + 2\zeta\omega_n s + {\omega_n}^2\right)}$
26.	$\begin{Bmatrix} f(t-a) & \text{where } t > a \\ 0 & \text{where } t < a \end{Bmatrix}$	$e^{-as} F(s)$

FE/EIT

FE: PM Mechanical Engineering Exam

CHAPTER 3

Computers

CHAPTER 3

COMPUTERS

While the mainframe was the primary computing environment in the past, today the personal, or desktop computer is becoming the dominant computing tool. Regardless of the rapid advancements and changes in technology, the operations of the computer remain basically unchanged.

ARCHITECTURE

The basic computer architecture is composed of input/output (I/O), memory storage, and data processing functions. The basic computer system architecture is represented as follows:

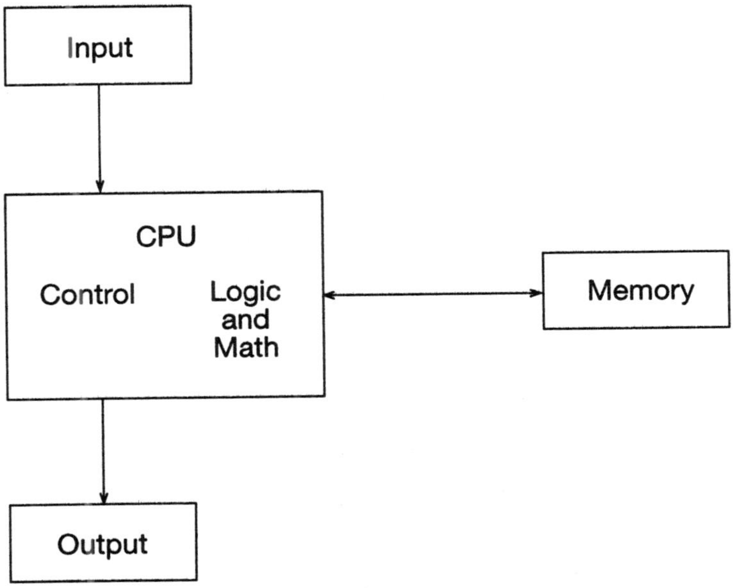

Figure 1. Basic Computer Architecture

The heart of the computer is the *central processing unit (CPU)*, which receives *input* material provided mechanically, electronically, and orally. Material is input by keyboards, tape/disk drives, CD ROMs, telephones, and facsimile machines. The CPU provides electronic, visual, audio and written *output* to memory and/or through line printers, plotters, paper, disks and magnetic tapes, hard and floppy drives, facsimile machines, electronic telexes, and speakers. Today, the media receiving computerized information and commands are endless.

MEMORY

Memory ranges from internal hard drives to tape backups, magnetic core, or drums. In most cases, it is recommended that more than one type of memory device is used. At present, the recordable CD ROMs are about to hit the market and be made available to personal computer users at an affordable cost. In order to take advantage of this great breakthrough, PC owners will have to have CD ROMs that not only read data, but will also have the ability to write data to disk. This change will more than likely end the use of floppy disks which at present are used for data storage and programming.

The hard core memory has the largest capacity for storage and operates and offers the fastest access to data and programs. All calculations and operations use the data and programs stored in the core memory. Temporary work and data storage is done on the disk memory which can be used later in the program. Some data stored on the disk memory can be reference files and standard programs which are available for use by other programs—I/O routines and math subroutines. Interface between input and output devices is normally done in drum memory. The drum memory serves as a buffer for input programs and data and output data.

The hard core memory units are nondestructible devices and store the data even when power is lost. Registers, temporary memory units (Random Access Memory, or RAM), lose the data stored when power is turned off. These units are used to control the computer and perform the logical and math functions.

The following are registers named after their functions: an accumulator accumulates results; a multiplier/quotient stores math results; storage holds information sent to or from storage; address holds locations of data or device; and instruction holds operational data being done.

The heart of the computer is the *central processing unit* (*CPU*) which manages the computer system and performs all calculations and makes the logical decisions on the data it processes. The control sector directs and implements all operations required by instructions, including control of input and output devices, entry or removal of information from memory, and directing information between storage and the mathematic and logic sector.

The logic and math sector performs all mathematical and logical operations. It calculates, shifts numbers, selects the correct mathematical sign of the answers, rounds off, and compares results. The logic part makes all of the decisions that change the order of instruction operation.

BINARY NUMBERING

Computers primarily manipulate numbers. Inside the computer there are many two-state electrical devices, each one representing one binary number, either a zero or a one. These binary numbers are called *bits.* Often, three bits are grouped together and called an octal number, for example, ($2^3 = 8$). Just as often, four bits are grouped together and called a hexadecimal or hex number, such as, ($2^4 = 16$). Most often, eight bits are grouped together and called a *byte.* A byte contains two hex numbers. Bytes are grouped together and called a *word.* Microcomputers use one-byte words, minicomputers use two-byte or 16-bit words, and the IBM 360/370 computer uses four-byte or 32-bit words. *Constants and variables* are stored within the computers as binary words, with anywhere from one half a word, up to four words being allocated to each constant or variable. The components of the computer that perform mathematical functions are binary in nature. The memory of the computer is a series of "0's" and "1's" called binary numbers. Mathematical calculations are very easy to do in the binary numbering system.

The *binary system* is a base-2 system which uses either a 0 or a 1 for all data. The binary system functions the same as the decimal system, base 10, utilizing only two numbers:

DECIMAL	BINARY	DECIMAL	BINARY
0	0	5	101
1	1	6	110
2	10	7	111
3	11	8	1000
4	100	9	1001

Binary numbers are converted to base 10 using the positional numbering method:

$$(A_N A_{N-1} \ldots A_2 A_1 A_0) = A_N B^N + \ldots + A_2 B^2 + A_1 B^1 + A_0 B^0$$

PROBLEM 1:

Convert $(1111)_2$ to base 10.

SOLUTION:

$$(1)2^3 + (1)2^2 + (1)2^1 + (1)2^0 = 15$$

To convert from base 10 to binary, use the remainder method:

Convert $(150)_{10}$ to base 2: $(10010110)_2$.

$150/2 = 75$	remainder	0
$75/2 = 37$	remainder	1
$37/2 = 18$	remainder	1
$18/2 = 9$	remainder	0
$9/2 = 4$	remainder	1
$4/2 = 2$	remainder	0
$2/2 = 1$	remainder	0
$1/2 = 0$	remainder	1

Binary mathematics follows the same rules as decimals, whereby bits are added singly and the carry over is added to the next pair of bits. There are only three rules for binary mathematics, compared to 55 for the decimal system.

1.	0	plus	0	equals	0	
2.	0	plus	1	equals	1	
3.	1	plus	1	equals	0	with a carryover of 1 to the next column to the left

PROBLEM 2:

Add 1 to 703 in binary (0010 1011 1111).

SOLUTION:

DECIMAL	BINARY			
		111	1110	carry
703	0010	1011	1111	
+ 1	+0000	0000	0001	
704	0010	1100	0000	

The same process can be performed for subtraction of binary numbers as follows:

1.	0	from	1	is	1	
2.	1	from	1	is	0	
3.	1	from	0	is	1	and borrow 1 from the next column to the left

PROBLEM 3:

Subtract 57 from 359.

SOLUTION:

DECIMAL	BINARY			
		001		borrow
359	0001	0110	0111	
– 57	– 0000	0011	1001	
302	0001	0010	1110	

As in decimal math, special steps have to be taken in binary arithmetic when subtracting a larger number from a smaller number in order to have the correct quantity with the correct sign.

The following is an example of the results of subtracting 590 from 220:

DECIMAL	BINARY			
	11111	101		borrow
220	0000	1101	1100	
– 590	– 0010	0100	1110	
– 370	(1) 1111	0111	0010	

When a larger number is subtracted from a smaller number, a “1” must be borrowed from the left-most column, even though the position does not exist.

PROBLEM 4:

Subtract 220 – 590.

SOLUTION:

$$\begin{array}{rr} & 220 \\ & -\underline{590} \\ (-1) & 630 = -1{,}000 + 630 = -370 \end{array}$$

The (–1) in the fourth column equals –1,000. We have borrowed 1,000 and added it to 220 to equal 1,220. From 1,220, we subtracted 590 and then subtracted the 1,000 borrowed.

How does the computer solve this problem with the negative sign? The computer resolves this problem by representing a negative number with the negative sign in its description and identification. Most digital computers use the 2s Complement (Inverse plus 1) method to represent negative numbers in 16-bit format.

A complement of the binary number is created—a 1 is substituted for each 0 and vice versa—then a 1 is added to generate the 2*s* complement.

PROBLEM 5:

Subtract 1*s* complement from 220.

SOLUTION:

220	0000	0000	1101	1100
1*s* complement	1111	1111	0010	0011
				+1
2*s* complement	1111	1111	0010	0100

Add the 2*s* complement of 220 to 590:

	1111	11	1	1
590	(1)0000	0010	0100	1110
2*s* complement of 220	+1111	1111	0010	0100
$(370)_{10}$	(1)0000	0001	0111	0010

The 2*s* complement is the negative of the number. To subtract that number from another number, you actually add the 2*s* complement to the number. The sign of the number is represented in the 16th bit by a "0" for the positive sign and a "1" for the negative sign.

Multiplication and division are actually accomplished by repeated additions and subtractions, which are really a result of shifting the partial products as done in regular multiplication of decimals.

There are three rules of multiplication:

1. Add the multiplicand to the least significant bit of the multiplier if it is 1, then shift the multiplier to the right one column.
2. Repeat above, if least significant bit is 0.
3. Add partials.

PROBLEM 6:

$10 \times 15 = 150$

SOLUTION:

```
    1010
    1111
    ----
    1010
   1010
  1010
 1010
--------
10010110
```

PROBLEM 7:

Division is also simple. Solve:

$$\frac{130}{10} = 13$$

SOLUTION:

```
          0000 1101
        ___________
1010   )1000 0010
         101 0
         -----
          11 00
          10 10
             1010
             1010
             ----
                0
```

So, digital computers actually add and never subtract. Since multiplication and division are basically addition and subtraction functions, these functions are accomplished by the computer using special hardware or instructions.

The *octal system* (base 8) is used to work with large computer output, allowing the numbers zero to seven. The same rules for addition of decimal numbers apply except that eight and nine do not exist.

PROBLEM 8:

$(4)_8 + (3)_8 =$

SOLUTION:

Since 4 + 3 in base 10 equals seven and seven is an octal digit, then:

$(4 + 3)_8 = (7)_8$

PROBLEM 9:

$(5)_8 + (9)_8 =$

SOLUTION:

The base 10 sum of 5 + 9 equals 14 and has to be converted to base 8. Use the remainder method to convert.

$\frac{14}{8} = 1$	remainder	6
$\frac{1}{8} = 0$	remainder	1

The answer is $(16)_8$

PROBLEM 10:

Convert (65) to base 8.

SOLUTION:

$\frac{65}{8} = 8$	remainder	1
$\frac{8}{8} = 1$	remainder	0
$\frac{1}{8} = 0$	remainder	1

The answer is $(101)_8$

PROBLEM 11:

Convert $(13)_8$ to base 10.

SOLUTION:

Using the positional numbering method:

$$(1)(8)^1 + (3)(8)^0 = (11)_{10}$$

Since the octal system is related to the binary system, conversion from one to the other is simple. Start with the least significant bit (right) and group the bits in threes which correspond to an octal digit.

PROBLEM 12:

Convert $(6430)_8$ to base 2.

SOLUTION:

$$(6)_8 = (110)_2$$

$$(4)_8 = (100)_2$$

$$(3)_8 = (011)_2$$

$$(0)_8 = (000)_2$$

The answer is $(110100011000)_2$

PROBLEM 13:

Convert (1110110) to octal (base 8).

SOLUTION:

Group from the right the bits into groups of threes:

1 110 110

Convert the groups to octal:

$$(1)_2 = (1)_8$$

$$(110)_2 = (6)_8$$

$$(110)_2 = (6)_8$$

The answer is $(166)_8$.

The hexadecimal (base 16) system allows the representation of four binary digits at once by combining numbers and letters. Letters A through F are used to represent numbers 10 through 15. Conversions to the base 10 are done using the equation for the positional numbering method.

Decimal	Hex
0	0
1	1
2	2
⋮	⋮
9	9
10	A
11	B
12	C
13	D
14	E

PROBLEM 14:

Convert $(5E4)_{16}$ to base 10.

SOLUTION:

(E)16 is $(14)_{10,}$

$5(16)^2 + 14(16)^1 + 4(16)^0 = (1508)_{10}$

PROBLEM 15:

Convert $(1{,}473)_{10}$ to base 16.

SOLUTION:

Use the remainder method:

$$\frac{1,473}{92} = 92 \qquad \text{remainder} \qquad 1$$

$$\frac{92}{16} = 5 \qquad \text{remainder} \qquad 12$$

$$\frac{5}{16} = 0 \qquad \text{remainder} \qquad 5$$

Since $(12)_{10}$ equals $(C)_{16}$, the answer is $(5C1)_{16}$.

Conversion from hexadecimal to binary is accomplished by grouping the binary bits into fours from the right.

PROBLEM 16:

Convert $(1111101100111101)_2$ to hexadecimal.

SOLUTION:

1111 1011 0011 1101

$(1111)_2 = (F)_{16}$

$(1011)_2 = (B)_{16}$

$(0011)_2 = (3)_{16}$

$(1101)_2 = (D)_{16}$

The answer is $(FB3D)_{16}$.

INTEGER CONSTANTS

Integer constants are fixed point numbers, whole numbers without commas or decimal points; they are positive, negative, or zero and, in general, cannot exceed the magnitude specified by the computer manufacturer (i.e., 2.15×10^9). *Real constants,* a basic real constant or integer constant followed by a decimal exponent) contain decimal points or floating point exponential numbers with a decimal point and an "E" to denote the exponent, or a floating point double precision exponential number with a decimal point and a "D" to denote the exponent. Floating point numbers are real numbers. For example:

Fixed point:	1, 2, 3, 40, 500, –15
Floating point:	1.0, 2.0, 3.0, 3.14, –15.
Exponential:	0.100E+0, .2689E–12, 30.2056E+6
Double precision:	.010D+0, –0.258975D–12

A plus sign following the E or D exponents is normally omitted. A minus is always printed. The power number is always a fixed point integer, without a decimal point.

Real numbers enclosed in a parenthesis and separated by a comma, signed or unsigned, are *complex constants*. The first number is real; positive, negative, or zero; and does not exceed the allowable magnitude. The second number is imaginary.

For example:

(4.5, –2.9)

(–10.1E05, .123E + 02)

COMPUTER PROGRAMMING

Today, there are many high-level programming languages (FORTRAN, BASIC, COBOL Q-BASIC, C+, Algol, PL/1, APL, Applesoft, etc.) used by computers to perform the many tasks they are designed to do. Most exam questions focus on the knowledge and understanding of Fortran and BASIC programming languages.

The object program is the language of the computer. Fortran commands are translated into the object program language by the compiler, the processor which translates into the machine language the analyzed source program statements. Errors in the source program are identified by the processor and relayed to the operator through error and diagnostic messages.

FORTRAN

FORTRAN was derived to write programs to solve mathematical and technical problems requiring the manipulation of numerical data. The name comes from the acronym formed by combining the FORmula TRANslator (FORTRAN).

Constants and variables are stored within the computer as binary words (one or more bytes: 8 bits for a byte)—half a word to four words are allocated for each.

Constants are fixed quantities. There are four types of constants:

numerical—integers, and real or complex numbers which do not include commas

logical (truths)—True. Or. False.

literal—strings of alphanumeric or special characters ($#&)

hexadecimal—16-bit numbers (character Z followed by numbers 0 thru 9, and letters A to F used for data initialization values only)

A string of alphabetic, numeric, or special characters form *literal constants,* which require one byte of storage to store from seven to 255 characters composing it. They are used in argument lists of CALL statements, data initialization values, or FORTRAN statements, delimited by *w*H, where *w* equals the number of string characters. For example:

20H Deposit/Withdraw Ten

13H 'Remember Me'

A *variable* is a name to which a numerical value is assigned. The name begins with a letter. Numbers can be added later, but a variable does not include punctuation or mathematical symbols. A special instance is a pair of parentheses with subscripted variable names. FORTRAN compilers accept names containing one to six characters.

Fixed point variables are always assigned integer values without decimal points. The first letter of a fixed point variable is a I, J, K, L, M, or N.

Integer Variables: IGOR, JACK, KELP, LOW, MAL, NAT; IGOR = 5, JACK = 7

Subscripted Variables: JACK(1,1), IGOR(5) = 7, JACK(1,1) = 0

Floating point variables are always assigned values with a decimal point. The first letters are A to Z, not including I to N: ALPHA, BETA, CAR, X, Y, Z; X = 10.5, Y = 9.008, Z = 1.0.

Subscripted Variables: ALPHA(1), BETA(5), CAR(22), X(1) = 13.3, Y(9) = 180.22

Exponential Numbers–Real Numbers (E Format): 3.0E + 15, 2.14E – 12, 40.E – 08

WRITING PROGRAMS

In order to get the computer to carry out calculations and operations, programs have to be written in the correct FORTRAN format. In some versions of FORTRAN the first character that would normally be printed is used for print spacing. A space causes output to be single spaced; a zero causes double spacing; a one causes output to be written on the next page. The following are examples of program formats:

```
          Write (1, 1000)
          Write (1, 1001)
          Write (1, 1003) Ncount, Nage, Salary
1000           Format (16X, ‘Yearly Salary By Age’ ///)
1001           Format (5X, “Run Number’, 10X, ‘Age’, 5X
               ‘Yearly Salary’ //)
1003           Format (15X,I5 5X,I3, 5X, F16.2)
```

The following is the resultant output: for Formats 1000, 1001, 1003:

(15x)	Yearly Salary By Age	
(4x) Run Number	(5x) Age	(5x) Yearly Salary
1	30	40,000.00
2	40	45,000.00
3	50	52,000.00

Note: x denotes a space

The slashes in the program terminate the reading of a punched card.

Expressions, Operations, and Simple Programs

The equal sign (=) in a FORTRAN expression does not mean equal. It means that the symbol on the right of the equal sign is evaluated first and then stored in a location referenced by the symbol on the left of the equal sign. Standard mathematical symbols are used for addition, subtraction, multiplication, and division.

Mathematical Functions

Math Operation	Math Symbol	FORTRAN Symbol	FORTRAN Sample
Addition	+	+	$A + B$
Subtraction	–	–	$A - B$
Multiplication	× or •	*	$A * B$
Division	/	/	A / B
Exponentiation	A^b	**	$A ** B$

Flow Diagrams

In order to visualize the different operations which take place in a written program, charts are used to show the flow between commands, operations, nodes, loops, etc. Figure 2 is an example of a flow diagram. The written program for the flow diagram follows.

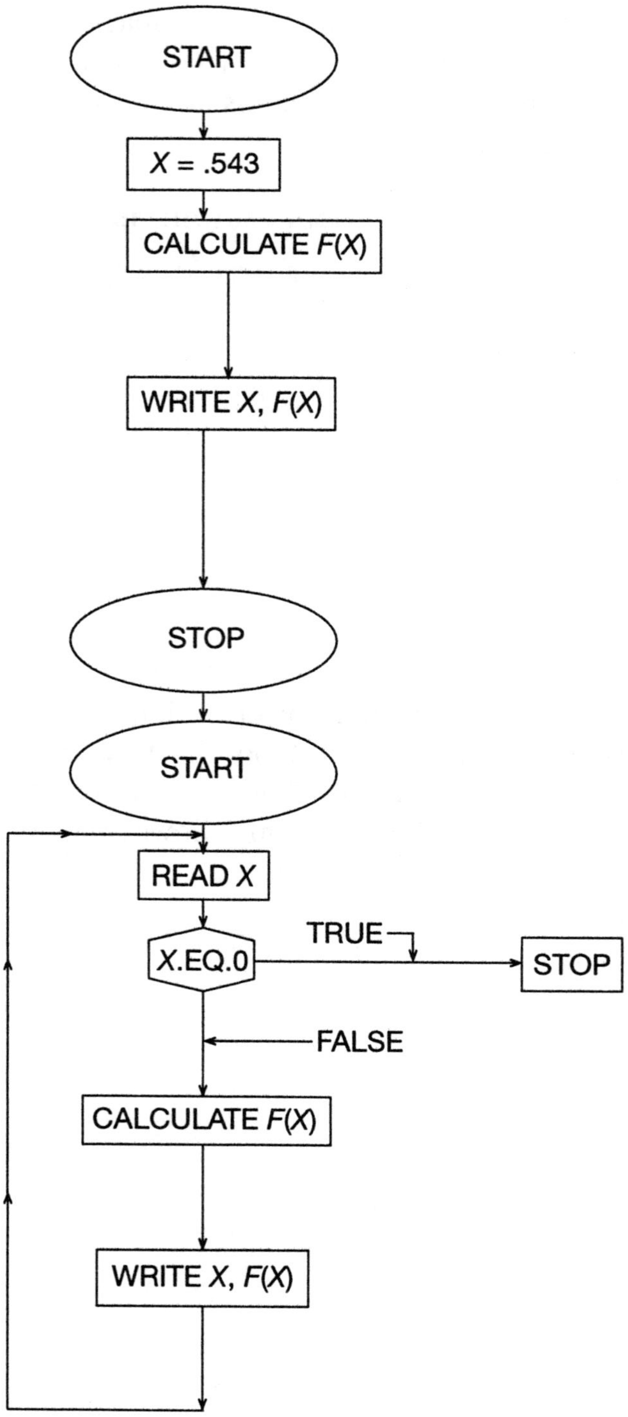

Figure 2. Flow Diagram

```
        X = 0.543
        FUNX = 1.0 + (X**2)/2.0+(X**3)/6.0+(X**4)/24.0
        WRITE (5,1000)X, FUNX
1000    FORMAT (1X,’X=’,F20.8/1X
               ‘F(X)=’,F20.8)
        STOP
        END
   1    READ (8,300) X
 300    FORMAT (F80.0)
        IF (X.EQ.0) GO TO 2
        X2=X*X
        X3=X*X2
        X4=X*X3
        ANSX=1+X+X2/2+X3/6+X4/24
        WRITE (5, 3000)
3000    FORMAT (1H1////10X’FOURTH
        ORDER APPROXIMATION TO
   1    EXPONENTIAL FUNCTION//)
        WRITE (5,3010) X
3010    FORMAT (1X. ‘VALUE OF
        EXPONENT X =’, F16.7/)
3020    FORMAT (1X,’E TO THE X
        POWER=’ E20.8)
        GO TO 1
   2    CONTINUE
        STOP
        END
        0.543
        0.0
```

BASIC

BASIC stands for Beginner's All-purpose Symbolic Instruction Code. It is very similar to FORTRAN. BASIC processes three types of data: *constants*, *variables*, and *strings*. Both *constants* and *variables* are real numbers, differing only in the fact that the values of variables can change where that of constants remain the same. *Variables* have to be one or two characters long, the first character a letter from A to Z and the second, a single digit 0 to 9 (B, B0, C1, D1, E2). *Strings* are used with Print statements. Characters enclosed in quotation marks of a print statement form a string. They provide information. For example:

Statement #	Variable name	Operator	Constant
10	LET A	=	12
20	LET B	=	16
30	LET C	=	A + B
40	PRINT "A = "A, "B = "B, "A – B = "C		
50	END		
Output:	A = 12 B = 16 A + B = 28		
	Stop at line 50		

Mathematical Operators:

Exponentiation	^
Multiplication	*
Division	/
Addition	+
Subtraction	–
Equals	=

PROBLEM 17:

Express in program $X = 2B + C^5$.

SOLUTION:

```
10      LET X = 2*B + C^5
```

A statement number (10) is always necessary.

Arithmetic calculations are done left to right, except where "hierarchy of operation" governs. Exponentiation, multiplication, division, then addition and subtraction occur in the order listed. Note: The "LET" is optional in most BASIC languages. LET B = 1 is equivalent to B = 1.

PROBLEM 18:

Determine value of A in the following BASIC statement.

```
10      LET A = 2 + 4*6+1
```

SOLUTION:

The computer would yield the following solution:

2 + (4*6) + 1 = 27 versus

(2 + 4) * 6 + 1 = 37.

Use of parenthesis is always advised to clarify the order of operation desired.

IF and GO TO Statements

The following program will find the square roots of numbers from 1 to 3.

```
10      PRINT "Table of Square Roots From 1 to 3"
15      LET B = 1
20      LET A = SQR(B)
30      PRINT B, A
40      LET B = 2
50      LET A = SQR(B)
60      PRINT B, A
70      LET B = 3
80      LET A = SQR(B)
90      END
```

The length of this program can be reduced by using an IF statement and a GO TO statement. The format of the statement is as follows:

IF (A) (Relational Operator) (B) THEN (Statement Number)

A and B are arithmetic expressions, constants or variables

If A and B fall into the relational operator, control of the program transfers to the statement number given. GO TO can be used in place of THEN.

IF 1 = 10 THEN 100

Not only can you use "=," you can use ">" greater than, "<" less than. ">=" greater than or equal to, "<=" less than or equal to, and "<>" not equal to.

If A and B do not stand in the relation given by the relational operator, then control is transferred to the next statement in the program.

A GO TO statement has the form:

```
10      GO TO (Statement Number)
```

It passes control of the program to the statement that is defined by the statement number. Solving the previous sample, using GO TO also:

```
10      PRINT "Table of Square Roots From 1 to 10"
15      LET B = 1
20      IF B > 10 THEN 90
30      LET A = SQR(B)
40      PRINT B, A
70      LET B = B + 1
80      GO TO 20
90      END
```

Statements 20 thru 60 form a loop. The value of B is set to 1, then the IF statement checks B to see if it is greater than 10. If B is less than 10, the program proceeds to the next statement, where the program calculates the square root of B and sets it equal to A. The next statement prints B and its square root, A. Next, the value of B is increased by 1 and then sent back to statement 60 and the loop repeats again until B equals 10. At that point, the program goes to statement 90 and stops.

```
10      IF A*B + 1 > 6 THEN 100
```

This IF statement evaluates the product of A and B plus 1 and checks to see if its value exceeds 6. If it does, control of the program goes to statement 100.

```
10      IF B > = 0 THEN 100
```

If B is greater than or equal to 0, control of the program goes to statement 100.

INPUT Statement

INPUT is defined as a list of variable names separated by commas.

This statement requests the input of an amount of data from the terminal equal to the number of variable names in the list. Compliance with this statement causes the printing of a question mark (?) on the terminal to tell the user to input the data. The INPUT statement makes programs more flexible.

```
10      PRINT “TYPE IN STARTING NUMBER”
20      INPUT S
30      PRINT “TYPE IN ENDING NUMBER”
40      INPUT E
50      IF S < 0 THEN 200
60      IF E < S THEN 200
70      PRINT “TABLE OF SQUARE ROOTS FROM “S” TO “E”
80      IF S > E THEN 300
90      LET A = SQR(S)
100     PRINT S, A
110     LET S = S + 1
120     GO TO 80
200     PRINT “INPUT ERROR”
210     GO TO 10
300     END
```

Statements 50 to 120 form the loop. Statement 80 checks if the program should END. To continuously loop back and repeat the program many times, BASIC uses the "FOR" statement.

FOR (running variable) = (A) TO (B) STEP (C)

where:

A is the "FROM" element

B is the "TO" element

C is the "STEP" element

FOR starts the statements that you want to repeat and end with the NEXT statement.

NEXT (running variable)

This statement tells the program to return to the FOR statement.

```
10    PRINT "TABLE OF SQUARE ROOTS FROM 1 TO 10"
20    FOR B = 1 TO 10 STEP 1
30    LET A = SQR(B)
40    PRINT B, A
50    NEXT B
60    END
```

DIM Statement

The DIMension statement sets up arrays and associates the array with the single variable name. DIM (variable) (number of elements in the array) DIM A(20). For example:

```
10    DIM A (20)
20    FOR N = 1 TO 20 STEP 1
30    PRINT "TYPE IN TEMP FOR TIME UNIT" N
40    INPUT A(N)
50    NEXT N
60    LET T = 0
70    FOR N = 1 TO 20 STEP 1
```

```
80    LET T = T + A(N)
90    PRINT N, A(N)
100   NEXT N =
110   LET B = T/20
120   PRINT "THE AVERAGE TEMP IS" B
130   END
```

Twenty readings are taken. Statements 20 to 50 are the input loop. A new variable, T, is introduced to accumulate the total sum of all of the temperatures. Statement 60 sets T initially to a value of 0 and the loop 70 to 100, performs the addition, and prints out a table of time units and their respective temperatures. At the end of the later loop, B is calculated as the average of 20 numbers and is printed.

Elementary Functions

Function names consist of three letters and an argument in parenthesis. They are:

Function	Interpretation
SIN(X)	Trigonometric sine of X (radians)
COS(X)	Trigonometric cosine of X (radians)
TAN(X)	Trigonometric tangent of X (radians)
ATN(X)	Angle in (radians) whose tangent is X
EXP(X)	Exponential function (e^x) (power of 1.71828)
LOG(X)	Natural logarithm of X (lnX)
ABS(X)	Absolute value of X
INT(X)	Integer part of X
SGN(X)	+1, 0 – 1 depending if $X > 0$, $X = 0$, $X < 0$
SQR(X)	Square root of X

For example:

Integer Function

INT(X)	INT(.98643)	= 0
	INT(2.8)	= 2
	INT(1.99876)	= 1
	INT(–2.8)	= –3

Integer functions drop any fraction beyond the real number.

SPREADSHEETS

Spreadsheets are among the most widely used and practical programs available for personal computers. Although developed primarily for bookkeeping, they are also useful for scientific calculations, data manipulation, and graph production.

A spreadsheet program allows you to type in numbers and formulae to represent any system that can be described numerically. For instance, a model of the cash flow of a business might include figures for the expected monthly costs and income of the business over a period of time, yielding as output an estimate of how much cash the business would have at the end of each month. Similarly, you could build a simple model of a scientific experiment showing the temperature along an extended surface as a function of position along the extended surface.

The reasons for modeling such a situation are as follows:

1. To make repeated calculations on a large number of figures, and automatically recalculate when values change.
2. To analyze a situation and discover any significant patterns in the results.
3. To make a projection based on the change of input data.
4. To present numerical results in a graphical form making them easier to understand.

A spreadsheet is a matrix of rows and columns called *cells*. The columns are designated with capital letters while the rows have a numerical designation. The upper left cell is given the designation A1 while the 27th, 28th, and 29th cell in the first row have the designation of AA1, AB1, and AC1, respectively.

Cells can hold a label, a number, or a formula. Labels are used to identify the information in adjacent cells. A cell can also contain a value or a number. A value or number can be directly entered into the spreadsheet cell or it can be extracted from a table stored in the spreadsheet by using functions. The ability to store and retrieve data stored in a columnar fashion is another unique property of the spreadsheet. Lastly, a cell can contain a formula. The formula is the most important part of the spreadsheet as it allows the construction of interrelationships between the various cells. The simple operations of addition, subtraction, multiplication, division, and exponentiation use the symbols +, –, *, /, and ^ together with cell designations. If the content of cell C1 is to contain the product of cell A1 and B1, then the correct formula for cell C1 would be: A1 * B1. The formula can also contain functions. Suppose that cell B4 is to contain the square root of cell B1. If the function for square root is sqrt( ), then the correct formula for cell B4 would be: sqrt(B1).

To illustrate its scientific use, a spreadsheet will be constructed to find the Reynold's Number and friction factor for water flowing through a circular pipe.

Referring to Figure 3, six columns and 14 rows of a spreadsheet are shown. The first column, A, contains the names of the data in the third column. The second column, B, contains the units required for the data in the third column. The third column, C, contains the original data. Cells C3 to C6 contain the required values of the flow rate, nominal diameter, length, and roughness of the pipe while cell C8 contains the temperature of the water. From this initial data, the spreadsheet will look up values for the actual inside diameter in C16 and the density of the water based on the temperature in C15. The spreadsheet now calculates via formula the absolute temperature in K in cell C13, the viscosity of the water in C14, and the relative roughness of the pipe in C17.

The velocity is calculated using the GPM and the actual inside diameter. The Reynold's Number is calculated from the density, velocity, actual inside diameter, and the viscosity. Finally, the friction factor is calculated from the relative roughness and the Reynold's Number. Spreadsheets are designed to recalculate as they go along, so let's see what happens when you change one of the original numbers. The fourth column, D, is the same as column C except that the pipe diameter has been changed from 14 to 8 inches. Note how the values adjust themselves. The fifth column is the same as the third with the temperature changed to 120°F. In the sixth column, the GPMs are changed.

Figure 3. Spreadsheet to Determine Reynold's Number and Friction Factor

	UNITS	VALUE	VALUE	VALUE	VALUE
(A)	(B)	(C)			
GPM	GPM	5400	5400	5400	4000
NOMINAL DIAMETER	inches	14.0	8.0	14.0	14.0
PIPE LENGTH	feet	100	100	100	100
ROUGHNESS		1.50E-04	1.50E-04	1.50E-04	1.50E-04
REYNOLDS No.		1.14E+06	1.90E+06	2.29E+06	8.47E+05
TEMPERATURE	degrees F	60.0	60.0	120.0	60.0
FRICTION *f*		0.0139	0.0146	0.0134	0.0142
FRICTION *Ft*		0.013	0.014	0.013	0.013
VELOCITY	ft/sec	12.549	34.589	12.549	9.296
TEMPERATURE	degrees K	288.6	288.6	321.9	288.6
VISCOSITY	cP	1.124	1.124	0.556	1.124
DENSITY	lbm/ft^3	62.4	62.4	61.7	62.4
INSIDE DIAMETER	inches	13.250	7.981	13.250	13.250
RELATIVE ROUGH		1.36E-04	2.26E-04	1.36E-04	1.36E-04

The Copy Command

Once the formulas for column three are created, the formula for the remaining columns can be filled in using the Copy command. There are two types of copying, *absolute* and *relative*. With absolute copying, the exact content of the cell is copied in its new destination. With relative copying, the formulas are modified to reflect the new relative position in the spreadsheet.

Normally, copying is performed in a relative fashion. On occasion, such as when data is to be extracted from a table, that component of the formula must be treated in an absolute fashion. This is accomplished by placing "$" around particular cell elements. Suppose data exists in a table that occupies cells C23:E32. In order to copy this cell designation without any changes, it would be written as $C$23:$E$32.

PROBLEM 19:

Cells A1...A5 of a spreadsheet contains the values of 1, 2, 3, 4 and 5, respectively, which represents the value of the variable x. Cells B1...B5 are to contain the corresponding values of variable $y = x^2 + 1$. What would be the correct content for cell B3?

SOLUTION:

The data for cell B3 must come from cell A3. The correct answer is A3 * A3 + 1.

PROBLEM 20:

In a spreadsheet, the number in cell A10 is set to 4. Then B10 is set to A10 + A10. The formula is now copied to cells C10 and D10. The number in cell D10 is most nearly ____ ?

SOLUTION:

The correct answer is 32. Since the cells copied were references, cell B10 would contain 4 + 4 = 8, cell C10 would contain 8 + 8 = 16, and cell D10 would contain 16 + 16 = 32.

CHAPTER 4

Dynamic Systems

CHAPTER 4

DYNAMIC SYSTEMS

Dynamic systems are ubiquitous components of engineering systems. Essentially, a dynamic system is a device that moves or changes with time. For example, systems such as spring-mass-damper interconnections are dynamic systems. Also, thermodynamic cycles and heat transfer situations can be viewed as dynamic systems with pressure, temperature enthalpy, and entropy acting as the varying state components. Fluid pipe flow is another example, with pressure and fluid quantity acting as the variables. Finally, electrical systems with voltages and currents and magnetic fluxes acting as varying components are also examples of dynamic systems.

TIME DOMAIN ANALYSIS

Consider the engineering system described by the Laplace domain transfer function:

$$\frac{X(s)}{U(s)} = G(s) = \frac{B(s)}{A(s)} = \frac{b_m s^m + b_{m-1} s^{m-1} + \ldots + b_0}{s^n + a_{n-1} s^{n-1} + \ldots + a_0}.$$

We can note the following:

System Descriptions

The function description shown is known as a transfer function form. It involves a denominator polynomial of order n, and a numerator of order m. The coefficient of the s^n term in the denominator is always one. All of the other coefficients are arbitrary and some of them may be zero.

It is noted that $n \geq m$ always, and for real physical systems, $n > m$. This implies that inputs cause changes in the outputs of systems and not the other way around. We call systems that obey this rule *causal* systems and the mathematical descriptions of such systems, that is $G(s)$, are strictly proper. The physical reasoning for the need for causal systems is also seen in examining the frequency response of such systems.

PROBLEM 1:

For a given mechanical system, give a physical reason for its causality in terms of a mathematical model description.

SOLUTION:

Most mechanical systems are described by Newton's Law:

$$F = ma$$

which can also be written as:

$$\ddot{x} = \frac{1}{m} F$$

where F, the external forces, is a function of x and $\dot{x}$, the system displacement and velocity. Thus, the model can be rewritten:

$$\ddot{x} = \frac{1}{m} f(x, \dot{x}).$$

Normally, the output of the system is its displacement or velocity. According to the model, the system computes higher derivatives of the output in order to find a solution. This implies causality.

Alternative descriptions to the transfer function form shown are also useful.

Pole-Zero Form

$$G(s) = \frac{K \prod_{i=1}^{m} (s + z_i)}{\prod_{i=1}^{n} (s + P_i)} = \frac{K(s + z_1)(s + z_2) \ldots (s + z_m)}{(s + p_1)(s + p_2) \ldots (s + p_n)}$$

This form has the advantage that all of the roots of both the denominator and numerator are explicitly identified. The terms *pole* and *zero* will be explained in the next section.

Time Constant Form

$$G(s)=\frac{K\prod_{i=1}^{q}(\tau_i s+1)\times\prod_{k=1}^{\frac{m-q}{2}}\left(\left(\frac{s}{\omega_{nk}}\right)^2+2\xi_k\frac{s}{\omega_{nk}}+1\right)}{\prod_{i=1}^{r}(\tau_i s+1)\times\prod_{k=1}^{\frac{n-r}{2}}\left(\left(\frac{s}{\omega_{nk}}\right)^2+2\xi_k\frac{s}{\omega_{nk}}+1\right)}$$

where there are q purely real roots of the numerator, and

$$\frac{m-q}{2}$$

pairs of complex conjugate roots, and there are r purely real roots of the denominator and

$$\frac{n-r}{2}$$

pairs of complex conjugate roots in the denominator. This form can be used to describe the various first order *time constants, natural frequencies,* and *damping ratios* of the system.

Note that all of the above forms are entirely equivalent in that they can be used to describe exactly the same underlying physical system. Therefore, they will all share exactly the same mathematical properties. The only difference is in the way the system is described. This allows for some insight into the characteristics of the system.

PROBLEM 2:

Given:

$$G(s)=\frac{15(3s+1)\left(\left(\frac{s}{2}\right)^2+2(0.7)\frac{s}{2}+1\right)}{(10s+1)(65s+1)\left(\left(\frac{s}{0.5}\right)^2+2(0.1)\frac{s}{0.5}+1\right)\left(\left(\frac{s}{6}\right)^2+2(0.9)\frac{2}{6}+1\right)}$$

in time constant form, find the description in pole-zero form and in transfer function form.

SOLUTION:

The system can also be represented in pole-zero form as:

$$G(s) = \frac{0.156(s+0.33)\,(s+1.4+1.43j)\,(s+1.4-1.43j)}{(s+0.1)\,(s+0.015)\,(s+5.4+2.62j) \\ +(s+5.4-2.62j)\,(s+0.05+0.5j)\,(2+0.05-0.5j)}$$

and in transfer function form as:

$$G(s) = \frac{0.156s^3 + 0.488s^2 + 0.769s + 0.208}{s^6 + 11.015s^5 + 38.589s^4 + 10.624s^3 + 9.784s^2 + 1.048s + 0.014}.$$

All forms of $G(s)$ are entirely equivalent.

POLES AND ZEROS

Given a system described in *pole-zero* form:

$$G(s) = \frac{K(s+z_1)\,(s+z_2)\ldots(s+z_m)}{(s+p_1)\,(s+p_2)\ldots(s+p_n)}$$

if we choose to evaluate this transfer function, $G(s)$, for particular (possibly complex) values of $s = s^*$, then $G(s^*)$ will simply be a complex number.

PROBLEM 3:

Given:

$$G(s) = \frac{1}{s+1},$$

find the complex valued solution at $s = 2 + j$.

SOLUTION:

$$G(s)\big|_{s=2+j} = \frac{1}{2+j+1} = \frac{1}{3+j} = \frac{1}{3+j} \times \frac{3-j}{3-j}$$

$$= \frac{1}{3^2+1^2}(3-j) = \frac{3}{10} - \frac{1}{10}j.$$

If, for example, we evaluate $G(s)$ at $s = z_i\ i = 1, \ldots, m$, then:

$$G(s)\big|_{s=-z_i} = \frac{K(-z_i+z_1)\ldots(-z_i+z_i)\ldots(-z_i+z_m)}{(-z_i+p_1)\ldots(-z_i+p_n)}$$

$$= \frac{K(-z_i+z_1)\ldots(0)\ldots(-z_i+z_m)}{(-z_i+p_1)\ldots(-z_i+p_n)} = 0.$$

Therefore, when $G(s)$ is evaluated at each

$$s = -z_i$$

$$i = 1, \ldots, m,$$

the function is zeroed. Thus, each

$$s = -z_i$$

$$i = 1, \ldots, m,$$

is called a *zero* of $G(s)$.

Similarly, evaluating $G(s)$ at each $s = -p_i$

$$i = 1, \ldots, n, \text{ gives:}$$

$$G(s)\big|_{s=-p_i} = \frac{K(-p_i+z_1)\ldots(-p_i+z_m)}{(-p_i+p_1)\ldots(-p_i+p_i)\ldots(-p_i+p_n)}$$

$$= \frac{K(-p_i+z_1)\ldots(-p_i+z_m)}{(-p_i+p_1)\ldots(0)\ldots(-p_i+p_n)} = \pm\infty.$$

Therefore, the function $G(s)$ gets large in magnitude when evaluated at:

$$s = -p_i,\ i = 1, \ldots, n.$$

If we consider a graph of $G(s)$ versus s, the graph will go to zero at the zeros of the transfer function and will tend to infinity at the values:

$$s = -p_i,\ i = 1, \ldots, n.$$

In a colloquial way, one can view this as if someone stuck a tent pole under the graph and pushed it up to infinity at these points. Thus, we call the points:

$$s = -p_i,\ i = 1, \ldots, n$$

poles of the transfer function.

PROBLEM 4:

Let:

$$G(s) = \frac{(s+2)(s-12)}{(s+8)(s-6)}$$

and evaluate $G(s)$ for various inputs of s.

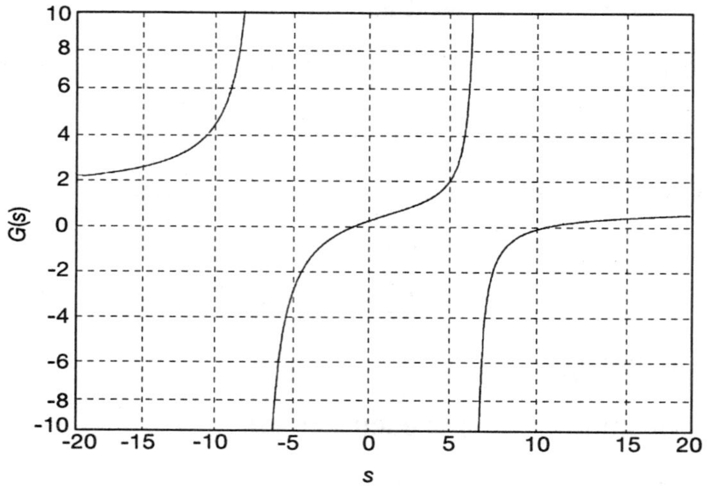

Figure 1. Plot of *G*(*s*) Showing Poles and Zeros

SOLUTION:

The results are shown in Figure 1. Note that the function goes to zero at the zeros of $G(s)$ and the plot goes to infinity at the poles of $G(s)$.

It is noted that there are exactly *m zeros* and *n poles* of the transfer function $G(s)$, which had a numerator of order m and a denominator of order n.

The *relative order* of the system is $RD = n - m$. This is sometimes called the *pole excess* of the system. *Causal* systems are those with more poles than zeros, and so consequently another way of describing the collection of all causal systems is those systems with $RD > 0$.

System Stability

Stability is, for obvious reasons, a key issue in engineering systems. Thus, we should have a clear and precise definition of stability and a methodology for determining if a given system is stable.

For our purposes, it suffices to consider a form of stability analysis called *bounded-input-bounded-output stability*. This implies that there is a guarantee that the output of a system will be bounded (that is, will not blow up to infinity) at all times if there are bounds on the size of the input at all times. A test for bounded-input-bounded-output (BIBO) stability is as follows.

Consider a system described by:

$$\frac{X(s)}{U(s)} = G(s)$$

where $X(s)$ is the output, $U(s)$ is the input, and $G(s)$ is the system transfer function. The transfer function contains poles and zeros as in:

$$G(s) = \frac{K(s+z_1)(s+z_2)\ldots(s+z_m)}{(s+p_1)(s+p_2)\ldots(s+p_n)}.$$

Further, $U(s)$ is an input signal that has the Laplace domain description:

$$U(s) = \frac{K_u(s+z_{u1})(s+z_{u2})\ldots(s+z_{um})}{(s+p_{u1})(s+p_{u2})\ldots(s+p_{un})}.$$

For example, if $u(t)$ is a unit step input then:

$$U(s) = \frac{1}{s}$$

and if $u(t) = \sin \omega t$ then:

$$U(s) = \frac{\omega}{s^2+\omega^2} = \frac{\omega}{(s+j\omega)(s-j\omega)}.$$

Given this system and input, the output can be described by:

$$X(s) = \frac{K(s+z_1)(s+z_2)\ldots(s+z_m)}{(s+p_1)(s+p_2)\ldots(s+p_n)} \times \frac{K_u(s+z_{u1})(s+z_{u2})\ldots(s+z_{um})}{(s+p_{u1})(s+p_{u2})\ldots(s+p_{un})}.$$

Now, taking the inverse Laplace transform, the time domain solution for the output is:

$$x(t) = c_1 e^{-p_1 t} + c_2 e^{-p_2 t} + \cdots + c_n e^{-p_n t} + c_{u1} e^{-p_{u1} t} + c_{u2} e^{-p_{u2} t} + \cdots + c_{un} e^{-p_{un} t}.$$

Each of the poles of both the system and input are, in general, complex valued. If they are real valued, then the complex part is zero. Further, if any poles are complex valued, they will appear as *complex conjugate pairs* and imply an oscillating or trigonometric time function solution. This trigonometric function will have an exponential term in front of it whose exponent is given by the real part of the pole.

In summary, a portion of the terms in the sum solution for the output will look like:

$$x_{ri}(t) = c_i e^{-a_i t}$$

where $p_i = a_i = \in \Re$ or is a purely real number. The remaining terms will appear in pairs and will look like:

$$x_{ck}(t) = c_k e^{-a_k t}(e^{-jb_k t} + \rho_k e^{jb_k t})$$

where:

$$p_k = a_k + jb_k \text{ and } p_{k+1} = a_k - jb_k.$$

The important point to recognize here is that all of the terms $e^{jb_k t}$ for all values of b_k are purely oscillatory with constant unity magnitude. They are unit vectors in the complex plane that rotate about the origin.

Note that if any of the exponential terms with real exponents (that is, all of the $e^{-a_j t}$, $j = 1, \ldots, n$) are growing with time, then the output will grow in magnitude with time. Therefore, for stability, we will require that none of the terms in the sum grow with time. This is sufficiently satisfied if:

$$a_j \geq 0 \ \forall \ j = 1, \ldots, n.$$

That is, all of the real parts of the exponents are positive for all of the system and input poles. However, by definition of BIBO stability, the input poles will necessarily meet this condition since $u(t)$ is a bounded function of time by construction and so all poles of $U(s)$ will have negative real parts.

Thus, we only need to check the poles of the system and a precise definition of BIBO stability is that for all poles, $p_i = -a_i + jb_i$ of $G(s)$, if:

$$real(p_i) \leq 0$$

then the system is BIBO stable.

A slightly stronger sense of stability is that of *asymptotic stability*, which implies that the output of the system tends to zero for energy bounded input signals. Therefore, for asymptotic stability, the condition on the poles changes to:

$$real(p_i) < 0 \text{ for all } i = 1, \ldots, n.$$

PROBLEM 5:

Evaluate the stability of

$$G_1(s) = \frac{2}{s+4}$$

and

$$G_2(s) = \frac{2}{s-4}.$$

SOLUTION:

Using a step input for each case, the time solutions are:

$$y_1(t) = L^{-1}\left(\frac{2}{(s(s+4)}\right) = \frac{1}{2}(1 - e^{-4t})$$

and:

$$y_2(t) = L^{-1}\left(\frac{2}{(s(s-4)}\right) = \frac{1}{2}(-1 + e^{4t}).$$

By letting time get large, it can be seen that the output y_1 stays bounded while y_2 does not. Therefore, we say that system G_1 is stable while system G_2 is unstable.

Characteristic Equations

From the above discussion, we have seen that the *system poles* play a major role in determining system stability. In fact, all of the dynamic modes of the system response are entirely determined by the system poles.

The *zeros* of the system influence the relative weighting of each of the modes. That is, the system poles determine the a_i in the exponents, and the zeros (along with the poles) determine the c_i in the partial fraction expansion description of the output and also in the sum of the modes found in the output.

It is noted that the system poles are the roots of the denominator of the transfer function:

$$G(s) = \frac{B(s)}{A(s)}.$$

That is, the roots of the equation $A(s) = 0$. We will label this special polynomial used in determining the dynamic response of systems, $A(s)$, as the *characteristic polynomial*. Further, we will label the equation $A(s) = 0$ as the *characteristic equation* of the system transfer function $G(s)$.

PROBLEM 6:

State the characteristic equation of the example system in Problem 4.

SOLUTION:

The system was:

$$G(s) = \frac{(s+2)\,(s-12)}{(s+8)\,(s-6)}$$

and therefore the characteristic equation is:

$$(s + 8)(s - 6) = 0.$$

TIME DOMAIN PERFORMANCE

In order to analyze the time domain performance of systems, we will examine the characteristics of the output of the system when a particular input is applied. The input we will choose is a unit step input, that is:

$$u(t) = \begin{cases} 1 & t \geq 0 \\ 0 & t < 0 \end{cases}.$$

The Laplace transform of this input is:

$$U(s) = \frac{1}{s}.$$

This input is selected because it is simple; it is relatively easy to experimentally implement to test systems and to implement on system simulations; and it contains a rich frequency content in that the Fourier transform of the step function contains all frequency signals. Thus, the step input is a signal rich input potentially capable of providing a wealth of information about the system.

This unit step input will be applied to first, second, and higher order systems in order to determine the general behavior of these types of processes.

First Order Systems

Consider the system:

$$\frac{X(s)}{U(s)} = G(s) = \frac{k}{\tau s + 1}$$

and using:

$$U(s) = \frac{1}{s}$$

gives:

$$X(s) = G(s)\, U(s) = \frac{k}{s(\tau s + 1)}.$$

Constructing a partial fraction expansion of the output:

$$X(s) = \frac{c_1}{s} + \frac{c_2}{s + \frac{1}{\tau}}$$

where:

$$c_1 = sX(s)\Big|_{s=0} = \frac{k}{\tau s + 1}\Big|_{s=0} = k$$

$$c_2 = \left(s + \frac{1}{\tau}\right)X(s)\Big|_{s=-\frac{1}{\tau}} = \frac{k}{\tau s}\Big|_{s=-\frac{1}{\tau}} = -k.$$

Thus:

$$x(t) = k\left(1 - e^{-t/\tau}\right).$$

It is quickly verified that the dynamic response of this system is entirely governed by τ or equivalently the pole $s = -\,{}^{1}/_{\tau}$.

We will label τ the time constant of this first order system.

By evaluating the dynamic part of the response, the term $e^{-t/\tau}$, we can compute that:

t	$e^{-t/\tau}$
0	0
τ	0.37
3τ	0.05
4τ	0.02

The initial value of the response is zero and, when $t = \tau$, 63% of the dynamics have dissipated. Also, by the time $t = 3\tau$, 95% of the dynamics have occurred. Finally, when $t = 4\tau$ has elapsed, 98% of the dynamic part of the response has disappeared.

We can also note that:

$$\dot{x}(t) = \frac{k}{\tau} e^{-t/\tau}.$$

Therefore, the initial slope of the response is k/τ and the slope monotonically decreases at the same rate as the dynamics fade. Eventually, the slope becomes very small and asymptotically approaches zero. This implies that the system eventually reaches a *steady state.*

It is important to recognize that as the time constant gets smaller, the pole location, $s = -1/\tau$, moves more to the left on the real number line. That is, the pole becomes more negative. Further, as τ gets smaller, the system response effectively gets faster. Thus, there is a fundamental connection between fast system response and system poles that are more to the left in the complex plane.

PROBLEM 7:

Find the step responses of two first order systems:

$$G_1(s) = \frac{1}{s+1}$$

and

$$G_2(s) = \frac{1}{0.33s+1}.$$

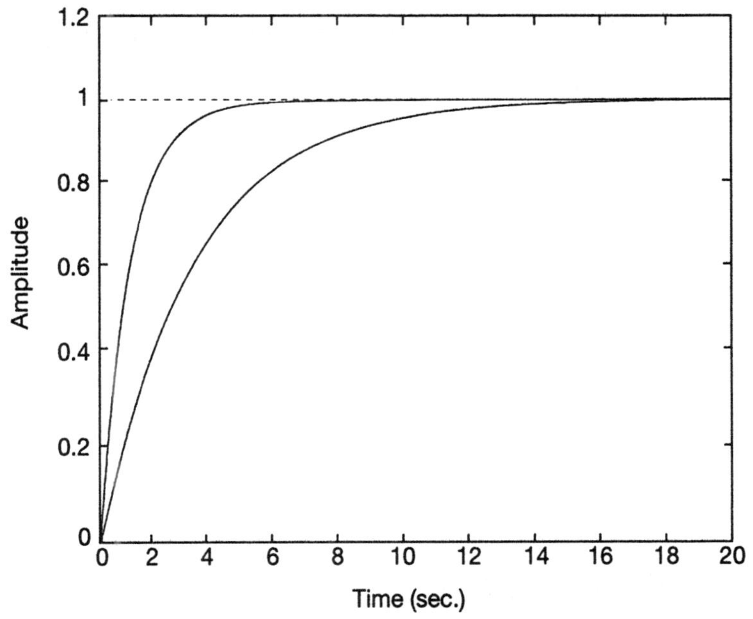

Figure 2. Step Reponse of Two First Order Systems

SOLUTION:

Figure 2 shows the response to a step input of the two first order systems with different time constants. The first and faster responding system, G_1, has a time constant of 1.0 seconds. The slower system, G_2, has a time constant of 0.33 seconds. It is noted that this system responds roughly three times slower than the first system.

Second Order Systems

Consider the engineering system:

$$\frac{X(s)}{U(s)} = G(s)$$

where:

$$G(s) = \frac{k\omega_n^2}{s^2 + 2\xi\omega_n s + \omega_n^2}$$

The variable ω_n is usually called the system *natural frequency* and the variable ξ is often referred to as the *damping ratio*. This standard form case can be used to cover essentially all possible alternatives or variations on particular second order system cases. These variations include the following:

Underdamped, Overdamped, and Critically Damped Systems

The poles of the system in standard form are the roots of the characteristic equation:

$$s^2 + 2\xi\omega_n s + \omega_n^2 = 0$$

which are:

$$s_{1,2} = \frac{-2\xi\omega_n \pm \sqrt{(2\xi\omega_n)^2 - 4\omega_n^2}}{2}$$

$$= -\xi\omega_n \pm \sqrt{\omega_n^2\left(\xi^2 - 1\right)} = -\xi\omega_n \pm j\omega_n\sqrt{1-\xi^2}.$$

From this it is seen that there are three basic possibilities:

1. $\xi < 1$ **Underdamped case**. Here there will be two complex conjugate poles with equal real parts and complex parts that are the negative inverse of each other.
2. $\xi = 1$ **Critically damped case**. Here there will be two poles that are identical since the discriminant in the quadratic form is precisely zero. The two poles will be $s_{1,2} = -\xi\omega_n$.
3. $\xi > 1$ **Overdamped case**. Here there will be two real poles. The poles will both be on the negative real axis (for stable systems) and they will be distinct. This case can be treated as the cascaded or series combination of two first order systems.

Addition of a Zero

If the actual system is:

$$G(s) = \frac{as + b}{s^2 + 2\xi\omega_n s + \omega_n^2},$$

this can be written as:

$$G(s) = \frac{as}{s^2 + 2\xi\omega_n s + \omega_n^2} + \frac{b}{s^2 + 2\xi\omega_n s + \omega_n^2}.$$

The second term is exactly as per the standard form, given that $b = k\omega_n^2$, and the first term is simply the Laplace version of the time derivative of the standard form (again with a correction for the gain). To verify this, recall that taking the derivative in the time domain is equivalent to multiplying by s in the Laplace domain. Thus, this system can be broken into two parts, one that is exactly equivalent to the standard form of the problem and one that is the derivative of the standard form in the time domain.

Thus, if the second term has an inverse transform that is a decaying sine wave, then the first term will be a decaying cosine wave with a different amplitude but the same decay rate and the same oscillation frequency.

Explicit Pole Form

If the actual system is written:

$$G(s) = \frac{c}{(s + p_1)\,(s + p_2)},$$

then this can be expressed as:

$$G(s) = \frac{c}{s^2 + (p_1 + p_2)s + p_1 p_2}.$$

Thus, using the relations:

$$\omega_n = \sqrt{p_1 p_2}, \quad \xi = \frac{p_1 + p_2}{2\sqrt{p_1 p_2}}, \quad k = \frac{c}{p_1 p_2}$$

the two forms are seen to be equivalent. Furthermore, the poles of the system can be evaluated as:

$$s = -p_1 = -\xi\omega_n + j\omega_n\sqrt{1-\xi^2}, \text{ and } s = -p_2 = -\xi\omega_n - j\omega\sqrt{1-\xi^2}.$$

It is quickly verified that the poles will always appear as either a set of real poles or as a pair of complex conjugates.

Real-Imaginary Pole Form

If the actual system is written:

$$G(s) = \frac{c}{(s + a)^2 + b^2}$$

then this can be expressed as:

$$G(s) = \frac{c}{s^2 + 2as + a^2 + b^2}.$$

Thus, using the relations:

$$\omega_n = \sqrt{a^2 + b^2}, \quad \xi = \frac{a}{\sqrt{a^2 + b^2}}, \quad k = \frac{c}{a^2 + b^2}.$$

the standard and this form are found to be equivalent. In this form, the poles of the system are computed to be:

$$s = -a + jb \text{ and } s = -a - jb.$$

Therefore, we can see that:

$$a = \xi\omega_n, \quad b = \omega_n\sqrt{1 - \xi^2}.$$

In other words, the real part of the system poles is $-a = -\xi\omega_n$ and the imaginary part is:

$$\pm b = \pm\, \omega_n\sqrt{1 - \xi^2}.$$

Pole-Zero Plots

The poles and zeros of a dynamic system can be plotted on a complex plane. From the locations of these poles and zeros, one can assess aspects of the dynamic performance of the underlying physical system.

For example, from the discussion of stability analysis, it was shown that for asymptotic stability of a system:

$$\text{Re}(p_i) < 0\ \forall\ i = 1, \ldots, n.$$

This says that the real parts of all of the system poles must be negative or lie in the closed, left-half plane of the complex plane.

Similarly, for BIBO stability, the poles of the system must lie in the open, left-half plane of the complex plane. The difference being that poles with zero real parts are allowed. Poles with real parts equal to zero are often called *marginally stable poles* since they imply dynamic system modes that are on the verge of instability.

PERFORMANCE REGIONS

Natural Frequency

For a second order system written in standard form with poles:

$$s=-p_1=-\xi\omega_n+j\omega_n\sqrt{1-\xi^2}, \text{ and } s=-p_2=-\xi\omega_n-j\omega\sqrt{1-\xi^2},$$

the magnitude of these poles is:

$$M=\sqrt{\operatorname{Re}(p_1)^2+\operatorname{Im}(p_1)^2}.$$

The magnitude is the same for both poles. Substituting in this data:

$$M=\sqrt{(-\xi\omega_n)^2+(\omega_n\sqrt{1-\xi^2})}=\sqrt{(\xi\omega_n)^2+\omega_n^2-(\xi\omega_n)^2}=\omega_n.$$

Thus, systems with equal values for ω_n (often called the *natural frequency* of the system) have equal magnitudes for the pairs of complex conjugate poles. In other words, the poles lie on circles centered at the origin and of magnitude $M = \omega_n$. We usually restrict discussion to *stable systems*, or those with poles only in the left-half plane. Thus, systems of equal *natural frequency* are those with poles on the semicircle centered at the origin and of magnitude $M = \omega_n$. Systems with natural frequencies higher than this value appear outside of the semicircle.

Damping Ratio

The angle of the poles is given by:

$$\phi_1=\cos^{-1}\left(\frac{|\operatorname{Re}(p_1)|}{M(p_1)}\right).$$

The angle of the other pole can be found to be $\phi_2 = -\phi_1$ since the real parts are precisely the same and the imaginary parts are negative inverses. On substitution:

$$\phi_1=\cos^{-1}\left(\frac{\xi\omega_n}{\omega_n}\right)=\cos^{-1}(\xi).$$

The parameter ξ is often called the *damping ratio* of a system. Thus, systems with equal damping ratios have poles that lie along two mirror (in the real axis) vectors emanating from the origin at angles given by

$\pm.\phi = \pm\cos^{-1}(\xi)$. Systems with higher damping ratios will appear inside the cone-shaped region bounded by the limiting vectors.

PROBLEM 8:

Plot lines of constant damping ratio and, separately, lines of constant natural frequency on the complex plane.

SOLUTION:

Lines of constant natural frequency and damping ratio are shown in Figure 3. We continue the time domain performance analysis of second order systems by examining the application of a unit step input to the second order system in standard form. That is, let:

$$U(s) = \frac{1}{s}$$

in:

$$X(s) = G(s)U(s) = \frac{\omega_n^2}{s^2 + 2\xi\omega_n s + \omega_n^2} \times \frac{1}{s}.$$

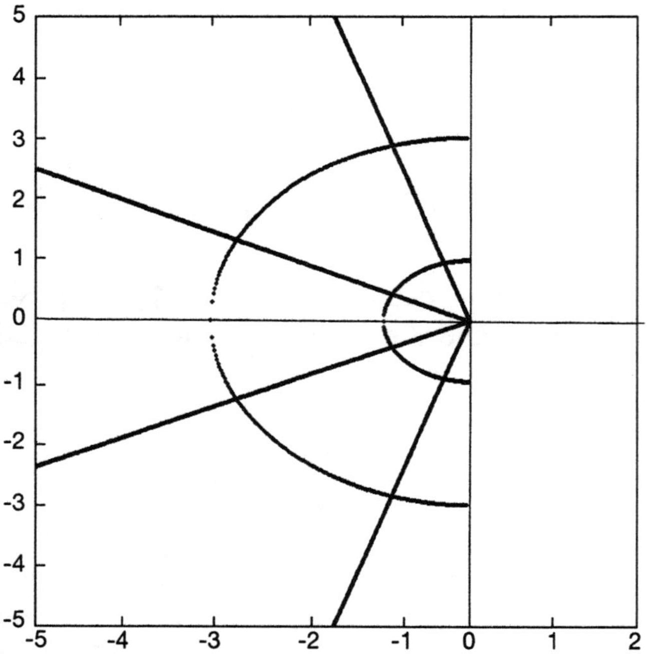

Figure 3. Performance Regions in the Complex Plane

Using inverse Laplace transforms, we can find that:

$$\dot{x}(t) = 1 - e^{-\sigma t}\left(\cos(\omega_d t) + \frac{\xi}{\sqrt{1-\xi^2}}\sin(\omega_d t)\right)$$

where $\omega_d = \omega_n\sqrt{1-\xi^2}$, which is often called the *damped natural frequency,* and $\sigma = \xi\omega_n$, which is equal to the magnitude of the real part of the system poles.

The rate of change of this function is given by:

$$\dot{x}(t) = \sigma e^{-\sigma t}\left(\cos(\omega_d t) + \frac{\xi}{\sqrt{1-\xi^2}}\sin(\omega_d t)\right)$$

$$-\omega_d e^{-\sigma t}\left(-\sin(\omega_d t) + \frac{\xi}{\sqrt{1-\xi^2}}\cos(\omega_d t)\right).$$

Collecting terms:

$$\dot{x}(t) = e^{-\sigma t}\left((-\sigma + \sigma)\cos(\omega_d t) + \left(\frac{\xi^2\omega_n + \omega_d}{\sqrt{1-\xi^2}}\right)\sin(\omega_d t)\right)$$

$$= \left(\frac{\omega_n\left(\xi^2 + \sqrt{1-\xi^2}\right)}{\sqrt{1-\xi^2}}\right)e^{-\sigma t}\sin(\omega_d t).$$

It is noted that the initial value of the system is:

$$x(0) = 1 - e^0\left(\cos(0) - \frac{\xi}{\sqrt{1-\xi^2}}\sin(0)\right) = 1 - 1(1-0) = 1 - 1 = 0.$$

The initial slope of the output is:

$$\dot{x}(0) = 0.$$

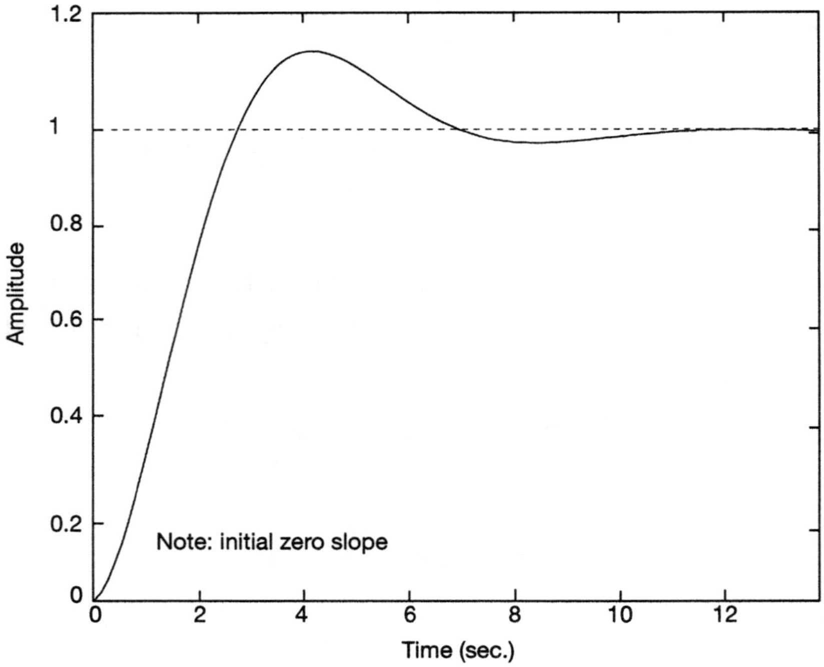

Figure 4. Step Response of a Second Order System

Also, the final value of the output is:

$$\lim_{t \to \infty} x(t) = x_f = 1 - 0 = 1 .$$

Beyond these basic calculated artifacts of the output, the shape of the response of the system to a unit step input is essentially defined by four parameters:

Rise Time—t_r, the first time the system crosses its eventual final value after the application of the step input. This is a measure of the speed of response of the system.

Peak Time—t_p, the time at which the system reaches its highest or peak value after a step input before returning back down and eventually settling at a final value. This is also a measure of the speed of response of a system and is usually only used in conjunction with finding the overshoot in a system.

Maximum Percentage Overshoot—M_p, how far the system goes past its eventual final value at the peak after a step input. This is a measure of how oscillatory the system is.

Settling Time—t_s the time beyond which the system stays within a pre-specified percentage bound of the final value after a step input.

This is a combined measure of the speed of response of the system and its oscillatory nature. It is a good practical measure of how fast the system can reach a new, steady, operating condition after a disturbing input.

The Rise Time (t_r)

This parameter can be defined in several ways, but, for our purposes, is defined as the time after a step input is applied until the system first crosses its eventual final value. Therefore, in order to compute the *rise time*, we need to find the first time at which $x(t) = x_f$. By observing the system output solution, this will occur when the trigonometric parts of the output are zero, leaving only the constant out front, which is precisely the final value. That is at the rise time:

$$\cos(\omega_d t) + \frac{\zeta}{\sqrt{1-\xi^2}} \sin(\omega_d t) = 0.$$

By rearranging this we get:

$$1 + \frac{\zeta}{\sqrt{1-\xi^2}} \tan(\omega_d t) = 0$$

$$\tan(\omega_d t) = \frac{\sqrt{1-\xi^2}}{\xi}$$

$$t = \frac{1}{\omega_d} \tan^{-1}\left(\frac{\sqrt{1-\xi^2}}{\xi} \right)$$

Thus, the rise time can be expressed as:

$$t_r = \frac{1}{\omega_n} \frac{1}{\sqrt{1-\xi^2}} \tan^{-1}\left(\frac{\sqrt{1-\xi^2}}{\xi} \right).$$

This function can be very roughly approximated by:

$$t_r = \frac{1.8}{\omega_n}$$

where the approximation takes into account the possibility of extra poles and zeros which may affect the response of the system. Thus, the approximation is only a conservative measure. However, the approximation is useful in that it makes clear that the rise time is essentially not a function of the damping ratio at all, but rather only a function of the system *natural frequency*.

PROBLEM 9:

Plot the rise time variation with the damping ratio and compare this with the approximation of no variation with the damping ratio.

SOLUTION:

Figure 5 indicates the value β in the function:

$$t_r = \frac{\beta}{\omega_n}.$$

where:

$$\beta = -\frac{1}{\sqrt{1-\xi^2}}\tan^{-1}\left(\frac{\sqrt{1-\xi^2}}{\xi}\right).$$

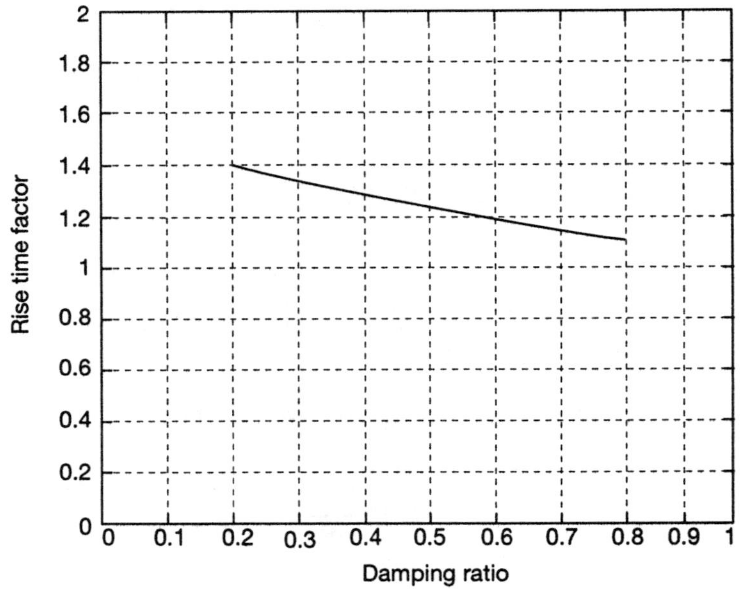

Figure 5. Rise Time Variation with Damping Ratio

It shows that β does not vary significantly over a wide range of damping ratios. Thus, considering this factor as a constant is a reasonable approximation. It is also noted that β appears to vary from 1.4 to 1.1. This is in contrast to the chosen value of 1.8 in the approximation, a difference that can be attributed to other factors such as extra poles and zeros in real systems.

The Peak Time (t_p)

The *peak time* is another measure of the speed of response of a system, much like the rise time. It is most often used in conjunction with a measure of the maximum overshoot of a system. The peak time is defined as the moment when the system reaches its maximum value. This is equivalent to the slope of the system reaching zero for the first time after the initial step is applied. That is, the first time after $t = 0$ when:

$$\dot{x}(t) = \left(\frac{\omega_n\left(\xi^2 + \sqrt{1-\xi^2}\right)}{\sqrt{1-\xi^2}} \right) e^{-\sigma t} \sin(\omega_d t) = 0.$$

This condition will occur when:

$$\sin(\omega_d t) = 0.$$

for the second time. The first time occurs when $t = 0$. Thus, the peak time, t_p, satisfies:

$$\omega_d t_p = \pi$$

or:

$$t_p = \frac{\pi}{\omega_d} = \frac{\pi}{\omega_n\sqrt{1-\xi^2}}.$$

It can be shown that this function has stronger variation with damping ratio than the rise time, but also it is seen that the variation with natural frequency is precisely the same as for rise time.

PROBLEM 10:

Given that, for a system, $\omega_n = 2$ rad/sec and $\xi = 0.6$, find the peak time.

SOLUTION:

$$t_p = \frac{\pi}{\omega_n\sqrt{1-\xi^2}} = \frac{\pi}{2\sqrt{1-0.6^2}} = 1.96 \text{ sec.}$$

The Maximum Percentage Overshoot (M_p)

The *maximum percentage overshoot* is the largest excess, expressed as a percentage, by which the output of the system exceeds its final value. This will occur at precisely the *peak time*. That is, the maximum overshoot will occur when the output is:

$$x(t_p) = -e^{-\sigma t_p}\left(\cos\left(\omega_d t_p\right) + \frac{\xi}{\sqrt{1-\xi^2}}\sin\left(\omega_d t_p\right)\right)$$

and given that:

$$t_p = \frac{\pi}{\omega_d} = \frac{\pi}{\omega_n\sqrt{1-\xi^2}},$$

$$x(t_p) = -e^{-\sigma t_p}\left(\cos\left(\omega_d t_p\right) + \frac{\xi}{\sqrt{1-\xi^2}}\sin\left(\omega_d \frac{\pi}{\omega_d}\right)\right)$$

$$= -e^{-\sigma t_p}(-1+0) = 1 + e^{-\sigma t_p}.$$

Also, noting that the final value of the output is $x_f = 1$, the excess above the final value, divided by the final value, and expressed as a percentage is:

$$M_p = \frac{e^{-\sigma t_p}}{1} \times 100\% = 100 e^{-\xi\omega_n \frac{\pi}{\omega_n\sqrt{1-\xi^2}}} = 100\, e^{\frac{-\pi\xi}{\sqrt{1-\xi^2}}}.$$

It is noted that the maximum percentage overshoot is a function of the damping ratio alone, and not one of natural frequency.

The *inverse function* for the damping ratio from overshoot can be found as:

$$\xi = \sqrt{\frac{\left(\ln\left(\frac{M_p}{100}\right)\right)^2}{\pi^2 + \left(\ln\left(\frac{M_p}{100}\right)\right)^2}}.$$

PROBLEM 11:

What is the damping ratio required for a second order system to have a maximum percentage overshoot of 35%?

SOLUTION:

$$\xi = \sqrt{\frac{\left(\ln\left(\frac{M_p}{100}\right)\right)^2}{\pi^2 + \left(\ln\left(\frac{M_p}{100}\right)\right)^2}} = \sqrt{\frac{(\ln(0.35))^2}{\pi^2 + (\ln(0.35))^2}} = \sqrt{\frac{1.102}{\pi^2 + 1.102}} = 0.32$$

The Settling Time (t_s)

The *settling time* is a measure of the time that it takes before the output of a system enters a region near its final value and no longer leaves that region. From examination of the output equation:

$$x(t) = 1 - e^{-\sigma t}\left(\cos(\omega_d t) + \frac{\xi}{\sqrt{1-\xi^2}}\sin(\omega_d t)\right)$$

it is seen that the trigonometric terms in the bracket will continuously oscillate with a constant amplitude. Therefore, if the system is to settle to some steady state condition, the exponential term must decay to zero. To find the settling time, it is sufficient to find the time at which the exponential term has decayed to smaller than the bound required for the settling region. Let us define the settling region bound as b about the final value, x_f. Thus the settling time occurs at:

$$t = t_s$$

such that

$$e^{-\sigma t} \le b.$$

At the limit of the condition:

$$e^{-\sigma t_s} = b$$
$$-\sigma t_s = \ln(b)$$
$$t_s = \frac{-\ln(b)}{\sigma} = \frac{-\ln(b)}{\xi \omega_n}.$$

Typical values for the bound are one percent, two percent, and five percent depending on designer preferences. Using these conditions, the settling time formula can be explicitly stated as:

Bound, c	Settling time formula, t_s
1 %	$\frac{3}{\xi \omega_n}$
2 %	$\frac{4}{\xi \omega_n}$
5%	$\frac{4.6}{\xi \omega_n}$

PROBLEM 12:

A system of general form is constructed with:

$$X(s) = G(s)\,U(s), \quad G(s) = \frac{1}{\left(\frac{s}{\omega_n}\right)^2 + 2\xi \frac{s}{\omega_n} + 1}$$

and a unit step input or $U(s) = 1/s$. Evaluate the time responses for the general system as the damping ratio and natural frequency are varied.

SOLUTION:

The time responses of four versions of this system are shown in Figure 6. Table 1 summarizes the time domain performance properties of the systems as predicted by the design equations.

System	Natural frequency, ω_n	Damping ratio, ξ	Rise time, t_r	Peak time, t_p	Maximum percentage overshoot, M_p	Settling time, t_s
1	1	$\frac{2}{5}$	1.8	3.4	25	10
2	2	$\frac{2}{5}$	0.9	1.7	25	5
3	1	$\frac{1}{5}$	1.8	3.2	53	20
4	$\frac{5}{3}$	$\frac{2}{3}$	1.1	2.5	6	3.6

Table 1. Performance Measures for Several Second Order Systems

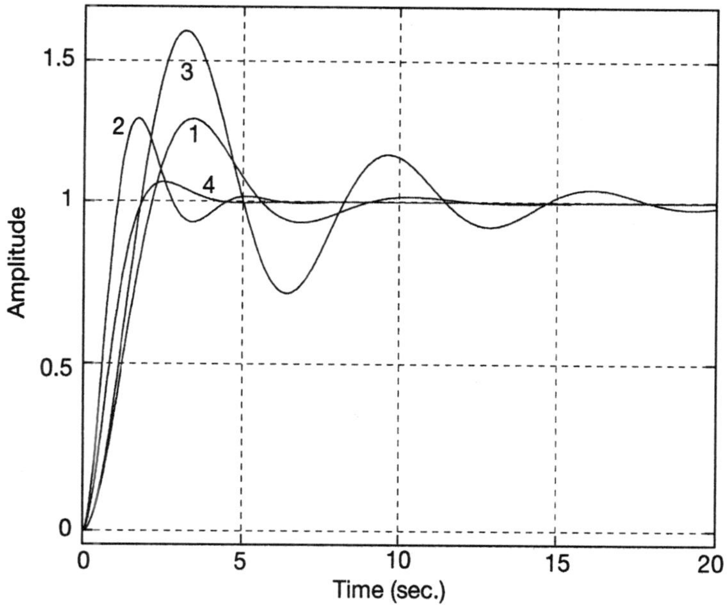

Figure 6. Step Response of Several Second Order Systems

FE/EIT

FE: PM Mechanical Engineering Exam

CHAPTER 5

Energy Conversion and Power Plants

CHAPTER 5

ENERGY CONVERSION AND POWER PLANTS

THE DIESEL CYCLE

The diesel cycle is the ideal cycle for compression-ignition engines. Air is assumed to be the working fluid and the cold air standard is assumed.

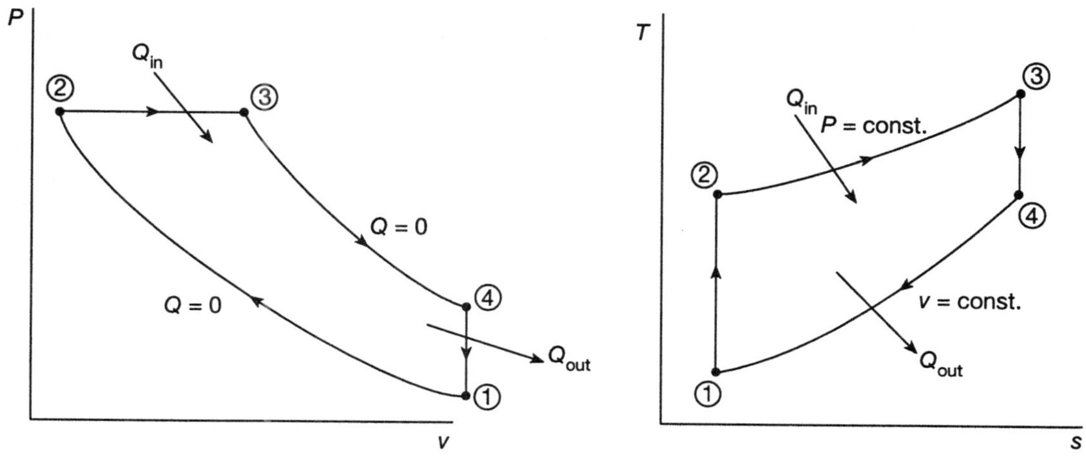

Figure 1. The Diesel Cycle; *P-v* and *T-s* Diagrams

Two important pressure ratios are used to describe the operation of the cycle. The compression ratio r is defined as follows:

$$r = \frac{V_1}{V_2}$$

while the cutoff ratio is defined as:

$$r_c = \frac{V_3}{V_2}$$

The fuel is idealized as being injected into the system at the beginning of the expansion stroke of the cycle. The simultaneous expansion and detonation of fuel maintains a constant pressure until point 3 where the fuel supply is cut off. The efficiency of the diesel cycle is given as:

$$\eta = 1 - \frac{1}{r^{k-1}}\left[\frac{r_c^{\,k} - 1}{k(r_c - 1)}\right]$$

where:

$$k = \frac{c_P}{c_V}$$

PROBLEM 1:

An air-standard diesel cycle with a compression ratio of 22.5 and a cutoff ratio of 3 is operating at a speed of 2,000 rpm. Assuming the heating value of diesel fuel to be 44,200 kJ/kg and the power output of the engine to be 74.6 kW at 2,000 rpm, determine:

(a) the fuel consumption rate needed to maintain the engine, and

(b) the mass flow rate of cooling water for a temperature rise of 10 K of the water.

SOLUTION:

The efficiency of the cycle will be found first knowing that $r = 22.5$ and $r_c = 3$.

(a)

$$\eta = 1 - \frac{1}{r^{k-1}}\left[\frac{r_c^{\,k} - 1}{k(r_c - 1)}\right] = 1 - \frac{1}{22.5^{1.4-1}}\left[\frac{3^{1.4} - 1}{1.4(3-1)}\right] = 0.6243$$

$$\eta = \frac{\dot{W}_{out}}{\dot{Q}_{in}}; \; \dot{Q}_{in} = \frac{\dot{W}_{out}}{\eta} = \frac{(74.6 \text{ kJ/s})}{0.6243} = 119 \text{ kJ/s}$$

$$\dot{m}_f = \frac{\dot{Q}_{in}}{HV} = \frac{(119 \text{ kJ/s})}{(44{,}200 \text{ kJ/kg})} = 0.0027 \text{ kg/s} = 9.73 \text{ kg/h}$$

(b)

$$\dot{Q}_{out} = \dot{Q}_{in} - \dot{W}_{out} = 119 - 74.6 = 44.4 \text{ kJ/s}$$

$$\dot{Q}_{out} = \dot{m}_{cw} c_w \; (\Delta T)_w \, c_w = 418 \text{ kJ/kg–K;}$$

from problem statement, $(\Delta T)_{cw} = 10$ K

$$\dot{m}_{cw} = \frac{\dot{Q}_{out}}{c_{cw} \Delta T_{cw}} = \frac{(44.4 \text{ kJ/s})}{(4.18 \text{ kJ/kg–K})(10K)}$$

$$= 1.06 \text{ kg/s} = 63.7 \text{ kg/min.}$$

BRAKE POWER

The brake horsepower of a reciprocating engine derives its name from the fact that the power output of an engine can be measured by means of a braking arrangement or dynameter as shown in Figure 2.

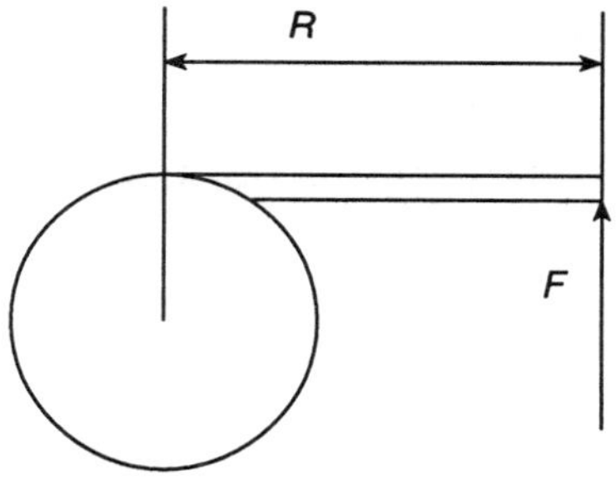

Figure 2. Braking Arrangement for Engine

$$\dot{W}_b = 2\pi TN = 2\pi FRN$$

where:

$\dot{W}_b$ = brake power, W

T = torque, N-m

N = rotational speed, rev/s

F = force at end of brake arm, N

R = length of brake arm, m

The brake power is related to the indicated power and the friction power by the following:

$$\dot{W}_i = \dot{W}_b + \dot{W}_f$$

where:

$\dot{W}_i$ = indicated power

$\dot{W}_f$ = friction power

The brake thermal efficiency and indicated thermal efficiency are defined as:

$$\eta_b = \frac{\dot{W}_b}{\dot{m}_f(HV)} \quad \text{and} \quad \eta_i = \frac{\dot{W}_i}{\dot{m}_f(HV)}$$

where:

η_i = indicated thermal efficiency

η_b = brake thermal efficiency

$\dot{m}_f$ = fuel consumption rate, kg/s

HV = heating value of fuel, kJ/kg

The mechanical efficiency, η_m, is defined as follows:

$$\eta_m = \frac{\dot{W}_b}{\dot{W}_i} = \frac{\eta_b}{\eta_i}$$

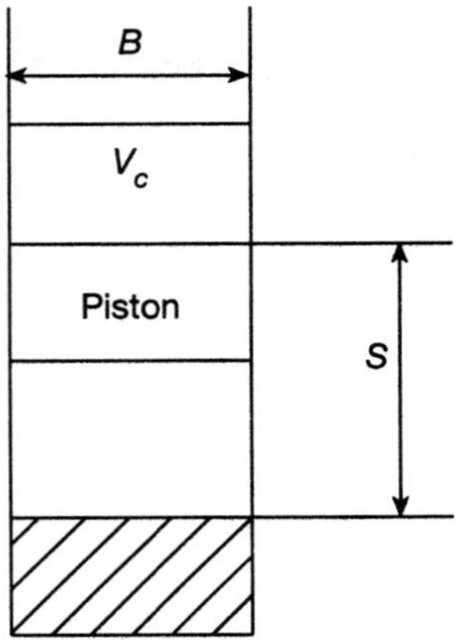

Figure 3: Cylinder Piston Arrangement

The following refers to the cylinder piston arrangement shown in Figure 3:

B = bore or width of cylinder

S = stroke or total distance piston moves

V_c = clearance volume, or volume remaining in cylinder when piston is at tdc (top dead center)

V_d = piston displacement per cylinder and is given by $(\pi/4)B^2S$

The total volume, $V_t = V_d + V_c$. The compression ratio, $r_c = V_t/V_c$. The mean effective pressure, mep, can be defined as follows:

$$\text{mep} = \frac{\dot{W}n_s}{V_d n_c N}$$

where:

n_s = number of crank revolutions per power stroke

n_c = number of cylinders

V_d = displacement volume of cylinder

The *mep* can be based on brake power (*bmep*), indicated power (*imep*), or friction power (*fmep*).

The volumetric efficiency is given by the following relation:

$$\eta_v = \frac{2\dot{m}a}{\rho_a V_d n_c N} \quad \text{(four-stroke cycles only)}$$

where:

$\dot{m}_a$ = mass flow rate of air into engine

p_a = density of air

The specific fuel consumption is defined as follows:

$$\text{sfc} = \frac{\dot{m}_f}{\dot{W}} = \frac{1}{\eta(HV)}$$

Use η_b and $\dot{W}_b$ for the brake specific fuel consumption (*bsfc*) and η_i and $\dot{W}_i$ for the indicated specific fuel consumption (*isfc*).

PROBLEM 2:

An automobile requires 26.1 kW at the engine flywheel to drive on a level road at 96.5 km/hr. The engine is a four-stroke-cycle and has a piston displacement of 5.73 liters. The wheels have a rolling radius of 0.319 m and the rear axle ratio is 3:1. At wide open throttle at 96.5 km/hr, it develops a maximum brake mean effective pressure of 793 kPa, has a mechanical efficiency of 85%, and a brake specific fuel consumption of 0.304 kg/kW-hr. Assume gasoline to be 2.72 kg/gal. Also assume that at constant rpm the friction power and indicated thermal efficiency of the engine are constant regardless of load. Determine:

(a) the brake power and friction power, and

(b) the fuel consumption under road conditions.

SOLUTION:

(a) Distance around wheel = (2π)(0.319 m) = 2.00 m per revolution.

At 96.5 km/hr, in one minute the car will be traveling at 96.5 × 1,000/60 = 1,608 m/min.

Wheel rpm = (1,608 m/min)/(2.00 m) = 804 rpm.

Engine rpm = 3 × wheel rpm = (3)(804) = 2,412 rpm.

The product of V_d and n_c is given in the problem (5.73 liters).

$\dot{W}_b$ is found from *bmep* as follows for full throttle conditions:

$$\dot{W}_b = \frac{\text{bmep}V_d n_c N}{n_s} = \frac{(793\ \text{kN/m}^2)(5.73 \times 10^{-3}\ \text{m}^3)(2{,}412\ \text{rev/min})}{(2\ \text{rev/stroke})(60\ \text{s/min})}$$

$$= 91.3\ \text{kW}$$

$$\dot{W}_i = \frac{\dot{W}_b}{\eta_m} = \frac{(91.3\ \text{kW})}{(0.85)} = 107\ \text{kW}$$

$$\dot{W}_f = \dot{W}_i - \dot{W}_b = 107\ \text{kW} - 91.3\ \text{kW} = 15.7\ \text{kW}$$

(b) Road conditions:

$$\dot{W}_i = \dot{W}_b + \dot{W}_f = 26.1\ \text{kW} + 15.7\ \text{kW} = 41.8\ \text{kW}$$

since friction power is constant. Also *isfc* at wide open throttle = *isfc* at road conditions.

$$isfc = bsfc \times \eta_m = (0.304\ \text{kg/kW-hr}) \times (0.85)$$

$$= 0.258\ \text{kg/ikW-hr}$$

$$\text{kg/hr of fuel} = isfc \times \dot{W}_i = (0.258\ \text{kg/ikW-hr}) \times (41.8\ \text{kW})$$

$$= 10.8\ \text{kg fuel/hr}$$

$$\text{gal/hr} = \frac{10.8\ \text{kg/hr}}{2.72\ \text{kg/gal}} = 3.97\ \text{gal/hr}$$

$$\dot{m}_f = \frac{96.5\ \text{km/hr}}{3.97\ \text{gal/hr}} = 24.3\ \text{km/gal}\ (15\ \text{mpg})$$

THE BRAYTON CYCLE

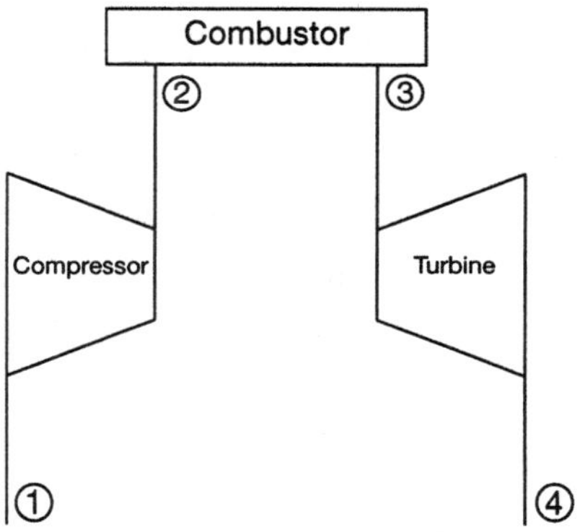

Figure 4. Brayton Cycle Components

The Brayton Cycle is an all-gas system. In the simple model, (Figure 4) air is assumed to be the working substance. Kinetic and potential energy effects are normally neglected. The basic equations for this cycle are as follows:

$$w_{12} = h_1 - h_2 = c_P\,(T_1 - T_2) = w_{\text{comp}}\ \text{(negative)}$$

$$w_{34} = h_3 - h_4 = c_P\,(T_3 - T_4) = w_{\text{turbine}}\ \text{(positive)}$$

$$w_{\text{net}} = w_{12} + w_{34}$$

$$q_{23} = h_3 - h_2 = c_P\,(T_3 - T_2) = q_{\text{comb}}\ \text{(positive)}$$

$$q_{41} = h_1 - h_4 = c_P\,(T_1 - T_4) = q_{\text{rej}}\ \text{(negative)}$$

$$q_{\text{net}} = q_{23} + q_{41}$$

$$\eta = \frac{w_{\text{net}}}{q_{23}}$$

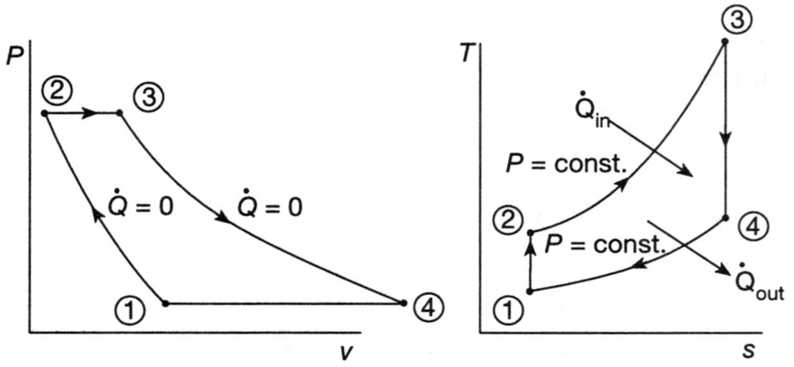

Figure 5. *P-v* and *T-s* Diagrams for Brayton Cycle

The P-v and T-s diagrams for the simple Brayton Cycle are shown in Figure 5. The efficiency of this cycle is affected by the large compressor work that is required. The compressor work can be reduced by using multiple-stage compressors together with intercooling. Note that the outlet temperature of the turbine (4) is higher than the inlet temperature of the combustor. This allows the use of regeneration where some of the wasted turbine energy is used to preheat the air entering the combustion chamber. The regenerative Brayton Cycle is pictured in Figure 6. If we perform a first law energy balance on the regenerator, which is merely a heat exchanger, then:

$$h_3 - h_2 = h_5 - h_6 \quad \text{or} \quad T_3 - T_2 = T_5 - T_6$$

Since the air into the combustor is at a higher temperature, the amount of energy input is reduced.

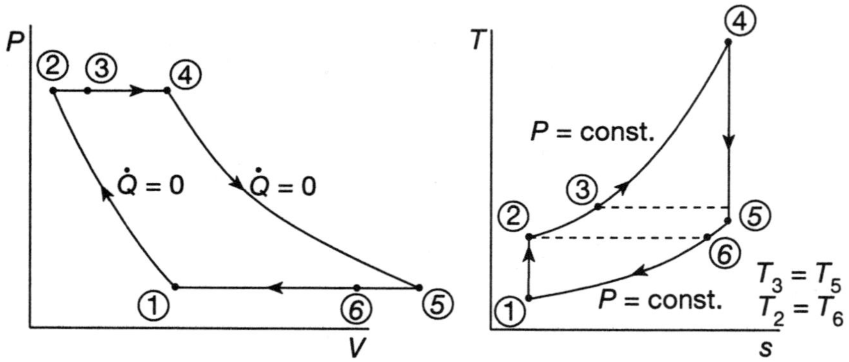

Figure 6. Regenerative Brayton Cycle

$$q_{34} = h_4 - h_3 = c_P\,(T_4 - T_3)$$

$$\eta = \frac{w_{\text{net}}}{q_{34}}$$

The amount of heat transfer to the air before entering the combustion chamber and the final value of T_3 (the greater T_3 the less energy is required in combustor) is dependent on the efficiency of the regenerator, η_{reg}.

$$\eta_{reg} = \frac{h_3 - h_2}{h_5 - h_2} = \frac{T_3 - T_2}{T_5 - T_2}$$

$$h_3 = h_2 + \eta_{reg}\,(h_5 - h_2) \text{ or}$$

$$T_3 = T_2 + \eta_{reg}\,(T_5 - T_2)$$

$\eta_{reg} = 0$ means that no regenerator is present in which case $T_3 = T_2$. $\eta_{reg} = 1$ means that $T_3 = T_5$. This is the maximum value allowed by the second law for T_3.

RANKINE CYCLE

Here we shall investigate modifications to the simple Rankine cycle, in particular the regenerative Rankine cycle. In the regenerative cycle, multiple turbines are used. In addition, part of the steam between turbine stages is bled-off and diverted to the feedwater heater. The purpose of the

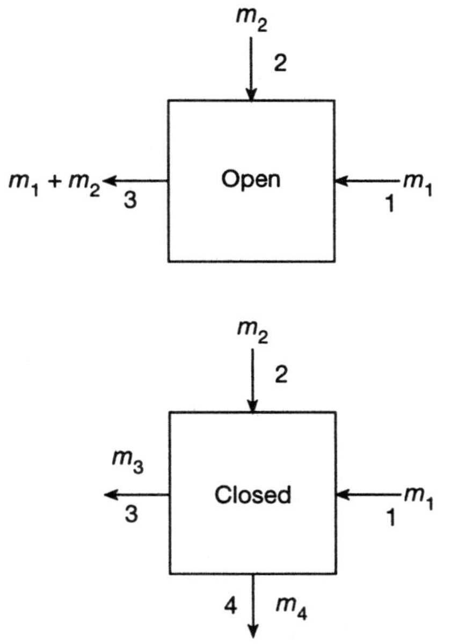

Figure 7. Open and Closed Feedwater Heaters

feedwater heater is to lessen the amount of heat input required by the boiler by increasing the temperature of the water prior to entry into the boiler. In an *open feedwater heater*, the steam from between the turbine stages are mixed with the high pressure water from the pump in a mixing chamber. This requires a second pump to raise the pressure to the required boiler pressure. In a *closed feedwater heater*, the condensed steam is routed to a trap that allows the liquid to be throttled to a lower pressure region but traps the vapor. Another possibility is using a pump and mixing chamber to route the condensate to the feedwater line.

The mass and energy equations for the open feedwater heater are as follows:

$$\text{mass} = \dot{m}_1 + \dot{m}_2 = \dot{m}_3$$

$$\text{energy} = \dot{m}_1 h_1 + \dot{m}_2 h_2 = \dot{m}_3 h_3 = (\dot{m}_1 + \dot{m}_2) h_3$$

The mass and energy equations for the closed feedwater heater are as follows:

$$\text{mass} = \dot{m}_1 + \dot{m}_3 \text{ and } \dot{m}_2 = \dot{m}_4$$

$$\text{energy} = \dot{m}_1 h_1 + \dot{m}_2 h_2 = \dot{m}_3 h_3 + \dot{m}_4 h_3$$

The following equations apply to the pump of the Rankine Cycle:

$$w = h_1 - h_2 = \frac{h_1 - h_{2s}}{\eta_P}$$

$$h_{2s} - h_1 = v(P_2 - P_1)$$

$$w = \frac{v(P_2 - P_1)}{\eta_P}$$

PROBLEM 3:

Consider a regenerative Rankine Cycle with one feedwater heater as shown in Figure 8. Steam enters the turbine at 8 MPa, 480°C, and expands to 0.7 MPa where some of the steam is extracted and diverted to an open feedwater heater. The remaining steam is allowed to expand through a second turbine stage to the condenser pressure of 0.008 MPa. The fluid exiting the feedwater heater is assumed to be saturated vapor at 0.7 MPa. The isentropic efficiency of each turbine stage is 85%. If the net power output of the cycle is 10 MW, determine:

(a) h_2 and h_3, the enthalpies leaving each turbine stage,

(b) the fraction of steam extracted and sent to the feedwater heater,

(c) the thermal efficiency of the cycle, and

(d) the mass flow rate of steam entering the first turbine stage.

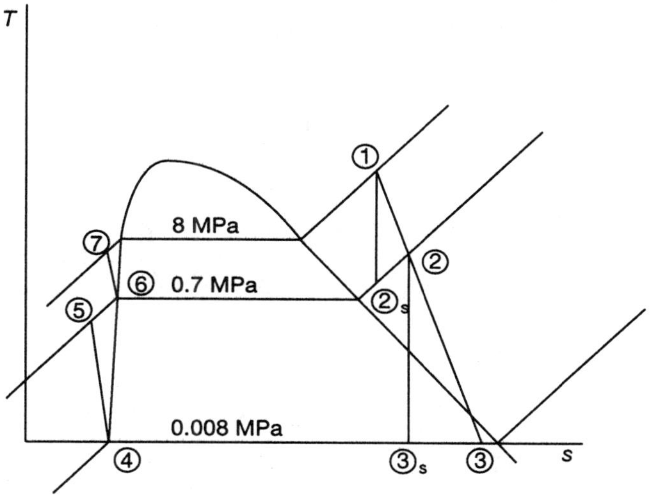

Figure 8. *T-s* Diagram for Problem 3

The following properties have been obtained from the steam table:

$h_1 = 3{,}348.4$ kJ/kg $\quad h_{2s} = 2{,}741.8$ kJ/kg

$h_{3s} = 2{,}146.3$ kJ/kg $\quad h_4 = 173.88$ kJ/kg

SOLUTION:

(a) $$\eta_T = \frac{h_1 - h_2}{h_1 - h_{2s}} = \frac{3{,}348.4 - h_2}{3{,}348.4 - 2{,}741.8} = 0.85$$

$h_2 = 2{,}833$ kJ/kg

$$\eta_T = \frac{h_2 - h_3}{h_2 - h_{3s}} = \frac{2{,}833 - h_3}{2{,}833 - 2{,}146.3} = 0.85$$

(b) At state 4 on the diagram, the saturated liquid is at 8 kPa and

$h_4 = 173.88$ kJ/kg; $v_4 = 1.0039 \times 10^{-3}$ m^3/kg

Likewise for state 6, the saturated liquid is at 0.7 MPa and

$h_6 = 697.22$ kJ/kg; $v_6 = 0.001108$ m^3/kg

$w_p = v_4\,(P_5 - P_4) = 1.0039 \times 10^{-3}$ m^3/kg[(700 – 8) kPa]

$$w_{p1} = 0.695 \text{ kJ/kg}$$

$$h_5 = h_4 + w_P = 173.88 \text{ kJ/kg} + 0.695 \text{ kJ/kg} = 174.6 \text{ kJ/kg}$$

For each 1 kg of steam leaving the boiler, y kg expands partially in the turbine and is extracted at state 2. Thus, $(1 - y)$ kg are expanded through the second turbine stage, pass through the condenser and pump 1, and then finally into the feedwater heater. The energy balance on the feedwater heater becomes:

$$yh_2 + (1 - y)h_5 = h_6$$

Solving for y:

$$y = \frac{h_6 - h_5}{h_2 - h_5} = \frac{697.22 - 174.6}{2{,}883 - 174.6} = 0.197.$$

(c) $h_7 = h_6 + w_{P2}$

$$w_{P2} = v_6(P_7 - P_6) = (0.001108 \text{ m}^3\text{/kg})[(8{,}000 - 700)\text{kPa}]$$

$$= 8.088 \text{ kJ/kg}$$

$$h_7 = 697.22 \text{ kJ/kg} + 8.088 \text{ kJ/kg} = 705.3 \text{ kJ/kg}$$

$$\eta_{th} = 1 - \frac{q_{out}}{q_{in}} = \frac{(1-y)(h_3 - h_4)}{(h_1 - h_7)} = 1 - \frac{(1 - 0.197)(2{,}249 - 173.88)}{(3{,}338.4 - 705.3)}$$

$$= 0.369 \text{ or } 36.9\%$$

(d)

$$w_{net} = q_{in} - q_{out} = 2{,}643.1 - 1{,}667 = 976 \text{ kJ/kg}$$

$$\dot{W} = \dot{m}_1 w_{net}$$

$$\dot{m}_1 = \frac{\dot{W}}{w_{net}} = \frac{(10 \text{ MW})(1000 \text{ MW/kW})}{976 \text{ kJ/kg}} = 10.2 \text{ kg/s}$$

PROBLEM 4:

An ideal Brayton-Cycle gas turbine shown in Figure 9 with regeneration, but no reheat or intercooling, has an inlet air temperature of 17°C and an inlet turbine temperature of 827°C. The compressor ratio is 6.0 and the regenerator is 80% efficient. The output of the system is to be 10 MW. The heating value of the fuel is 44,000 kJ/kg. The combustion process is 80% efficient. Using the cold-air assumption for ideal gases, determine:

(a) the work input to compressor,

(b) the work output from turbine,

(c) the cycle efficiency,

(d) the mass flow rate of air required, and

(e) the rate of fuel consumption.

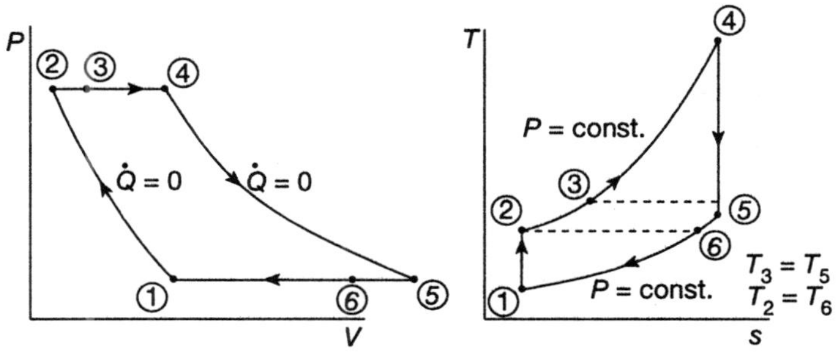

Figure 9. *P-v* and *T-s* Diagram for Problem 4

SOLUTION:

$$T_1 = 17°\text{C} = 290\text{ K};\ T_4 = 827°\text{C} = 1{,}100\text{ K}$$

$$P_2/P_1 = 6;\ = P_4/P_5;\ P_5/P_4 = 1/6$$

Assume processes 1-2 and 4-5 to be isentropic and that $k = 1.4$ and $c_P = 1.00$ kJ/kg-K, thus:

$$\frac{T_2}{T_1} = \left(\frac{P_2}{P_1}\right)^{\frac{k-1}{k}}$$

$$T_2 = (290\text{K})(6)^{0.286} = 484\text{K}$$

$$\frac{T_4}{T_5} = \left(\frac{P_4}{P_5}\right)^{\frac{k-1}{k}}$$

$$\text{T}_5 = T_4\left(\frac{P_5}{P_4}\right)^{\frac{k-1}{k}} = (1{,}100\text{ K})(1/6)^{0.286} = 659\text{ K}$$

$$\eta_{reg} = \frac{T_3 - T_2}{T_5 - T_2} = \frac{T_3 - 484}{659 - 484} = 0.80$$

$$T_3 = 484 + (0.80)(659 - 484) = 624 \text{ K}$$

With the above information, the required calculations can be made:

(a) $w_{comp} = c_P\,(T_2 - T_1) = (1.0 \text{ kJ/kg-K})[(484 - 290)\text{K}] = 194 \text{ kJ/kg}$

(b) $w_{turb} = c_P\,(T_4 - T_5) = (1.0 \text{ kJ/kg-K})[(1{,}100 - 659)\text{K}] = 441 \text{ kJ/kg}$

(c) $q_{in} = c_P\,(T_4 - T_3) = (1.0 \text{ kJ/kg-K})[(1{,}100 - 624)\text{K}] = 476 \text{ kJ/kg}$

$$w_{net} = w_{turb} - w_{comp} = 441 - 194 = 247 \text{ kJ/kg}$$

$$\eta = w_{net}/q_{in} = (247 \text{ kJ/kg})/(476 \text{ kJ/kg}) = 0.519 \text{ or } 51.9\ \%$$

(d) $$\dot{m} = \frac{\dot{W}_{net}}{w_{net}} = \frac{(10 \times 10^3 \text{ kg/s})}{(247 \text{ kJ/kg})} = 40.5 \text{ kg/s}$$

(e) $$\dot{Q}_{in} = \dot{m}_{air}\,\dot{q}_{\text{in}} = (40.5 \text{ kg/s})(476 \text{ kJ/kg}) = 19{,}300 \text{ kJ/s}$$

$$\dot{m}_{fuel,\ ideal} = \frac{\dot{Q}_{in}}{HV} = \frac{(19{,}300 \text{ kg/s})}{44{,}000 \text{ kJ/kg}} = 0.438 \text{ kg/s}$$

$$\dot{m}_{fuel,\ actual} = \frac{\dot{m}_{fuel,ideal}}{\eta_{comb}} = \frac{0.438}{0.80} = 0.547 \text{ kg/s}$$

FE/EIT

FE: PM Mechanical Engineering Exam

CHAPTER 6

Fans, Pumps, and Compressors

CHAPTER 6

FANS, PUMPS, AND COMPRESSORS

Fans are available in a wide range of sizes and are used for many purposes. Fans are generally classified by their blade shape and their use. Uses include removing contaminants (ventilation) and supplying air. Ventilation includes general ventilation where large volumes of air are used to remove contaminants, such as fumes or smoke.

When selecting a fan, consider the fan's environment and the volume of air to be moved. Size and performance must be specified based on the velocity to be maintained through the fan, the blade tip speed, and the power consumed by the drive motor.

FAN PERFORMANCE

The following terms are used to describe the pressure a fan operates against: *velocity pressure*, *static pressure*, and *total pressure*.

Velocity pressure is developed by the airstream as a result of airflow. If you hold your hand in an air duct (away from moving fan blades), you will feel velocity pressure. At standard air conditions (59°F and 29.921 in. Hg), the equation for velocity pressure is:

$$P_v = \rho_a \left(\frac{V}{1{,}097} \right)^2 \qquad \text{for non-standard air pressure}$$

$$P_v = \left(\frac{V}{4{,}005} \right)^2 \qquad \text{for standard air pressure}$$

where:

$V = \text{ft/min}$

$P_v = \text{inches of water gauge (Wg)}$

$1 \text{ in. Wg.} = 0.0365 \text{ psi}$

For SI units the velocity pressure is given by the following:

$$P_v = \rho_a\left(\frac{V^2}{2}\right) \quad [P_v] = \text{Pa};$$

where:

$V = \text{m/s}^2$

$\rho_a = \text{kg/m}^2$

Note: $\rho_a = 1.2 \text{ kg/m}^3$ for standard air.

Static pressure is exerted by a fluid against the walls of the container holding the fluid. In a fan system, this is the pressure felt by the ductwork. Sufficient static pressure must be maintained in a duct to push the air out of the diffusers. Because static pressure is the difference between total pressure and velocity pressure, static pressure is increased in a duct by reducing airspeed (e.g., enlarging the duct size).

Total pressure is the sum of velocity pressure and static pressure. Total pressure is a true measure of airflow energy. Velocity and static pressures can be interchanged by varying air speed in the duct. Total pressure is a measure of the duct losses resulting from friction, turbulence, or other factors.

TYPES OF FANS

Fans fall into two general categories—axial flow, where the air enters and leaves the fan with no change in direction (propeller, tube-axial, vane-axial), and centrifugal flow, where the flow changes direction twice—once when entering and once when leaving (forward curved, backward curved or inclined, and radial).

Propeller fans usually run at low speeds (under 12,000 feet per minute tip speed) and moderate temperatures (under 100°F). They experience a large change in airflow with small changes in static pressure. Propeller fans are often used indoors as exhaust fans. Outdoor applications include air-cooled condensers and cooling towers.

Tube axial fans operate in a higher pressure range and are not sensitive to small changes in static pressure. Fan blade tip can reach speeds of 15,000 feet per minute and temperatures of 150°F.

Vane axial fans are similar to tube axials, but have stator vanes installed downstream of the fan blades to convert the swirl of the air into useful static pressure. As a result, they have a higher static pressure with less dependence on the duct static pressure. Vane axials are typically the most energy-efficient fans available and should be used whenever possible.

Forward-curved fans are used in clean environments and operate at lower temperatures. They are well suited for low tip speed and high-airflow work.

Backward-inclined fans are more efficient than forward-curved fans. Backward-inclined fans reach their peak power consumption and then power demand drops off well within their usable airflow range. Backward-inclined fans are known as "non-overloading" because changes in static pressure do not overload the motor.

Radial fans are an industrial workhorse because of their high static pressures and ability to handle heavily contaminated airstreams. Because of their simple design, radial fans are well suited for high temperatures and medium blade tip speeds.

FAN EFFICIENCY

Fan efficiency is a comparison of the shaft horsepower and air horsepower. Static efficiency is calculated using static pressure, and total efficiency is calculated using total pressure. Total efficiency is preferred because static efficiency may be changed significantly by simply changing the duct size and, thereby, the static pressure. The equation for fan efficiency is:

$$\eta_s = \text{static efficiency} = \frac{\dot{Q}(P_1 - P_2)}{6{,}356\dot{W}}$$

$$\eta_t = \text{total efficiency} = \frac{\dot{Q}(P_{t1} - P_{t2})}{6{,}356\dot{W}}$$

where:

= cfm (ft^3/min)

P = in. Wg.

= hp (horsepower) = shaft power input

For SI systems

η_s = static efficiency =

η_t = total efficiency =

where:

= m^3/sec

P = Pa (N/m^2)

$\dot{W}$ = watts

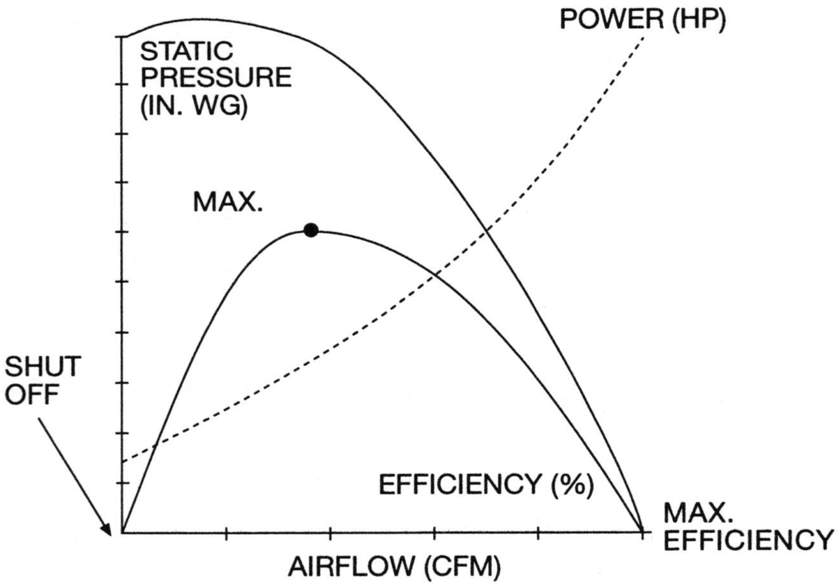

Figure 1. Fan Operating Characteristics

The operating characteristics of a typical fan running at a fixed speed is shown in Figure 1.

Terms regularly used with fan airflow include *shutoff*, *free delivery*, and *system effect*. At *shutoff*, the fan is running but airflow is reduced to zero by blocking the duct. This is represented by the left side of Figure 1. At *free delivery*, the outlet resistance is zero and the fan flow is at maximum. This is represented by the right side of Figure 1. Both conditions correspond to zero efficiency. A fan should be selected as close as possible to the point of maximum efficiency.

Fans are typically tested and rated in a laboratory under ideal conditions with uniform airflow. *System effect* is an estimate of the effect of field-installed conditions on fan performance. Obstructions include restrictions in the fan inlet or outlet. Airflow changes may be caused by duct fittings or an elbow. System effect factors are published for duct fittings, coils, dampers, and other elements.

	1" SP		2" SP	
CFM	RPM	BHP	RPM	BHP
2,900	751	0.58	962	1.14
3,190	780	0.65	984	1.24
3,480	811	0.73	1,008	1.35
3,770	844	0.82	1,032	1.48
4,060	878	0.92	1,058	1.61
4,350	914	1.03	1,086	1.75
4,640	951	1.14	1,116	1.91
4,930	989	1.27	1,148	2.08
5,220	1027	1.41	1,181	2.25

Table 1. Typical "Multi-Rating" Table

Fan performance is usually shown in the manufacturer's literature in "multi-rating" tables (as shown in Table 1). These tables show the fan speed (rpm) and power (horsepower) required for a desired airflow (cfm) at a given pressure rise.

The performance of a fan is altered by the speed at which it rotates. Since many fans are coupled to a drive motor via a belt and pulley system, changing the diameter of one or both pulleys will change the fan speed.

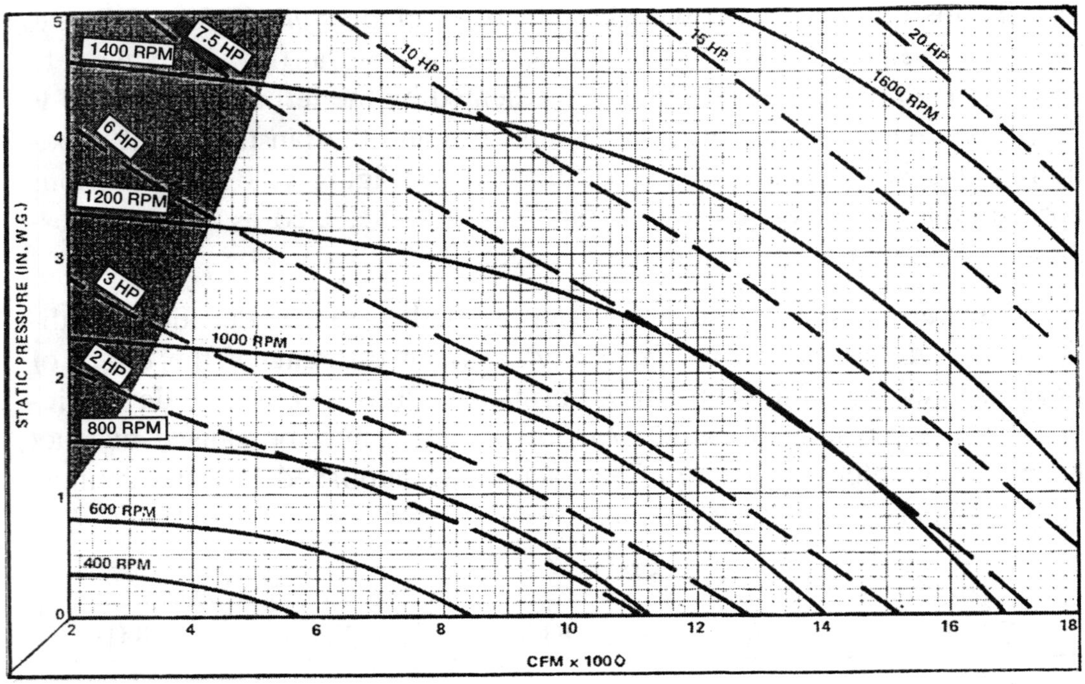

Figure 2. Fan Characteristic of 24-inch Backward-Fan

Pictured in Figure 2 are the characteristics of a 24-inch backward-curved supply air fan with an exit area of 3.14 ft^2. The y-axis is the static pressure (in inches of water gauge), the x-axis is the flow rate (in thousands of cubic feet per minute), the solid lines correspond to various rpm's of the fan, and the dashed lines correspond to the shaft hp. From the air horsepower and shaft horsepower the efficiency of the fan can be determined.

PROBLEM 1:

Referring to the "multi-rating" table, what would be the static efficiency of the fan if it produced a static pressure of 1 in. Wg. at a flow rate of 3,770 cfm?

SOLUTION:

For the stated conditions, the shaft power, $\dot{W}$, equals 0.82 hp.

$$\eta_s = \frac{\dot{Q}(P_1 - P_2)}{6,356\dot{W}} = \frac{(3,770)(1)}{(6,356)(0.82)} = 0.72 \text{ or } 72\%$$

PROBLEM 2:

Using the same conditions as in Problem 1, determine the exit velocity pressure P_{v2} if the exit area is 3.0 ft^2.

$$\dot{Q} = AV; \quad V = \frac{\dot{Q}}{A} = \frac{3,770}{3.0} = 1,260 \text{ ft/min}$$

SOLUTION:

$$P_{v2} = \left(\frac{V}{4,005}\right)^2 =, \left(\frac{1,260}{4005}\right)^2 = 0.0984 \text{ in.Wg.}$$

PROBLEM 3:

Referring to the fan characteristic chart in Figure 2, assume the fan is rotating at a speed of 1,000 rpm's and the volume flow rate is 10,000 cfm. Determine the static efficiency of the fan under these conditions.

SOLUTION:

Locating the intersection of 10,000 cfm and 1000 rpm, the fan's static pressure is 1.5 in. Wg. By interpolating the data on the chart, the shaft work, $\dot{W}$, equals 4.4 hp.

$$\eta_s = \frac{\dot{Q}(P_1 - P_2)}{6,356\dot{W}} = \frac{(10,000)(1.5)}{(6,356)(4.4)} = 0.536 \text{ or } 54\%$$

THE FAN LAWS

The fan laws can be deduced from the following similitude equations.

$$\left(\frac{Q}{ND^3}\right)_2 = \left(\frac{Q}{ND^3}\right)_1$$

$$\left(\frac{\dot{m}}{\rho ND^3}\right)_2 = \left(\frac{\dot{m}}{\rho ND^3}\right)_1$$

$$\left(\frac{H}{N^2D^2}\right)_2 = \left(\frac{H}{N^2D^2}\right)_1$$

$$\left(\frac{P}{\rho N^2 D^2}\right)_2 = \left(\frac{P}{\rho N^2 D^2}\right)_1$$

$$\left(\frac{\dot{W}}{\rho N^3 D^5}\right)_2 = \left(\frac{\dot{W}}{\rho N^3 D^5}\right)_1$$

where:

Q = volumetric flow rate

$\dot{m}$ = mass flow rate

H = head

P = pressure rise

$\dot{W}$ = power

ρ = fluid density

N = rotational speed

D = impeller diameter

Subscripts 1 and 2 refer to geometrically similar fans or to the same fan operating under different conditions. The first set, and perhaps most important set of the fan laws, involves the variation of Q, P, and $\dot{W}$ (the dependent variables), with rotational speed (N), impeller diameter (D), and density (r) (the independent variables). Solving the first equation for the Q ratio, the fourth equation for the P ratio, and the fifth equation for the $\dot{W}$ ratio results in the following:

$$\frac{Q_2}{Q_1} = \left(\frac{N_2}{N_1}\right)\left(\frac{D_2}{D_1}\right)^3 \tag{1a}$$

$$\frac{P_2}{P_1} = \left(\frac{\rho_2}{\rho_1}\right)\left(\frac{N_2}{N_1}\right)^2\left(\frac{D_2}{D_1}\right)^2 \tag{1b}$$

$$\frac{\dot{W}_2}{\dot{W}_1} = \left(\frac{\rho_2}{\rho_1}\right)\left(\frac{N_2}{N_1}\right)^3\left(\frac{D_2}{D_1}\right)^5 \tag{1c}$$

The remaining fan laws are determined by changing the dependent and independent variables. In the second set of laws, Q, N, and $\dot{W}$ are now the dependent variables and D, P, and r are the independent variables. The equations are derived by replacing the N_2/N_1 of Equation (1a) with N_2/N_1 from Equation (1b). Following this procedure, the second set of fan laws are derived:

$$\frac{Q_2}{Q_1} = \left(\frac{\rho_1}{\rho_2}\right)^{1/2}\left(\frac{P_2}{P_1}\right)^{1/2}\left(\frac{D_2}{D_1}\right)^{2} \tag{2a}$$

$$\frac{N_2}{N_1} = \left(\frac{\rho_1}{\rho_2}\right)^{1/2}\left(\frac{P_2}{P_1}\right)^{1/2}\left(\frac{D_2}{D_1}\right)^{-1} \tag{2b}$$

$$\frac{\dot{W}_2}{\dot{W}_1} = \left(\frac{\rho_1}{\rho_2}\right)^{1/2}\left(\frac{P_2}{P_1}\right)^{3/2}\left(\frac{D_2}{D_1}\right)^{2} \tag{2c}$$

In the third set of equations, N, P, and $\dot{W}$ are the dependent variables and D and Q are the independent variables. Following the procedure from above, the following equations are derived:

$$\frac{N_2}{N_1} = \left(\frac{Q_2}{Q_1}\right)\left(\frac{D_2}{D_1}\right)^{-3} \tag{3a}$$

$$\frac{P_2}{P_1} = \left(\frac{\rho_2}{\rho_1}\right)\left(\frac{Q_2}{Q_1}\right)^{2}\left(\frac{D_2}{D_1}\right)^{-4} \tag{3b}$$

$$\frac{\dot{W}_2}{\dot{W}_1} = \left(\frac{\rho_2}{\rho_1}\right)\left(\frac{Q_2}{Q_1}\right)^{3}\left(\frac{D_2}{D_1}\right)^{-4} \tag{3c}$$

PROBLEM 4:

Using the data from Problem 3, determine the new fan characteristics if the fan speed is increased from 1,000 rpm's to 1,200 rpm's. Compare these results with the data from the chart. For the new operating conditions, determine the static efficiency.

SOLUTION:

The required independent variable in this case is rotational speed N, which means that the first set of fan equations apply. The initial data is as follows:

$\dot{Q}_1 = 10{,}000$ cfm $\qquad P_{s1} = 1.5$ in. Wg. $\qquad \dot{W}_1 = 4.4$ hp

$$\frac{\dot{Q}_2}{\dot{Q}_1} = \frac{N_2}{N_1}$$

$$\dot{Q}_2 = \frac{(1,200)}{,(1000)}(10,000) = 12,000 \text{ cfm}$$

$$\frac{P_{s2}}{P_{s1}} = \left(\frac{N_2}{N_1}\right)^2$$

$$P_{s2} = 0.15\left(\frac{1,200}{1,000}\right)^2 = 2.16 \text{ in. Wg}$$

$$\frac{\dot{W}_2}{\dot{W}_1} = \left(\frac{N_2}{N_1}\right)^3$$

$$\dot{W}_2 = (4.4)\left(\frac{1,200}{1,000}\right)^3 = 7.60 \text{ hp}$$

These values match the values obtained from the chart at 12,000 cfm and 1,200 rpm. Solving for the new value of static efficiency:

$$\eta_s = \frac{\dot{Q}(P_1 - P_2)}{6,356\dot{W}} = \frac{(12,000)(2.16)}{(6,356)(7.60)} = 0.536 \text{ or } 54\%$$

The fan efficiency is the same as in the previous problem. This demonstrates an important characteristic of the fan selection process. Once a fan has been selected for an application, the efficiency will never change. The only way to get a better operating efficiency is to select a different fan.

SYSTEM CHARACTERISTICS

Combining both the system and fan characteristics on one plot is very useful in matching a fan to a system. This will ensure that the fan operation is at the desired condition. Once the head loss has been determined for a given flow rate, the head loss for any other flow rate can be calculated approximately by the following:

$$\frac{\Delta P_{s2}}{\Delta P_{s1}} = \left(\frac{\dot{Q}_2}{\dot{Q}_1}\right)^2$$

The system characteristic is a parabola which begins at the origin. The operating point of a fan and system will always be the intersection of the system characteristic and fan characteristic. If the intersection is at a lower than required volume flow rate, the fan speed must be increased. If the intersection is at a higher than required volume flow rate, the fan speed must be lowered or the system resistance must be increased. The former is preferred to the latter since increasing resistance will waste power.

PROBLEM 5:

It is desired to select a fan for a system that requires a volume flow rate of 10,000 cfm and a static pressure drop of 2.0 in. Wg. Assume that the fan in Figure 2 is to be used and the fan is initially set to rotate at 1,200 rpm. What will be the operating conditions of the system under these conditions? What modifications should be made to allow the system to operate properly?

SOLUTION:

Inspection of the fan graph shows the static pressure to be 2.65 in. Wg at 10,000 cfm. Since the static pressure is higher than required, the system will flow at a higher volume flow rate. To determine the actual flow rate, the system characteristic must be determined and plotted on the fan characteristic curve. The intersection of this curve and the 1200 rpm curve will determine the actual operating point. Try using 11,000 cfm and substituting into the equation:

$$\frac{\Delta P_{s2}}{\Delta P_{s1}} = \left(\frac{\dot{Q}_2}{\dot{Q}_1}\right)^2$$

$$\Delta P_{s2} = (2.0)\left(\frac{11,000}{10,000}\right)^2 = 2.42 \text{ in. Wg}$$

This meets the 12200 rpm curve. From the graph, the shaft power is read as 7.5 hp.

$$\eta_s = \frac{\dot{Q}(P_1 - P_2)}{6,356\dot{W}} = \frac{(11,000)(2.42)}{(6,356)(7.50)} = 0.558 \text{ or } 56\%$$

To properly flow the system, the speed of the fan must be reduced. By interpolating between the 1,000 rpm and 1,200 rpm curves at the desired flow rate of 10,000 cfm, the desired fan speed is found to be 1,080 rpm. The shaft power is likewise found by interpolation and found to be 6 hp.

Final conditions:

$$N = 1{,}080 \text{ rpm}$$

$$\dot{W} = 6 \text{ hp}$$

$$\Delta P_s = 2.0 \text{ in. Wg}$$

It is easy to verify that under these operating conditions the static efficiency is 53%. The system and fan characteristics are shown in Figure 2.

PUMPS

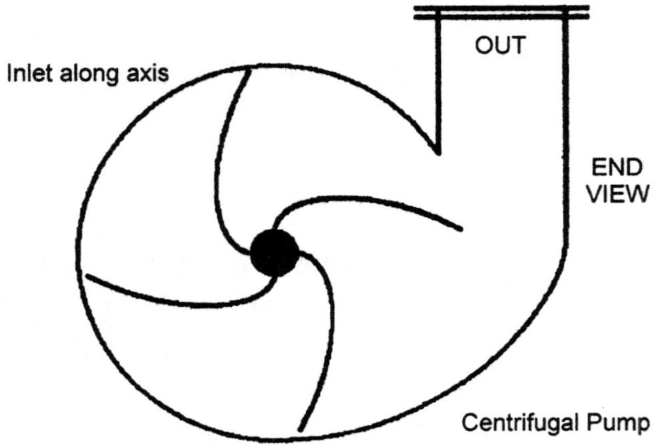

Figure 3. Centrifugal Pump

A centrifugal pump is a device that converts driver energy to kinetic energy in a liquid by accelerating it to the outer rim of a revolving device known as an impeller. The important concept here is that the energy created is kinetic energy. The amount of energy given to the liquid corresponds to the velocity at the edge or vane tip of the impeller. The faster the impeller revolves or the bigger the impeller is, then the higher will be the velocity of the liquid at the vane tip and the greater the energy imparted to the liquid.

The kinetic energy of a liquid coming out of an impeller is harnessed by creating a resistance to the flow. The first resistance is created by the pump volute (casing), which catches the liquid and slows it down. When the liquid slows down in the pump casing, some of the kinetic energy is converted to pressure energy. It is the resistance to the pump's flow that is read on a pressure gauge attached to the discharge line.

Head

In Newtonian fluids (non-viscous liquids like water or gasoline) the term *head* is used to measure the kinetic energy that a pump creates. Head is a measurement of the height of a liquid column that the pump would develop resulting from the kinetic energy the pump gives to the liquid. The main reason for using head instead of pressure to measure a centrifugal pump's energy is that the pressure from a pump will change if the specific gravity (weight) of the liquid changes, but the head will not change. A pump's performance on any Newtonian fluid, whether it's heavy (sulfuric acid) or light (gasoline), can be described by using the term *head.*

To convert head to pressure, the following formula applies:

$$H_P(\text{ft}) = P(\text{psi}) \times 2.31/\text{S.G.}$$

$$H_P(\text{m}) = P(\text{Pa})/(\text{SG} \times 9800)$$

PROBLEM 6:

The outlet pressure of a pump that is pumping water is 32.5 psi. Determine the head produced by the pump.

SOLUTION:

Using the equation:

$$H_p = \frac{(32.5)(2.31)}{1.00} = 100 \text{ ft}$$

PROBLEM 7:

A pump is rated at an output head of 10 m. Determine the outlet pressure of the pump if the working fluid is gasoline (S.G. = 0.75).

SOLUTION:

Solving for P:

$$P(\text{Pa}) = (0.75)(9{,}800)(10) = 73{,}500 \text{ Pa} = 73.5 \text{ kPa}$$

Newtonian liquids have specific gravities typically ranging from 0.5 (e.g., light hydrocarbons) to 1.8 (heavy, like concentrated sulfuric acid). Water is a benchmark, having a specific gravity of 1.0.

The two main factors in determining how much head a pump creates are the impeller diameter, and the rpm of the impeller.

Impeller Diameter

If the speed (revolutions per minute) of the impeller remains the same, then the larger the impeller diameter the higher the generated head. Note that as you increase the diameter of the impeller the tip speed at the outer edge of the impeller increases commensurately. However, the total energy imparted to the liquid as the diameter increases goes up by the square of the diameter increase. This can be understood by the fact that the liquid's energy is a function of its velocity and the velocity accelerates as the liquid passes through the impeller. A wider diameter impeller accelerates the liquid to a final exit velocity greater than the proportional increase in the diameter.

RPM (Revolutions Per Minute)

As the number of revolutions per minute of an impeller increases, the velocity (and head) imparted to the liquid passing through it increases as well. As the impeller revolves more rapidly the rate of increase in the liquid velocity is higher than the rate of rpm increase. In other words, an impeller spinning at 2,000 rpm generates more than twice the head of the same impeller spinning at 1,000 rpm.

Pump Data

Pump data is generally presented in the form of a chart for a particular series of pump. The chart is for a fixed speed, usually 3,500 or 1,750 rpm. Since pumps are normally close-coupled with electric motors, the speeds represent the normal operating speed of the motors. Figure 4 depicts the characteristic pump curves for a typical pump. The operating speed is 3,500 rpm. The y-axis is for the head in feet and varies from 0 to 100. The x-axis represents the flow in gallons per minute (gpm) and varies from 0 to 180. There are five curves on the graph that correspond to five rotor diameters for this series of pump. They vary from 3.00 in. to 4.94 in. The elliptical type lines are lines of constant efficiency. These vary from 40 to 60%. At the bottom of the chart near the x-axis are five lines which correspond to each rotor diameter.

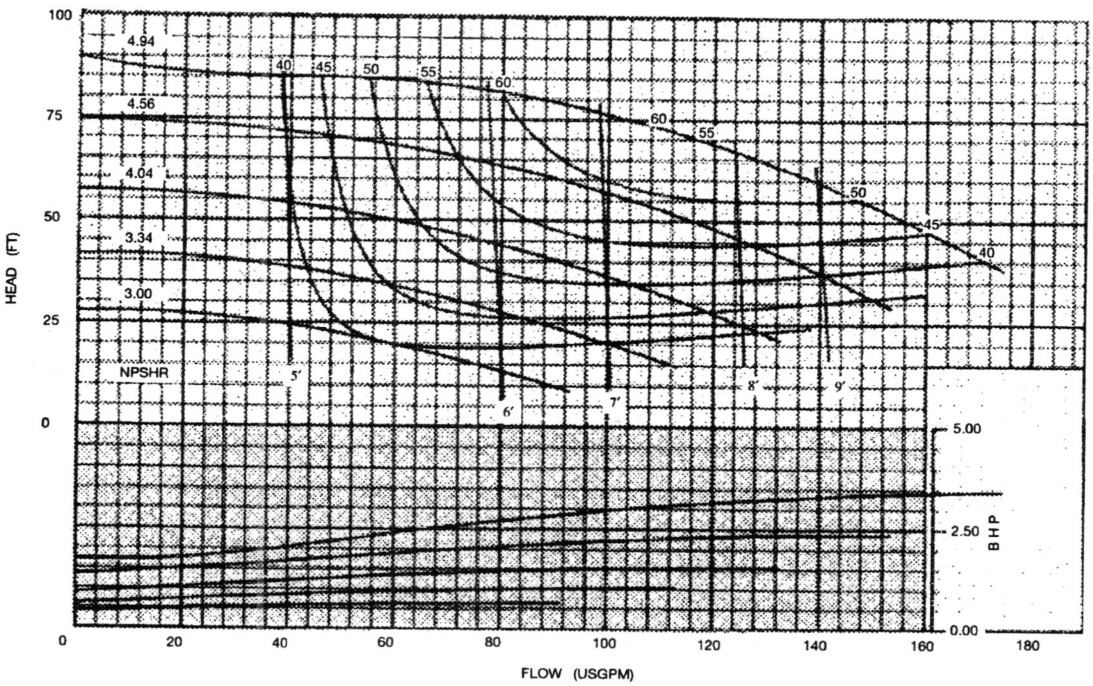

Figure 4. Pump Characteristics

These are used to find the brake horsepower (BHP). The last set of lines, which are almost vertical, are used to find the required net positive suction head.

Cavitation

Cavitation is a phenomenon that occurs when a liquid vaporizes as it passes through a pump and then quickly turns back into a liquid. The collapse of the vapor bubbles creates destructive microjets of liquid strong enough to damage the pump. Vaporization occurs if the pumped liquid drops below its vapor pressure. As a liquid accelerates through a pump it loses pressure (Bernoulli's Principle). If the pressure drops below the vapor pressure of the liquid, then gas bubbles will form instantly as the liquid vaporizes. These bubbles collapse just as quickly, causing cavitation. To prevent cavitation, the pressure (more correctly the head) of the liquid entering the pump must be high enough to prevent the subsequent pressure drop from reaching the liquid's vapor pressure.

NET POSITIVE SUCTION HEAD (NPSH)

A minimum amount of suction pressure (head) is needed for a pump to operate without cavitating. The term used to describe this suction pressure is *net positive suction head* (NPSH). The amount of NPSH the pump *requires* to avoid cavitation is called NPSHR. The amount of NPSH *available* to the pump from the suction line is termed NPSHA. The chart shown in Figure 4 has NPSHR lines running almost vertically. When selecting a pump it is necessary to see how much NPSH it requires at the duty point and make sure the NPSH available exceeds that amount.

$$\text{NPSHA} = \frac{P_B - P_v}{\gamma} + (S - h_\ell)$$

where:

P_B = pressure on the liquid surface in the suction tank (lbf/ft^2 or Pa)

P_v = vapor pressure of liquid at the liquid temperature (lbf/ft^2 or Pa)

S = vertical distance from pump centerline to level of liquid in the suction tank (ft or m). S is positive if level is above pump centerline and negative if below.

h_ℓ = total friction loss in pump suction pipe and fittings (ft or m)

γ = specific weight of fluid (lbf/ft^3 or N/m^3)

PROBLEM 8:

Suppose the pump in Figure 4 is installed in the system shown in Figure 5. The pump is operating at 3,500 rpm with the 4-in. impeller and delivering 100 gpm. Compute the NPSHA and compare with the NPSHR. Assume $h_\ell = 7'$ and water to be at 60°F.

SOLUTION:

Assume that $P_B = 14.7$ psi. P_v is found from the saturated water tables at 60°F and found to be 0.2562 psi, $\gamma = 62.4$ lbf /ft^3.

$$\text{NPSHA} = \frac{P_B - P_v}{\gamma} + (S - h_\ell)$$

$$= \frac{(14.7 - 0.2562) \times 144}{62.4} + (-10 - 7) = 16 \text{ ft}$$

From the pump chart, the NPSHR is 7 feet which means the pump will operate safely.

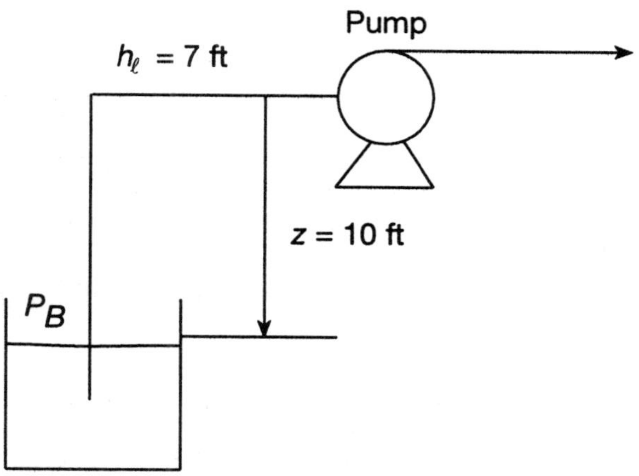

Figure 5. (Diagram for Problem 8)

Power Requirements

The power required to operate a pump is given by the following equation:

$$\dot{W} = \frac{\rho g H \dot{Q}}{\eta} = \frac{\gamma H \dot{Q}}{\eta} = \frac{\dot{m} g H}{\eta}$$

where:

$\dot{W}$ = pump power (W, kW, hp)

H = head increase of pump (m or ft)

$\dot{Q}$ = volumetric flow rate (m^3/s or ft^3/min)

$\dot{m}$ = mass flow rate of fluid through pump (kg/s or lbm/hr)

ρ = density of pump fluid (kg/m^3 or lbm/ft^3)

γ = specific weight of pump fluid (N/m^3 or lbf/ft^3)

η = pump efficiency

PROBLEM 9:

Water at a temperature of 40°C and a pressure of 100 kPa enters a pump at the rate of 50,000 kg/hr. The water leaves the pump at a pressure of 14 MPa and a temperature of 40°C. Determine the input power to the pump assuming the pump efficiency to be 80%.

SOLUTION:

The input power to the pump is given by the following:

$$\dot{W} = \frac{\rho g H \dot{Q}}{\eta} = \frac{\gamma H \dot{Q}}{\eta} = \frac{\dot{m} g H}{\eta}$$

To solve this equation, H, the head increase of the pump, must be determined.

$$P_1 = 100 \text{ kPa} = 100 \times 10^3 \text{ N/m}^2$$

$$P_2 = 14 \text{ MPa} = 14{,}000 \times 10^3 \text{ N/m}^2$$

For the SI system:

$$H = \frac{\Delta P}{(9{,}800) \times (\text{S.G.})} = \frac{13{,}900 \times 10^3 \text{ N/m}^2}{9{,}800 \text{ N/m}^3} = 1{,}418 \text{ m}$$

where S.G. is 1 for water.

Since the mass flow rate is given in the problem, the third form of the above equation should be used.

$$\dot{W} = \frac{(50{,}000 \text{ kg/h})(9.8 \text{ m/s}^2)(1{,}418 \text{ m})}{(3600 \text{ s/h})(0.8)} = 241{,}257 \text{ W} = 241 \text{ kW}$$

COMPRESSORS

A compressor is a device that increases the pressure of the gas or vapor that is flowing through the device. Compressors are classified as *positive displacement* or *dynamic*. *Positive displacement* compressors increase the pressure of the gas or vapor by reducing the volume. *Dynamic compressors* increase the pressure by a continuous transfer of angular momentum to the gas or vapor. The minimum work required to operate a compressor is found by integration of the reversible steady-flow work equation:

$$w_{rev} = -\int v dP - \Delta ke - \Delta pe$$

where Δke and Δpe are the changes in kinetic and potential energy. The Δpe term will be neglected in the following analysis. The Δke term is normally neglected but will be included for completeness. Three important cases will be investigated:

Isentropic (Pv^k = constant)

Polytropic (Pv^n = constant)

Isothermal (Pv = constant)

For the isentropic case, integration of the above equation results in:

$$w_{comp} = \frac{kR(T_1 - T_2)}{k-1} = \frac{kRT_1}{k-1}\left[1 - \left(\frac{P_2}{P_1}\right)^{\frac{(k-1)}{k}}\right]$$

For the polytropic case:

$$w_{comp} = \frac{nR(T_1 - T_2)}{n-1} = \frac{nRT_1}{n-1}\left[1 - \left(\frac{P_2}{P_1}\right)^{\frac{(n-1)}{n}}\right]$$

For the isothermal case:

$$w_{comp} = RT \ln \frac{P_1}{P_2}$$

PROBLEM 10:

Fifteen lbm/min of air are compressed in a low-pressure, water-jacketed, steady-flow compressor. The inlet conditions are 14.7 psia and 70°F, and the outlet conditions are 5 psig and 110°F. For a reversible and polytropic process and a water temperature rise of 6°F find:

(a) The work performed, and

(b) The mass flow rate of the circulating water.

SOLUTION:

$\dot{m}$ = 15 lbm/min; T_1 = 70°F = 530°R; T_2 = 110°F = 570°R;

P_1 = 14.7 psia; P_2 = 19.7 psia

Since the process is reversible and polytropic, the above equation can be used if n was known. To find n, we use the general polytropic relations:

$$\frac{T_2}{T_1} = \left(\frac{P_2}{P_1}\right)^{\frac{n-1}{n}}$$

$$\frac{570}{530} = \left(\frac{19.7}{14.7}\right)^{\frac{n-1}{n}}$$

$$n = 1.33$$

For the reversible, steady-flow process:

$$w_{comp} = \frac{nR(T_1 - T_2)}{n-1}$$

$$= \frac{(1.33)(0.06855)(530-570)}{(1.33-1)} = -11.05 \text{ Btu/lbm}$$

Work is negative since it is done on the system.

$$[R] = 0.06855 \text{ Btu/lbm} - {}^\circ\text{R}$$

$$\dot{W} = \dot{m}w = 15 \text{ lbm/min} \times (-11.05) \text{ Btu/lbm} = -166 \text{ Btu/min}$$

The heat transfer from the air as it passes through the compressor can be determined using a first law energy balance:

$$\dot{Q} = \dot{W} + \dot{m}(h_2 - h_1) = \dot{W} + \dot{m}c_p(T_2 - T_1)$$

$$= (-166) + 15 \times 0.24(570 - 530) = -22.0 \text{ Btu/min}$$

The negative sign indicates that heat is being removed from the compressor.

$$\dot{m}_{water} = \frac{\dot{Q}}{(\Delta T)_{water}} = \frac{22}{6} = 3.67 \text{ lbm/min}$$

PROBLEM 11:

Assume the process in the previous problem was irreversible and adiabatic. If the exit temperature of the air is 130°F, determine:

(a) the value of n for this process,

(b) the rate of work input required, and

(c) the significance of $-\int vdP$ in this case.

SOLUTION:

T_2 = 130°F = 590°R. The value of n is found by solving the polytropic relations:

$$\frac{T_2}{T_1} = \left(\frac{P_2}{P_1}\right)^{\frac{n-1}{n}} = \frac{590}{530} = \left(\frac{19.7}{14.7}\right)^{\frac{n-1}{n}}$$

$$n = 1.58$$

Since the process is adiabatic, $\dot{Q}$ = 0. The first law energy balance becomes:

$$\dot{W} = -\dot{m}(h_2 - h_1) = -\dot{m}\,c_p(T_2 - T_1)$$

$$= (-15) \times (0.240) \times (590 - 530) = -216 \text{ Btu/min}$$

Note that the rate of work input is greater in this case.

The term $-\int vdP$ has no significance in this problem since the process is irreversible.

PROBLEM 12:

Assume the compressor in the above problems was reversible and adiabatic. Determine the work required under these conditions.

SOLUTION:

Since the process is reversible and adiabatic and the air can be treated as an ideal gas, the outlet temperature of the compressor can be determined from:

$$\frac{T_2}{T_1} = \left(\frac{P_2}{P_1}\right)^{\frac{k-1}{k}}$$

$$T_2 = T_1\left(\frac{19.7}{14.7}\right)^{\frac{1.4-1}{1.4}} = 567°R = 116°F$$

$$\dot{W} = \dot{m}(h_2 - h_1) = -\dot{m}c_p(T_2 - T_1)$$
$$= (-15) \times (0.240) \times (576 - 530) = -166 \text{ Btu/min}$$

It should be clear from the above examples that cooling a gas as it is compressed is desirable since this reduces the work input to the compressor. When cooling through the casing of a compressor is not possible or practical, multistage compression with intercooling is employed where the gas is compressed in stages and cooled between each stage. Ideally, the cooling process takes place at constant pressure and the gas is cooled to the initial temperature T_1. The total work input for each stage of compression is as follows:

$$w_{comp} = w_{comp\ 1} + w_{comp\ 2}$$

$$= \frac{nRT_1}{n-1}\left[1-\left(\frac{P_x}{P_1}\right)^{\frac{(n-1)}{n}}\right] + \frac{nRT_1}{n-1}\left[1-\left(\frac{P_2}{P_x}\right)^{\frac{(n-1)}{n}}\right]$$

where P_x is the value of the intermediate pressure between P_1 and P_2. The value of P_x that results in the minimum amount of work is given as follows:

$$P_x = (P_1P_2)^{1/2} \text{ or } \frac{P_x}{P_1} = \frac{P_2}{P_x}$$

The minimum work occurs when the pressure ratio across each stage of the compressor is the same. When this condition is satisfied, $w_{comp\ 1} = w_{comp\ 2}$. The total work for both stages is merely twice the work for either stage.

PROBLEM 13:

Air is compressed steadily by a reversible compressor from an inlet state of 100 kPa and 300 K to an exit pressure of 800 kPa. Determine the compressor work per unit mass for (a) isentropic compression with $k = 1.4$, (b) polytropic compression with $n = 1.3$, (c) isothermal compression, and (d) ideal two-stage compression with intercooling and a polytropic exponent of 1.3.

SOLUTION:

$P_1 = 100$ kPa; $P_2 = 800$ kPa; $T_1 = 300$ K; $k = 1.4$; $n = 1.3$

(a)

$$W_{comp} = \frac{kRT_1}{k-1}\left[1-\left(\frac{P_2}{P_1}\right)^{\frac{(k-1)}{k}}\right]$$

$$= \frac{(1.4)(0.287)(300)}{1.4-1}\left[1-\left(\frac{800}{100}\right)^{\frac{(1.4-1)}{1.4}}\right] = -245 \text{ kJ/kg}$$

(b)

$$W_{comp} = \frac{nRT_1}{n-1}\left[1-\left(\frac{P_2}{P_1}\right)^{\frac{(n-1)}{n}}\right]$$

$$= \frac{(1.3)(0.287)(300)}{1.3-1}\left[1-\left(\frac{800}{100}\right)^{\frac{(1.3-1)}{1.3}}\right] = -230 \text{ kJ/kg}$$

c)

$$W_{comp} = RT\ln\frac{P_1}{P_2} = (0.287)(300)\ln\frac{100}{800} = -179 \text{ kJ/kg}$$

d) Ideal two-stage compressor with intercooling (n = 1.3). The value of P_x is found:

$$P_x = (P_1\, P_2)^{1/2} = [(100)(800)]^{1/2} = 282.8 \text{ kPa}$$

$$w_{comp} = 2w_{comp\,1} = 2\frac{nRT_1}{n-1}\left[1-\left(\frac{P_2}{P_1}\right)^{\frac{(n-1)}{n}}\right]$$

$$= \frac{2(1.3)(0.287)(300)}{1.3-1}\left[1-\left(\frac{283}{100}\right)^{\frac{(1.3-1)}{1.3}}\right]$$

$$- -202 \text{ kJ/kg}$$

FE/EIT

FE: PM Mechanical Engineering Exam

CHAPTER 7

Fluid Mechanics

CHAPTER 7

FLUID MECHANICS

FLUID MECHANICS

Fluids are substances that are capable of flowing and that conform to the shape of the containing vessel. While in equilibrium, fluids cannot sustain tangential or shear forces. Shear stresses occur in fluids in motion.

ρ = fluid density = mass per unit volume; $[\rho] = \text{kg/m}^3$

ρ_w = standard density of water = 1,000 kg/m^3

γ = specific weight = weight of substance per unit volume;

$$\gamma = \frac{W}{V} = \frac{mg}{Vg_c} = \rho \frac{g}{g_c}$$

g = gravitational acceleration = 9.80 m/s^2

$$g_c = \text{proportionality constant} = 1 \ \frac{\text{kg–m}}{\text{N–s}^2}$$

γ_w = 9,800 N/m^3

Specific Gravity (*SG*) is the ratio of the weight of a body to the weight of an equal volume of water.

$$SG = \frac{\gamma}{\gamma_w} = \frac{\rho}{\rho_w};$$ *SG* of mercury is 13.6;

ρ_{Hg} = 13,600 kg/m^3; γ_{Hg} = 133,300 N/m^3

Pressure (p) is the force exerted by a fluid per unit area.

$$[p] = \text{N/m}^3 = \text{Pascal (Pa)}$$

p or p_a = absolute pressure = pressure measured relative to zero datum

p_g = gage pressure

= pressure measured relative to atmospheric pressure

p_{atm} = standard atmospheric pressure

= 101,325 Pa = 101.3 kPa = 101,325 N/m^2

$$p_a = p_g + p_{atm}$$

p_g is useful for the solution of many fluids problems.

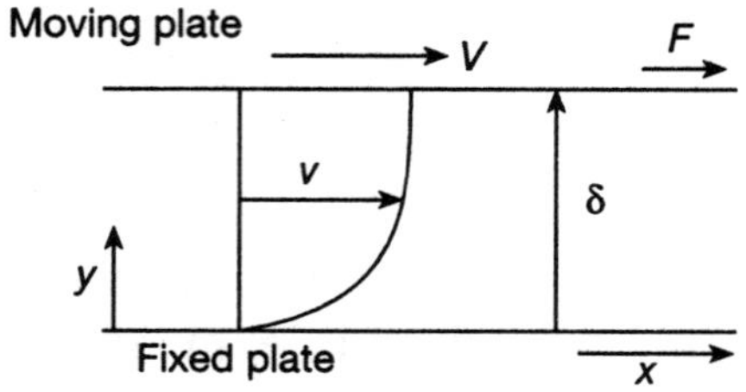

Figure 1. Velocity Distributing Between Two Plates

The *viscosity* of a fluid is its resistance to a shearing force. Figure 1 shows the distribution of velocity between two plates, one fixed and one moving at constant velocity V.

υ = fluid velocity at any position y

$\upsilon(0) = 0;\ \upsilon(y = \delta) = V$

$$\tau = \text{shear stress} = \frac{F}{A} = \mu \frac{dv}{dy}$$

$$\mu = \text{dynamic viscosity; } [\mu] = \frac{\text{kg}}{\text{m–s}} = \frac{\text{N–s}}{\text{m}^2} = \text{Pa–s}$$

Dynamic viscosity is often expressed in terms of Poise with 1 poise = 0.1 Pa-s; to convert from poise to Pa-s, multiply by 0.1.

$$\nu = \text{kinematic viscosity} = \frac{\mu}{\rho}; [\nu] = \frac{\text{m}^2}{\text{s}^2}$$

Kinematic viscosity is often expressed in terms of Stokes with 1 stoke $= 10^{-4}$ m^2/s; to convert from stokes to m^2/s, multiply by 10^{-4}.

To evaluate τ, the velocity distribution, $v(y)$ must be known or estimated using the following which assumes a linear velocity distribution:

$$\frac{dv}{dy} \cong \frac{\Delta v}{\Delta y} \cong \frac{V}{\delta}$$

where δ is the plate separation.

The *fluid pressure* is the same in all directions for a fluid at rest. It is dependent on the distance from the free surface. Assume h is the distance measured downward from the free surface of a fluid. This leads to the hydrostatic relation:

$$p_2 - p_1 = \gamma\,(h_2 - h_1) = \gamma\,\Delta h;\ (h_2 - h_1) = \Delta h$$

The pressures may be either absolute or gage. To find the gage pressure 1 meter below the surface of standard water, $p_2 - 0 = 9{,}800$ N/m^3 (1 m $-$ 0) $= 9{,}800$ N/m^2 $= 9.8$ kPa.

$\Delta p = 9.8$ kPa Δh(m) for water,

and

$\Delta p = 9.8$ kPa (SG) Δh(m) for other fluids.

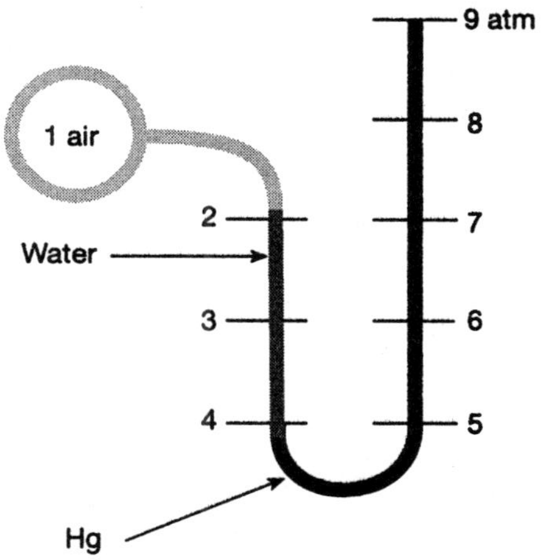

Figure 2. Manometer

Manometers, as shown in Figure 2, are devices that employ liquid columns to determine differences in pressure. According to hydrostatics laws, the pressure at the same level of a contiguous fluid is the same. When the manometer fluid is air, the pressure is approximately equal at all levels of the air.

$p_9 = p_{atm} = 101.3$ kPa (abs) = 0 kPa (gage)

$p_1 \neq p_8$ (fluid between 1 and 8 not the same)

$p_4 = p_5$ (same elevation and same fluid)

$p_3 \neq p_6$ (fluid between 3 and 6 not the same)

$p_2 \neq p_7$ (fluid between 2 and 7 not the same)

$p_1 = p_2$ (same at all levels for a gas)

For a particular level:

$$p_3 - p_2 = \gamma_w (h_3 - h_2) \Rightarrow p_3 = p_2 + \gamma_w (h_3 - h_2).$$

As we move down a fluid column, the pressure increases since Δh is positive; as we move up a fluid column, the pressure decreases since Δh is negative. To solve manometer problems, start at a known pressure point. If you move down the fluid column, you add pressure.

PROBLEM 1:

Referring to Figure 2, assume that $h_2 = h_7 = 1$ m, $h_4 = h_5 = 2$ m, and $h_9 = -0.5$ m. Assume the *SG* of mercury to be 13.6.

(a) Find the change of pressure between points 2 and 4 ($p_4 - p_2$)

(b) The gauge pressure of point 1 is closest to _______ .

SOLUTION:

(a) $p_4 - p_2 = \gamma (h_4 - h_2) = 9.8 \text{ kPa} \times (2 - 1) \text{ m} = 9.8 \text{ kPa} = 9{,}800 \text{ Pa}$

(b) $p_9 + \gamma_{HG} (h_5 - h_9) + \gamma_w (h_2 - h_4) = p_1$

$= 0 + 9.8 \text{ kPa} \times 13.6 \times (2 - (-0.5)) + 9.8 \text{ kPa} \times (1 - 2)$

$p_1 = 323$ kPa

FORCES ON SUBMERGED SURFACES

A plane surface in a horizontal position in a fluid at rest has a force acting on each surface equal to $F_H = p_H A$, where p_H is the pressure at the elevation of the horizontal surface and A is the area of the surface. If gauge pressure is used, then F_H is the force of water alone on the surface. If absolute pressure is used, then F_H is the force of water plus atmospheric pressure.

PROBLEM 2:

Assume that a horizontal surface is 10 m^2 and is at a depth of 0.5 m below the surface of standard water. Find the force of the water on the horizontal surface.

SOLUTION:

$$p_g = h\,\gamma_w$$

$$p_g = 0.5 \text{ m} \times 9{,}800 \text{ N/m}^3 = 4{,}900 \text{ N/m}^2$$

$$F_H = 4{,}900 \text{ N/m}^2 \times 10 \text{ m}^2 = 49{,}000 \text{ N}$$

For inclined surfaces, as pictured in Figure 3:

$$F = p_{CG}\,A = h_{CG}\,\gamma\,A$$

where:

p_{CG} = pressure at gravity center of surface

h_{CG} = fluid depth at gravity center

$h_{CG} = y_{CG} \sin\theta$

$F = y_{CG} \sin\theta\,\gamma\,A$

θ = Angle surface makes with horizontal

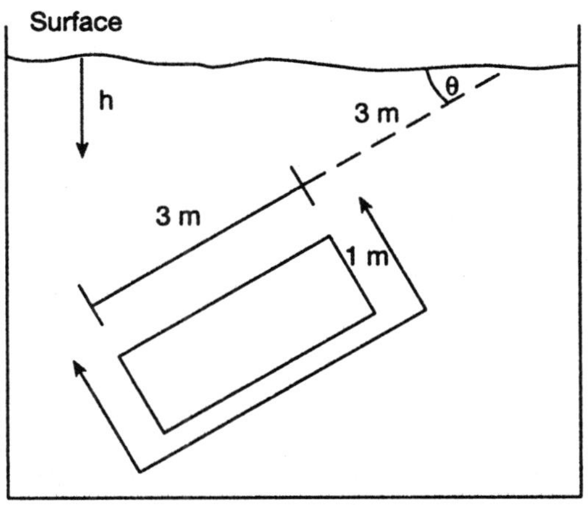

Figure 3. Forces on Submerged Surfaces

By way of an example for $\theta = 30°$, the force of fluid on the rectangle equals:

$$F = 4.5 \text{ m} \times \sin 30° \times 9{,}800 \text{ N/m}^3 \times (3 \times 1) \text{ m}^2$$

$$F = 66{,}150 \text{ N}$$

The line of action of the force is normal to the surface but does not pass through the center of gravity. I passes through a point called y_{CP}, the center of pressure, given by the following:

$$y_{CP} = \frac{I_{CG}}{y_{CG}A} + y_{CG}$$

Archimedes' Principle: Any body, floating or immersed in a liquid, is acted upon by a buoyant force equal to the weight of the liquid displaced.

FLOW OF FLUIDS

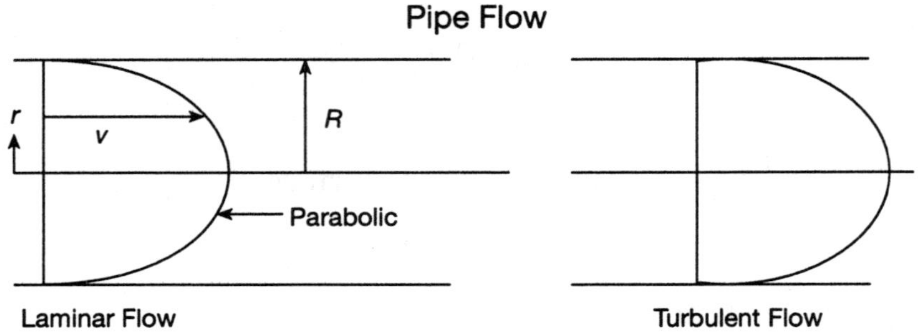

Figure 4. Laminar and Turbulent Pipe Flow

The continuity equation for pipe flow is given by the following:

$Q = A_1V_1 = A_2V_2 =$ volume flow rate; $[Q] = \text{m}^3/\text{s}$

$\dot{m} = \rho Q = \rho AV =$ mass flow rate; $[\dot{m}] = \text{kg/s}$

$A =$ cross section of area of flow

$V =$ average flow velocity

$\rho =$ the fluid density

Pipe flow may be classified as either *laminar* or *turbulent*. Laminar flow results in a parabolic velocity distribution with the average velocity equal to half of the maximum velocity. The maximum velocity occurs at the center of the pipe, V_0. Turbulent flow is characterized by a much flatter velocity distribution with the average velocity greater than V_0.

Flow is considered laminar when:

$Re =$ Reynold's Number $\leq 2{,}000$

$$Re = \frac{\rho VD}{\mu} = \frac{VD}{\nu}$$

When flow is laminar in a pipe of radius R, the velocity distribution is given by the following:

$$V = \mathrm{v}_{\max}\left[1 - \left(\frac{r}{R}\right)^2\right]$$

When the flow is turbulent, the velocity distribution is obtained experimentally.

PROBLEM 3:

Heavy fuel oil at 30°C is flowing in a 2.54 cm diameter pipe. Experimental data shows that the centerline velocity is 6 m/s. Find (a) V; (b) Q; (c) $\dot{m}$; (d) Re; (e) τ; and (f)

$$\frac{\Delta P}{I}$$

SOLUTION:

If the flow is laminar we can establish the velocity; therefore distribution g assume laminar flow. Also assume the following properties $n = 5.83 \times 10^{-5}$ m²/s; $SG = 0.899$.

(a) For laminar flow, the average velocity is half the centerline velocity. $V = v_{max}/2 = 6/2 = 3.0$ m/s

(b) $Q = AV$; $A = \pi/4\ \mathrm{D}^2 = \pi/4 \times (0.0254\ \mathrm{m})^2 = 5.07 \times 10^{-4}\ \mathrm{m}^2$

$Q = 5.07 \times 10^{-4}\ (\mathrm{m}^2) \times 3\ (\mathrm{m/s}) = 1.52 \times 10^{-3}\ \mathrm{m}^3/\mathrm{s}$

(c) $\dot{m} = \rho AV = \rho Q$; $\rho = \rho_w SG$; $\rho = 1{,}000\ \mathrm{kg/m}^3\ \dot{m}\ 0.899 = 899\ \mathrm{kg/m}^3$

$\dot{m} = 899\ \mathrm{kg/m}^3 \times 1.52 \times 10^{-3}\ \mathrm{m}^3/\mathrm{s} = 1.37\ \mathrm{kg/s}$

(d) $$Re = \frac{VD}{\nu} = \frac{3\ \mathrm{m/s} \times 0.0254\ \mathrm{m}}{5.83 \times 10^{-5}\mathrm{m}^2/\mathrm{s}} = 1{,}307;$$

laminar flow was confirmed:

(e) $$\tau = \mu \frac{dv}{dy}\Big)_w = -\mu \frac{dv}{dr}\Big)_{r=R}$$

Note: $dy = -dr$

$$v = v_{max}\,[1 - (r/R)^2]$$

$$\left.\frac{dv}{dr}\right|_{r=R} = -\frac{v_{\max} \times 2}{R};\ v_{\max} = 2 \times V;\ R = \frac{D}{2}$$

$$\tau = (8V\mu)\,/\,D;\ \mu = \rho \times \nu = 899\ \text{kg/m}^3 \times 5.83\ \text{x}\ 10^{-5}\ \text{m}^2\text{/s}$$

$$= 0.0524\ \text{kg/(m–s)} = 0.0524\ \text{N–s/m}^2$$

$$\tau = 8 \times 3\ \text{(m/s)} \times 0.0524\ \text{(N–s/m}^2\text{)/0.0254 (m)} = 49.5\ \text{N/m}^2$$

(f) The shear stress in the pipe causes the pressure to drop. $dpA_c = \tau A_l$, where A_c is the cross sectional area of the pipe and A_l is the lateral area of the pipe.

$$dp \times (\text{p}/4) \times D^2 = t \times (pDl).$$

Rearranging:

$$\frac{dp}{l} = \frac{\Delta p}{l} = \frac{4\tau}{D} = \frac{4 \times 49.5\ (\text{N/m}^2)}{0.0254\ \text{m}} = 7{,}800\ \text{Pa/m}$$

The above demonstrates that for fluid flowing in a constant cross-sectional pipe, the viscosity causes a decrease in pressure in the direction of flow.

$\dfrac{\Delta p}{\gamma} = h_f$ = head loss due to friction; $[h_f]$ = m. Similarly, $\Delta p = \gamma\, h_f$

Experimental evidence shows that h_f can be predicted from the Darcy's equation:

$$h_f = f \frac{L}{D} \frac{V^2}{2g}$$

$f = f\,(Re, e/D)$ the friction factor

D = diameter of pipe

L = length of pipe over which pressure drop occurs

e = roughness factor for pipe

$$\Delta p = \gamma h_f = \gamma f \frac{L}{D} \frac{V^2}{2g} = \rho \frac{g}{g_c} f \frac{L}{D} \frac{V^2}{2g} = f \frac{L}{D} \rho \frac{V^2}{2g_c}$$

$$f = \frac{64}{Re} \text{ for laminar flow}$$

$$f = \frac{0.316}{Re^{1/4}} \text{ turbulent flow; smooth tubes}$$

The energy equation for incompressible flow is:

$$\frac{p_1}{\gamma} + z_1 + \frac{V_1^2}{2g} = \frac{p_2}{\gamma} + z_2 + \frac{V_2^2}{2g} + h_f$$

Head losses also occur as the fluid flows through pipe fittings (i.e., elbows, valves, couplings, etc.) and sudden pipe contractions and expansions. These losses (also called minor losses) are given by:

$$h_{f,\,fitting} = C\frac{V^2}{2g}$$

where C is the characteristic head loss value for a particular fitting.

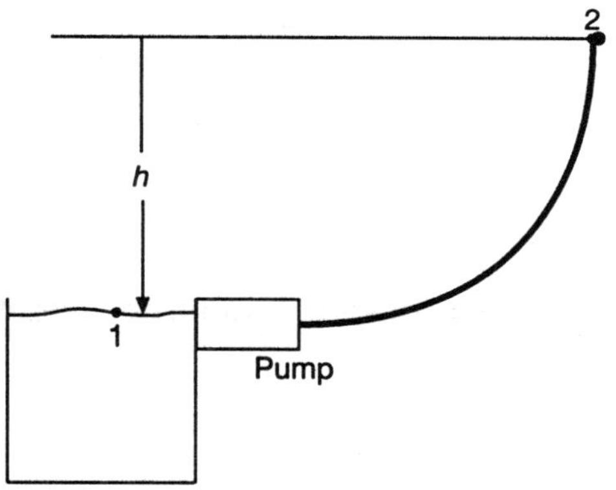

Figure 5. Pump and Reservoir System

Incorporating the minor loss term and E_m (defined below), the energy equation becomes:

$$\frac{p_1}{\gamma} + z_1 + \frac{V_1^2}{2g} = \frac{p_2}{\gamma} + z_2 + \frac{V_2^2}{2g} + h_f + h_{f,\,fittings} + E_m$$

If E_m is positive, a *turbine* exists between points 1 and 2.

If E_m is negative, a *pump* exists between points 1 and 2. In this case E_m represents the head that the pump must supply for the given flow conditions to continue.

The pump power equation is given by the following:

$$\dot{W} = \frac{Q\gamma E_m}{\eta}$$

where:

E_m = the total head loss the pump must overcome (m)

η = pump efficiency

$\dot{W}$ = power (watts)

The head loss at either an entrance or exit of a pipe from or to a reservoir is given by the $h_{f,\ fitting}$ equation. Values of C for various inlet conditions are shown in Figure 6 below.

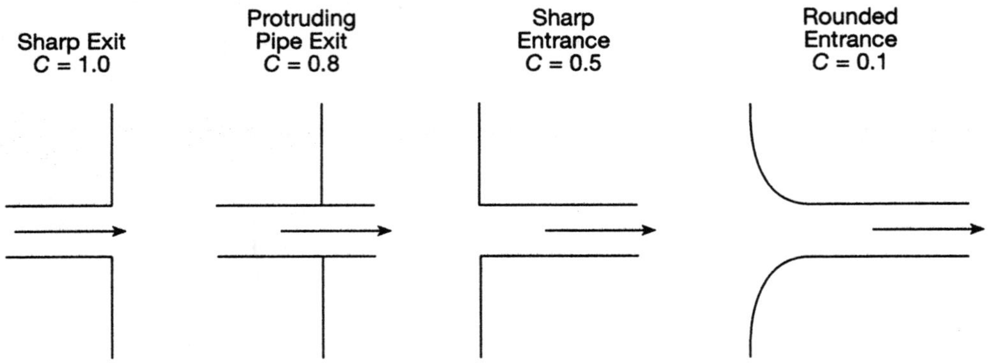

Figure 6. Entrance Loss Coefficients

With the loss terms omitted, the energy equation reduces to Bernoulli's equation:

$$\frac{p_1}{\gamma} + z_1 + \frac{V_1^2}{2g} = \frac{p_2}{\gamma} + z_2 + \frac{V_2^2}{2g}$$

The above is valid along a streamline in steady, frictionless, incompressible flow.

PROBLEM 4:

Medium fuel oil (SG = 0.861, $\nu = 5.2 \times 10^{-6}$ m^2/s) is pumped to reservoir C through 2,000 m of new riveted steel pipe (e = 0.00183 m) 40 cm in diameter. The gage pressure at A is 13.8 kPa when the flow rate is 0.2 m^3/s. Assume the pump efficiency, η, is 75%. Find:

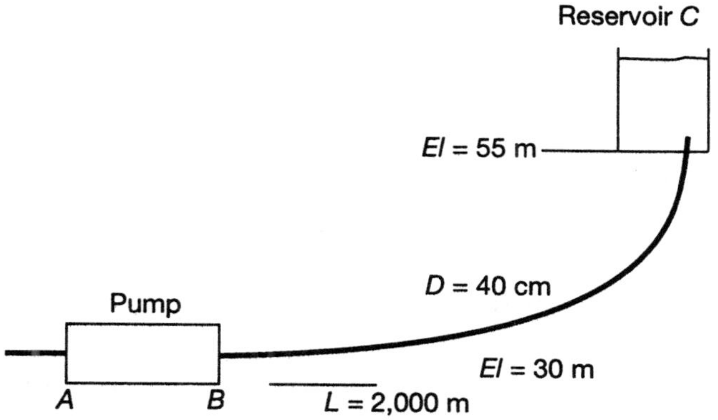

Figure 7. Sketch of Problem 4

(a) the pump power in watts, and

(b) the gage pressure at B.

SOLUTION:

(a) Write the energy equation between points A and C:

$$\frac{p_A}{\gamma} + z_A + \frac{V_A^2}{2g} = \frac{p_C}{\gamma} + z_C + \frac{V_C^2}{2g} + h_f + h_{f,\, fittings} + E_m$$

$$V_A = V_B = \frac{Q}{A} = \frac{0.2 \text{ m}^3/\text{s}}{(\pi/4)(0.4 \text{ m})^2} = 1.59 \text{ m/s};$$

$$Re = \frac{VD}{\nu} = \frac{(1.59 \text{ m/s}) \times (0.4 \text{ m})}{5.2 \times 10^{-6} \text{ m}^2/\text{s}} = 1.22 \times 10^5$$

Flow is turbulent. $e/D = 0.00183/0.4 = 0.0046$. $f = 0.030$.

$z_A = 30$ m; $z_c = 55$ m; $V_C = 0$ (velocity of a reservoir is always zero)

There is a fitting loss due to sudden enlargement from pipe to reservoir, $C = 1.0$.

$$H_{f,\,\text{fittings}} = 1.0\frac{V_B^2}{2g} = \frac{1.0 \times (1.59 \text{ m/s})^2}{2 \times 9.80 \text{ m/s}^2} = 0.129 \text{ m}$$

$$\frac{V_A^2}{2g} = 0.129 \text{ m}$$

$$h_f = fL/DV^2/2g = (0.03) \times (2000/.4) \times (1.59^2/2 \times 9.80) = 19.3 \text{ m}$$

$p_A = 13.8$ kPa $= 13{,}800$ N/m^2; $p_C = 0$; Substituting the above values into the energy equation:

$$1.41 + 30 + 0.129 = 0 + 55 + 0 + 19.3 + 0.129 + E_m;$$

$$E_m = -42.9 \text{ m}$$

which is the head input of the pump.

$$\dot{W} = \frac{\gamma Q E_m}{\eta} = \frac{(9{,}800 \text{ N/m}^3) \times (0.2 \text{ m}^3\text{/s}) \times (42.9m)}{0.75}$$

$$= 1.12 \times 10^5 \text{ W} = 112 \text{ kW}$$

(b) Write the energy equation from A to B:

$$\frac{p_A}{\gamma} + z_A + \frac{V_A^{\ 2}}{2g} = \frac{p_B}{\gamma} + z_B + \frac{V_B^{\ 2}}{2g} + E_m; \quad z_A = z_B; \; V_A = V_B$$

$$\frac{p_A}{\gamma} = E_m + \frac{p_B}{\gamma}; \; 1.41 \text{ m} = -42.9 \text{ m} + \frac{p_B}{\gamma}; p_B$$

$$= (44.31 \text{ m}) \times \left(9{,}800 \text{ N/m}^3\right)$$

$$= 4.34 \times 10^5 \text{ Pa} = 434 \text{ kPa (gage pressure)}$$

IMPULSE-MOMENTUM PRINCIPLE

When the path of a fluid stream is diverted, mechanics tells us that forces must be involved because the momentum of the system has been altered. For two-dimensional flow, this leads to the following set of scalar equations:

$$\Sigma F_x = \rho Q \, (V_{2x} - V_{1x})$$

$V_{1x} =$ initial x velocity

$V_{1y} =$ initial y velocity

$$\Sigma F_y = \rho Q \, (V_{2y} - V_{1y})$$

$V_{2x} =$ final x velocity

$V_{2y} =$ final y velocity

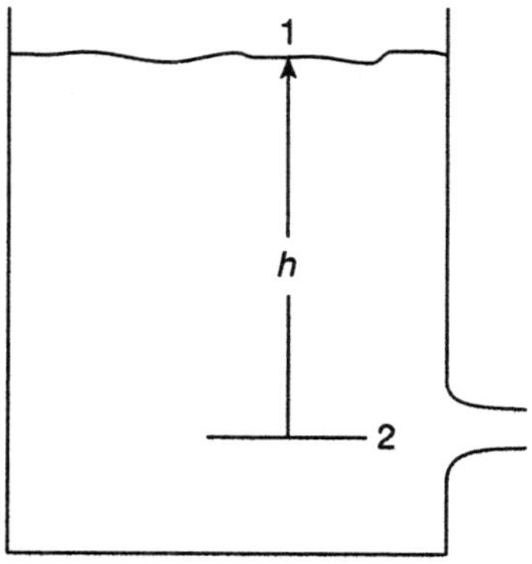

Figure 8. Sketch of Problem 5

PROBLEM 5:

For the smooth, 5 cm diameter nozzle shown in Figure 8, $h = 3$ m:

(a) What is the velocity of efflux?

(b) What is the volume flow rate Q?

SOLUTION:

Since the nozzle is "smooth" we can neglect losses and write Bernoulli's equation between points 1 and 2.

(a)

$$\frac{p_1}{\gamma} + z_1 + \frac{V_1^2}{2g} = \frac{p_2}{\gamma} + z_2 + \frac{V_2^2}{2g};$$

$$p_1 = p_2 = 0 \text{ gage}; \; V_1 = 0; \; z_1 = 3 \text{ m}; \; z_2 = 0$$

$$0 + 3 \text{ m} + 0 = 0 + 0 + \frac{V_2^2}{2g}; \; V_2 = \sqrt{2 \times g \times h}$$

$$= \sqrt{2 \times 9.80 \text{ m/s}^2 \times 3 \text{ m}} = 7.67 \text{ m/s}$$

(b) $Q = V_2 A_2$; $A_2 = (\pi/4)\, D_2^2$; $D_2 = 5$ cm $= 0.05$ m

$Q = (7.67 \text{ m/s}) \times (\pi/4) \times (0.05 \text{ m})^2 = 0.015 \text{ m}^3\text{/s}$

PROBLEM 6:

A jet of water 7.5 cm in diameter impinges on a flat plate held normal to its axis. For a velocity of 25 m/s what is the absolute value of the force required to keep the plate in equilibrium?

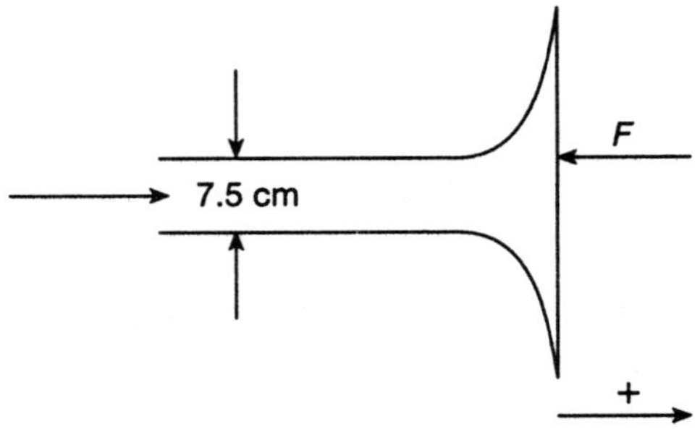

Figure 9. Sketch of Problem 6

SOLUTION:

F is the plate force; the force exerted on the fluid changing its momentum.

$$\Sigma F_x = \rho Q\,(V_{2x} - V_{1x});\; V_{1x} = \text{initial } x \text{ velocity}$$

where:

$$V_{2x} = \text{final } x \text{ velocity}$$

$$Q = AV = (\pi/4) \times (.075\text{ m})^2 \times (25\text{ m/s}) = 0.11\text{ m}^3/\text{s}$$

$$\rho = 1{,}000\text{ kg/m}^3;\; V_{2x} = 0$$

$$\Sigma F_x = (1{,}000\text{ kg/m}^3) \times (0.11\text{ m}^3/\text{s}) \times (0 - 25\text{ m/s}) = -2{,}750\text{ N}$$

The minus indicates that the force is opposite to the positive direction (to the right).

PROBLEM 7:

Water enters a pipe from a large reservoir and on issuing out of the pipe strikes a 90° deflector as shown in Figure 10. If a horizontal force of 890 N is developed by the deflector, find:

(a) V_2

(b) The power developed by the turbine

Assume $\eta = 1.00$

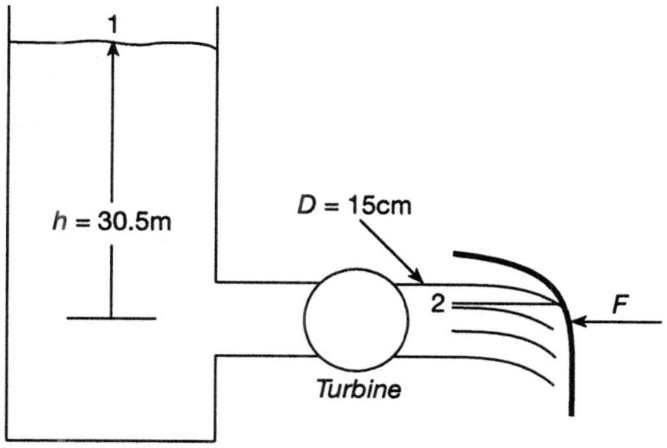

Figure 10. Sketch of Problem 7

SOLUTION:

(a) Write the energy equation between points 1 and 2 and neglect losses:

$$\frac{p_1}{\gamma} + z_1 + \frac{V_1^2}{2g} = \frac{p_2}{\gamma} + z_2 + \frac{V_2^2}{2g} + E_m$$

$p_1 = p_2 = 0$ kPa (gage); $V_1 = 0$; $z_1 = 30.5$ m; $z_2 = 0$

$$30.5\text{ m} = \frac{V_2^2}{2g} + E_m;$$

the momentum equation is used to find V_2

$$\Sigma F_x = \rho Q\,(V_{2x} - V_{1x}) = \rho A_2 V_2 (0 - V_{2x}) = -\rho A_2 V_2^2$$

$$-F = -\rho A_2 V_2^2$$

Cancelling minus signs on both sides of the equation:

$$890\text{ N} = (1{,}000\text{ kg/m}^3) \times (\pi/4) \times (0.15\text{m})^2 \times (V_2)^2;\ V_2 = 7.1\text{ m/s}$$

(b) $30.5\text{ m} = (7.1\text{ m/s})^2/(2 \times 9.80\text{ m/s}^2) + E_m$

$E_m = 27.9$ m; $Q = VA = 0.125\text{ m}^3/\text{s}$

$$\dot{W} = \frac{Q\,\gamma E_m}{\eta} = \frac{(0.125\text{ m}^3/s) \times (9{,}800\text{ N/m}^3) \times (27.9\text{ m})}{1}$$

$$= 3.43 \times 10^4\text{ W} = 34.3\text{ kW}$$

FE/EIT

FE: PM Mechanical Engineering Exam

CHAPTER 8

Heat Transfer

CHAPTER 8

HEAT TRANSFER

There are three fundamental modes of heat transfer: conduction, convection, and radiation.

CONDUCTION

A temperature gradient within a homogeneous substance results in an energy transfer within the medium which can be calculated by:

$$-kA\frac{\partial T}{\partial n}$$

where q is the rate of heat transfer (W) and $\partial T/\partial n$ is the temperature gradient in the direction normal to the area A. The thermal conductivity k is an experimental constant for the medium involved and depends on temperature and pressure. The units of k are W/m-K.

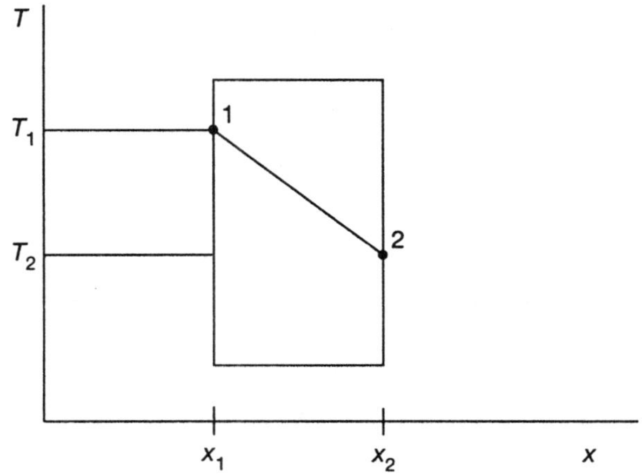

Figure 1. Linear Temperature Profile

If the temperature profile within the medium is linear, as shown in Figure 1, it is permissible to replace the temperature gradient with:

$$\frac{\Delta T}{\Delta x} = \frac{T_2 - T_1}{x_2 - x_1}$$

CONVECTION

When a solid body is exposed to a moving fluid having a temperature difference from that of the body, energy is carried or convected away by the fluid. If the upstream temperature of the fluid is T_∞ and the surface temperature of the solid is T_s, the heat transfer per unit time is given by:

$$q = hA\,(T_s - T_\infty)$$

This is known as Newton's law of cooling. The factor h is defined as the convective heat transfer coefficient and defines the ratio of heat transfer per unit area to overall temperature difference. The units of h are W/m^2-K.

RADIATION

The third mode of heat transmission is due to electromagnetic wave propagation, which can occur in a vacuum as well as in a medium. The fundamental Stefan-Boltzmann law is:

$$q = \sigma A T_\infty^{\ 4}$$

where T_∞ is the absolute temperature and σ is a constant equal to 5.67×10^{-8} W/m^2-K.

The ideal emitter or "black body" is one that gives off radiant energy according to the above equation. All real surfaces emit somewhat less than this amount and the thermal emission from such surfaces (gray bodies) can be represented by:

$$q = \varepsilon \sigma A T_\infty^{\ 4}$$

where ε, the surface emissivity, ranges from 0 to 1.

MATERIAL PROPERTIES

The thermal conductivity of substances are readily available in many reference books. If required, this information will be provided as part of the problem. As a general rule, k for a material increases both with increased temperature and increased density. Also available is the mass density ρ and the specific heat capacity c_p. A useful combination of terms is the thermal diffusivity, α, defined by the following:

$$\alpha = \frac{k}{\rho c_p}$$

the α is the ratio of the thermal conductivity to the thermal capacity of the material. Its units are m^2/s.

PROBLEM 1:

Determine the steady-state heat transfer per unit area through a 3.81 cm thick homogeneous wall with its two faces maintained at uniform temperatures of 311 K and 294 K. The thermal conductivity of the material is 0.193 W/m-K.

SOLUTION:

$$\frac{q}{A} = -k\frac{T_2 - T_1}{x_2 - x_1}$$

$$= -0.193 \text{ W/m–K}\left[\frac{294 - 311}{0.0381}\right]\frac{\text{K}}{\text{m}} = 86.1\frac{\text{W}}{\text{m}^2}$$

PROBLEM 2:

The forced convection coefficient for a hot fluid flowing over a cool surface is 120 W/m^2. The fluid temperature upstream from the cool surface is 394 K and the surface is held at 283 K. Determine the heat transfer per unit surface area from the fluid to the surface.

SOLUTION:

$$\frac{q}{A} = h(T_\infty - T_s) = 120 \text{ W/m}^2\text{–K}\,[394 - 283]\text{K} = 13{,}320\frac{\text{W}}{\text{m}^2}$$

PROBLEM 3:

Radiant energy can be sensed by a person standing near a brick wall. Suppose this wall had a surface temperature of 44°C and an emissivity value of 0.92. What would be the radiant thermal flux per square meter from a brick wall at this temperature?

SOLUTION:

$$\frac{q}{A} = \varepsilon\sigma T_s^4 = (0.92)(5.67\times10^{-8})(44+273)^4 = 527\frac{\text{W}}{\text{m}^2}$$

Note that absolute temperature must be used in all radiation calculations.

ONE-DIMENSIONAL STEADY-STATE CONDUCTION

Plane Wall: Fixed Surface Temperature

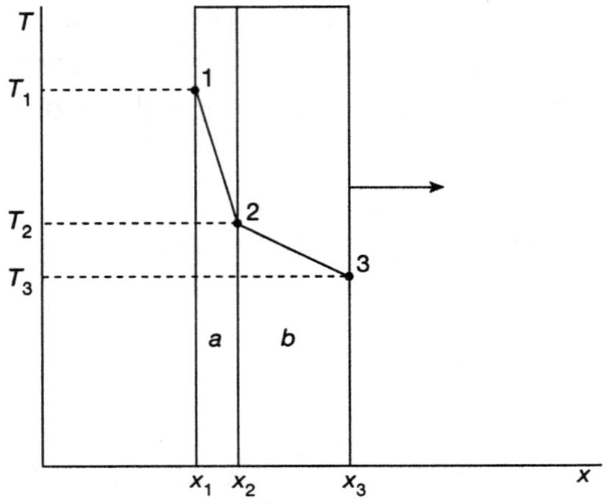

Figure 2. Composite Plane Wall

Integration of Fourier's equation yields the following:

$$q = -kA\frac{T_2 - T_1}{x_2 - x_1} = -kA\frac{T_2 - T_1}{\Delta x}$$

This equation can be rearranged as follows:

$$q = \frac{T_1 - T_2}{\dfrac{\Delta x}{kA}} = \frac{\text{thermal potential difference}}{\text{thermal resistance}}$$

Notice that the resistance to heat flow is directly proportional to the material thickness and inversely proportional to thermal conductivity and area. These principles are readily extended to the composite wall shown in Figure 2. Thus:

$$q = \frac{T_1 - T_2}{\dfrac{\Delta x_a}{k_a A}} = \frac{T_2 - T_3}{\dfrac{\Delta x_b}{k_b A}}$$

where:

k_a = thermal conductivity of section a

k_b = thermal conductivity of section b

Combining these equations gives:

$$q = \frac{T_1 - T_3}{\dfrac{\Delta x_a}{k_a A} + \dfrac{\Delta x_b}{k_b A}}$$

These equations illustrate the analogy between conduction heat transfer and electrical current flow. The terms in the denominator of the above represent the thermal resistance of an element.

$$\text{conductive heat flow} = \frac{\text{overall temperature difference}}{\text{total thermal resistance}}$$

Radial Systems: Fixed Surface Temperature

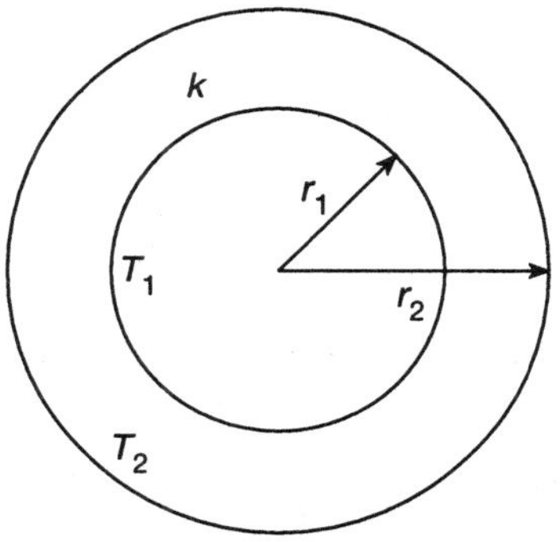

Figure 3. Single-Layer Cylindrical Wall

Figure 3 depicts a single-layer cylindrical wall of a homogeneous material with constant k and uniform surface temperature. Integration of Fourier's equation yields:

$$q = 2\pi k L \frac{T_1 - T_2}{\ln \dfrac{r_2}{r_1}}$$

where L is the length of the cylinder. From this equation, the thermal resistance of the single-cylinder layer is:

$$\frac{\ln \dfrac{r_2}{r_1}}{2\pi k L}$$

For a two-layered cylinder the heat transfer is:

$$q = \frac{T_1 - T_3}{\dfrac{\ln r_2 / r_1}{2\pi k_a L} + \dfrac{\ln r_3 / r_2}{2\pi k_b L}}$$

For a radial system in a spherical shape:

$$q = \frac{4\pi k (T_1 - T_2)}{\dfrac{1}{r_1} - \dfrac{1}{r_2}}$$

HEAT GENERATION SYSTEMS

Plane Wall

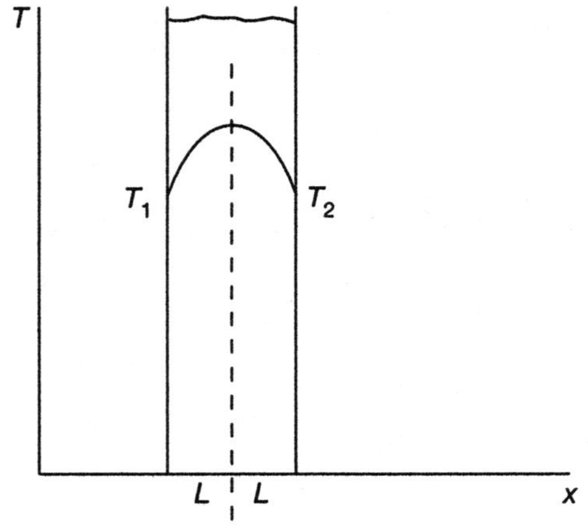

Figure 4. Plane Wall with Uniform Generation

Consider the plane wall in Figure 4 with uniform heat generation $\dot{q}$. If the faces are maintained at T_1 and T_2 respectively, then:

$$T = \left[\frac{T_2 - T_1}{2L} + \frac{\dot{q}}{2k}(2L - x)\right]x + T_1$$

For the case of $T_1 = T_2 = T_s$, the above reduces to:

$$T = T_s + \frac{\dot{q}}{2k}(2L - x)x$$

The heat flux out the left face is given by $q = -\dot{q}AL$. The minus sign indicates that the heat transfer is in the negative x-direction; the product AL is $^1/_2$ the plate volume.

Cylinder

For a long cylinder of constant k and uniform internal energy generation, $\dot{q}$, the temperature distribution:

$$T - T_s = \frac{r_s^{\,2}\dot{q}}{4k}\left[1 - \left(\frac{r}{r_s}\right)^2\right]$$

where r_s is the radius at the surface of the cylinder.

A convenient dimensionless form of the temperature distribution is given in terms of the centerline temperature T_c:

$$\frac{T - T_s}{T_c - T_s} = 1 - \left(\frac{r}{r_s}\right)^2$$

CONVECTION BOUNDARY CONDITIONS

Overall Heat-Transfer Coefficient

It is often useful to express the heat transfer rate for a combined conduction-convection problem in terms of U, the overall heat-transfer coefficient.

Plane Wall

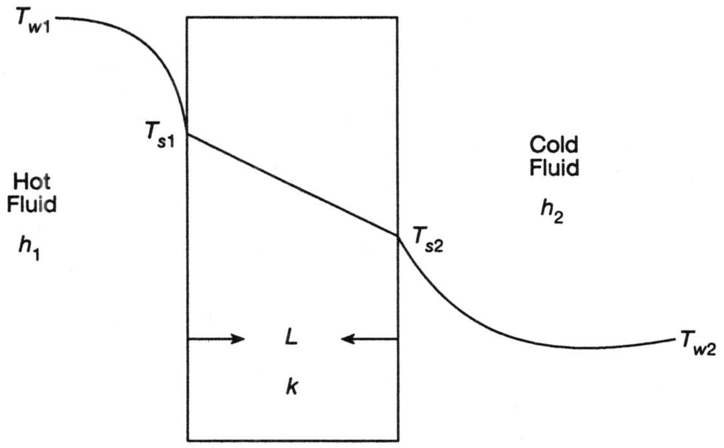

Figure 5. Plane Wall with Convective Boundary Conditions

Consider the plane wall in Figure 5 exposed to hot fluid 1 on its left face and cold fluid 2 on its right face. Applying the principles of convection:

$$q = h_1 A(T_{\infty 1} - T_1) = h_2 A(T_2 - T_{\infty 2})$$

or

$$q = \frac{(T_{\infty 1} - T_1)}{\frac{1}{h_1 A}} = \frac{(T_2 - T_{\infty 2})}{\frac{1}{h_2 A}}$$

In agreement with the electrical analogy, $1/hA$ can be thought of as a thermal resistance to the convective boundary. Combining this resistance factor with the wall's properties yields:

$$q = \frac{T_{\infty 1} - T_{\infty 2}}{\frac{1}{h_1 A} + \frac{L_a}{k_a A} + \frac{1}{h_2 A}} = \frac{T_{\infty 1} - T_{\infty 2}}{R_{tot}} = \frac{(\Delta T)_{overall}}{R_{tot}} = UA\Delta T$$

From the above it is obvious that $UA = 1/R_{\text{tot}}$. For a multilayer plane wall consisting of layers *a*, *b*, ... *U* can be expressed as follows:

$$U = \frac{1}{\frac{1}{h_1} + \frac{L_a}{k_a} + \frac{L_b}{k_b} + \ \dots \ + \frac{1}{h_2}}$$

For plane walls, both areas are considered the same. For other geometries, the surface on which the *U* is based must be specified.

Radial Systems

If we define U_1 as the overall heat transfer coefficient based on the inside area of a cylinder and r_1 as the inside radius of a cylindrical system of three composite layers, then:

$$U_1 = \frac{1}{\frac{1}{h_1} + \frac{r_1}{k_a}\ln\frac{r_2}{r_1} + \frac{r_1}{k_b}\ln\frac{r_3}{r_2} + \frac{r_1}{k_c}\ln\frac{r_4}{r_3} + \frac{r_1}{r_4}\frac{1}{h_4}}$$

Refer to Problem 4 for an exercise with U_o.

Critical Thickness of Insulation

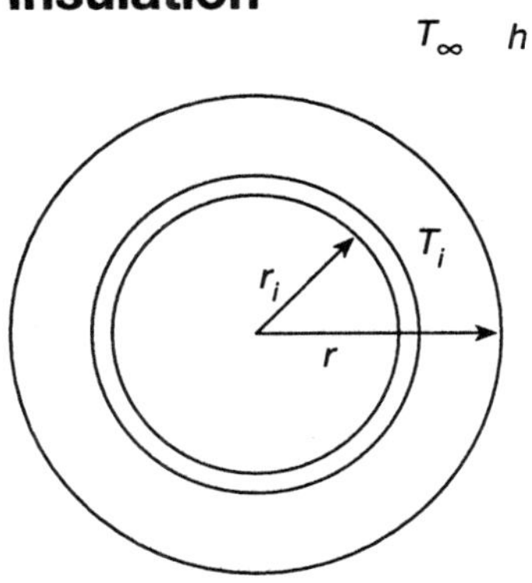

Figure 6. Critical Radius of Insulation

In many cases, the thermal resistance offered by a metal pipe or duct is small in comparison with that of the insulation. Also, the pipe wall temperature is often very nearly the same as that of the fluid inside the pipe. For a single layer of insulation material, the heat transfer rate is given by:

$$\frac{q}{L} = \frac{T_i - T_\infty}{\frac{\ln r / r_i}{2\pi k} + \frac{1}{2\pi h r}}$$

As a function of r, q/L has a minimum at $r = r_{crit} = k/h$. Thus for $r_i < r_{crit}$, the heat loss increases with the addition of insulation and then decreases with further addition of insulation.

HEAT TRANSFER FROM FINS

Extended surfaces or fins are used to increase the effective area for convective heat transfer in heat exchangers, internal combustion engines, heat sinks, etc. For straight fins of constant cross-sectional area the following equation can be used for convection boundary conditions:

$$q = \sqrt{hpkA_c}\,(T_b - T_\infty) \tan h(mL_c)$$

$$\frac{T - T_\infty}{T_b - T_\infty} = \frac{\cos hm(L - x)}{\cos hmL}$$

where:

p = exposed perimeter

A_c = cross-sectional area

z = fin width ($z > t$)

t = fin thickness

T_b = temperature at base of fin

T_∞ = fluid temperature

$m = \sqrt{hp/(kA_c)}$

$L_c = L + A_c/p = L + t/2$

For infinitely long fins:

$$q = \sqrt{hpkA_c}\,(T_b - T_\infty)$$

$$\frac{T - T_\infty}{T_b - T_\infty} = e^{-mx}$$

PROBLEM 4:

Steam at 121°C flows in an insulated pipe. The pipe is mild steel (k = 56.7 W/m-K) and has an inside radius of 5 cm and an outside radius of 5.7 cm. The pipe is covered with a 2.54-cm layer of 85% magnesia (k = 0.06 W/m-K). The inside heat transfer coefficient is 85 W/m^2-K and the outside coefficient is 12.5 W/m^2-K. Determine the overall heat-transfer coefficient U_o and the heat transfer rate per foot of pipe length if the surrounding air temperature is 18.3°C.

SOLUTION:

$$U_o = \frac{1}{\dfrac{r_3}{r_1 h_i} + \dfrac{r_3 \ln r_2 / r_1}{k_{steel}} + \dfrac{r_3 \ln r_3 / r_2}{k_{ins}} + \dfrac{1}{h_o}}$$

Changing cm to m and substituting:

$$U_o = \frac{1}{\dfrac{0.0824\text{ m}}{(0.05\text{ m})(85\text{ W/m}^2\text{–K})} + \dfrac{0.0824\text{ m}\ln(0.057/0.05)}{56.7\text{ W/m–K}} + \dfrac{0.0824\text{ m}\ln(0.0824/0.057)}{0.06\text{ W/m–K}} + \dfrac{1}{12.5\text{ W/m}^2\text{–K}}}$$

$U_o = 1.65\text{ W/m}^2\text{–K}$

$$\frac{q}{L} = \frac{U_oA(\Delta T)_{overall}}{L} = (1.65)(2)(\pi)(0.0824)(121 - 18.3) = 87.7 \text{ W/m}$$

Note that in the above problem, $A/L = 2\pi r_o L/L = 2\pi r_o$.

PROBLEM 5:

Determine the critical radius in cm for an asbestos-covered pipe (k_{abs} = 0.208 W/m-K) if the external heat transfer coefficient is 8.51 W/m^2-K.

SOLUTION:

$$r_{crit} = \frac{k}{h} = \frac{0.208 \text{ W/m-K}}{8.51 \text{ W/m}^2\text{-K}} = 0.0244 \text{ m} = 2.44 \text{ cm}$$

PROBLEM 6:

An aluminum fin (k = 200 W/m-K) is 0.3 cm thick, 3 cm wide, and 7.5 cm long. It protrudes from a wall with its base maintained at 300°C and an ambient temperature of 50°C. For h = 10 W/m^2-K, calculate the heat loss.

SOLUTION:

$$q = \sqrt{hpkA_c}\,(T_b - T_\infty)\tan h(mL_c)$$

$$A_c/p = (0.03)(0.003)/2(0.03 + 0.003) = 0.00136 \text{ m}$$

$$L_c = L + A_c/p = 0.075 \text{ m} + 0.00136 \text{ m} = 0.0764 \text{ m}$$

$$m = \sqrt{hp/(kA_c)} = \sqrt{\frac{10 \text{ W/m}^2\text{-K}}{(0.00136 \text{ m})(200 \text{ W/m-K})}} = 6.06 \text{ m}^{-1}$$

$$q = \sqrt{(10 \text{ W/m}^2\text{-K})(2 \times 0.033 \text{ m})(200 \text{ W/m-K})(0.03 \times 0.003 \text{ m}^2)}$$

$$\times (300 - 50)^\circ\text{C} \times \tan h[6.06 \times 0.0764]$$

$$= 11.8 \text{ W}$$

TIME-VARYING CONDUCTION

The Biot number (Bi) is a dimensionless parameter that can be expressed as:

$$Bi = \frac{\text{resistance to internal heat flow}}{\text{resistance to external heat flow}}$$

When the Bi is small, the internal temperature gradients are also small and the problem is solved using the "lumped heat-capacity" method that assumes the body temperature to be uniform throughout the body. In terms of physical parameters:

$$Bi = \frac{hL}{k}$$

For $Bi < 0.1$, the lumped heat-capacity method is valid.

In the above definition, L equals the characteristic dimension of the body.

$$L = \frac{V}{A_s}$$

where V is the volume of the body and A_s is the surface area. The characteristic length of a sphere is $r/3$. The characteristic length of a cylinder is $r/2$. When the lumped heat-capacity method is valid, the following equation is used to determine the temperature as a function of time:

$$\frac{T - T_\infty}{T_i - T_\infty} = e^{-\frac{hA_s}{\rho c V}t}$$

where:

T_i = initial temperature of body

T_∞ = ambient temperature

T = body temperature at a given time

ρ = density of body, kg/m^3

c = heat capacity of body, J/kg-K

t = time

PROBLEM 7:

A steel ball (c = 460 J/kg-K, k = 35 W/m-K) 5 cm in diameter is initially at a uniform temperature of 723 K. The ball is placed in an environment maintained at 373 K with a resulting heat transfer coefficient h = 10 W/m^2-K. Calculate the time required for the ball to attain a temperature of 423 K.

SOLUTION:

First we must check to see if the lumped heat-capacity method is appropriate.

$L_c = r/3$ for a sphere. $L_c = 0.025 \text{ m}/3 = 0.00833 \text{ m}$

$$Bi = \frac{hL}{k} = \frac{(10 \text{ W/m}^2\text{-K})(0.00833 \text{ m})}{(35 \text{ W/m-K})} = 0.0023 < 0.1$$

$$\frac{hA}{\rho c V} = \frac{h}{\rho c L} = \frac{10 \text{ W/m}^2\text{-K}}{(7{,}800 \text{ kg/m}^3)(460 \text{ J/kg-K})(0.00833 \text{ m})}$$

$$= 3.35 \times 10^{-4}\, s^{-1}$$

$$\frac{T - T_\infty}{T_i - T_\infty} = \frac{423 - 373}{723 - 373} = e^{-\frac{hA_s}{\rho c V}t} = e^{-3.35\times10^{-4}\,t}$$

$$t = 5{,}800$$

$$s = 1.6 \text{ hours}$$

CONVECTION HEAT TRANSFER

There are many parameters affecting convection heat transfer in a particular geometry. These include the system length scale L, the fluid thermal conductivity k, the fluid velocity V, density ρ, viscosity μ, specific heat c_p, and other factors related to the manner of heating (e.g., uniform wall temperature).

Dimensionless parameters are used to correlate data and in turn calculate the values of h. The Nusselt number is defined as:

$$Nu = \frac{hL}{k}$$

where L is the characteristic length for a given geometry. The Nu is usually correlated in terms of the Prandtl number and Reynold number defined as follows:

$$Pr = \frac{\mu c_p}{k}$$

$$Re = \frac{\rho VL}{\mu}$$

The Graetz number is used to correlate the entry problem and is given by:

$$Gz = Re\ Pr\ (D/L)$$

The Pr is indicative of the relative ability of the fluid to diffuse momentum and internal energy by molecular mechanisms. The Re is indicative of the relative importance of inertial and viscous effects in the fluid motion. At low Re, the viscous effects dominate and motion is laminar. At high Re the inertial effects give way to turbulence and dominate the momentum and energy-transfer processes.

CONVECTION CORRELATIONS

The following correlations are used to solve convection problems. For laminar flow ($Re < 2{,}000$) in a closed conduit:

$$Nu = 3.66 + \frac{0.19\mathrm{Gz}^{0.8}}{1 + 0.117\mathrm{Gz}^{0.467}}$$

For turbulent flow ($Re > 10^4$, $Pr > 0.7$) in a closed conduit the Sieder-Tate equation applies:

$$Nu = \frac{h_i D}{k_f} = 0.023 Re^{0.8} Pr^n;\ n = 0.4 \text{ for } T_w > T_b;\ n = 0.3 \text{ for } T_w < T_b$$

where:

T_b = local bulk temperature of fluid

T_w = local inside surface temperature of tube

k_f = thermal conductivity of the fluid

Re and Pr are evaluated at T_b.

For non-circular ducts, use the equivalent diameter defined as:

$$D_{eq} = \frac{4 \times \text{cross sectional area}}{\text{wetted perimeter}}$$

For liquid metals ($0.003 < Pr < 0.05$) in a closed conduit:

$Nu_D = 4.82 + 0.0185(Re_D Pr)^{0.827}$ (constant heat flux)

$Nu_D = 5.0 + 0.025(Re_D Pr)^{0.8}$ (constant surface temperature)

The heat transfer coefficient for condensation of a pure vapor on a vertical surface is:

$$Nu_l = \frac{h_L L}{k} = 0.943\left[\frac{\rho_l g(\rho_l - \rho_v)\lambda L^3}{\mu_l k_l (T_{sat} - T_s)}\right]^{1/4}$$

Note that $\rho_l(\rho_l - \rho_v) \approx \rho_l^2$.

where:

λ = heat of vaporization

T_{sat} = saturation temperature of vapor

T_s = temperature of surface

Properties other than λ are for the liquid and are evaluated at the average between T_{sat} and T_s.

For condensation outside horizontal tubes:

$$Nu_D = \frac{h_D D}{k} = 0.729\left[\frac{\rho_l g(\rho_l - \rho_v)\lambda D^3}{\mu_l k_l (T_{sat} - T_s)}\right]^{1/4}$$

HEAT EXCHANGER THEORY

The log mean temperature difference (*LMTD*) for counterflow in a tubular heat exchanger is:

$$\Delta T_{lm} = \frac{(T_{Ho} - T_{Ci}) - (T_{Hi} - T_{Co})}{\ln\left(\dfrac{T_{Ho} - T_{Ci}}{T_{Hi} - T_{Co}}\right)}$$

The log mean temperature difference (*LMTD*) for parallel flow in a tubular heat exchanger is:

$$\Delta T_{lm} = \frac{(T_{Ho} - T_{Co}) - (T_{Hi} - T_{Ci})}{\ln\left(\dfrac{T_{Ho} - T_{Co}}{T_{Hi} - T_{Ci}}\right)}$$

where:

ΔT_{lm} = log mean temperature difference

T_{Hi} = inlet temperature of hot fluid

T_{Ho} = outlet temperature of the hot fluid

T_{Ci} = inlet temperature of the cold fluid

T_{Co} = outlet temperature of the cold fluid

EXTERNAL HEAT TRANSFER

In all cases that follow, properties are evaluated at T_f, the film temperature, which is the average of the body surface temperature and the fluid temperature. The values listed are average values over the entire length of the surface.

Heat transfer for parallel flow to a constant-temperature flat plate is determined by:

$Nu_L = 0.648\ Re^{0.5}\ Pr^{1/3}$ ($Re < 10^5$; laminar flow)

$Nu_L = 0.0366\ Re^{0.8}\ Pr^{1/3}$ ($Re > 10^5$; turbulent flow)

Heat transfer for flow perpendicular to axis of a constant-temperature cylinder is:

$$Nu_D = cRe^n\ Pr^{1/3}$$

Re	*n*	*c*
1–4	0.330	0.989
4–40	0.385	0.911
40–4,000	0.466	0.683
4,000–40,000	0.618	0.193
40,000–250,000	0.805	0.0266

Heat transfer for flow past a constant-temperature sphere is:

$$Nu_D = 2.0 + 0.6\, Re^{0.5}\, Pr^{1/3} \qquad (1 < Re < 70{,}000,\ 0.6 < Pr < 400)$$

FREE CONVECTION

The following correlation can be used for a vertical flat plate (or a vertical cylinder of sufficiently large diameter) and a stationary fluid:

$$h_L = C\,(k/L)\,Ra_L{}^n$$

where:

L = length of plate in vertical direction

$$Ra_L = \text{Rayleigh Number} = \frac{g\beta(T_s - T_\infty)L^3}{\nu^2} Pr$$

T_s = surface temperature

T_∞ = fluid temperature

β = coefficient of thermal expansion = $2/(T_s + T_\infty)$ for an ideal gas. Temperatures are absolute.

Range of Ra_L	C	n
$10^4 - 10^9$	0.59	1/4
$10^9 - 10^{13}$	0.10	1/3

For free convection between a long horizontal cylinder and a stationary fluid:

$$h_D = C\,(k/D)\,Ra_D{}^n$$

where:

$$Ra_D = \frac{g\beta(T_s - T_\infty)D^3}{\nu^2} Pr$$

Range of Ra_D	C	n
10^{-3}–10^2	1.02	0.148
10^2–10^4	0.850	0.188
10^4–10^7	0.480	0.250
10^7–10^{12}	0.125	0.333

THERMAL RADIATION

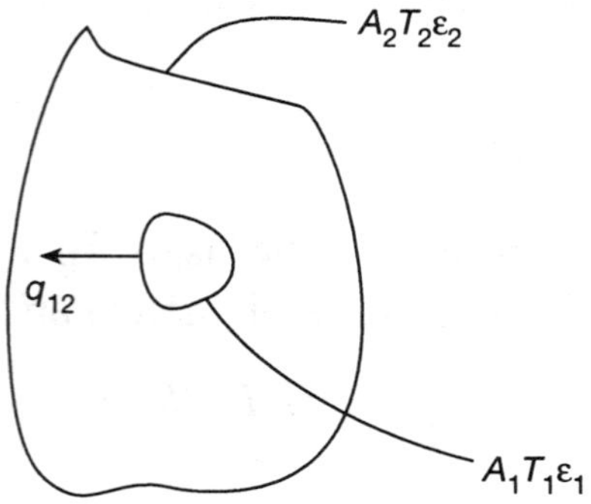

Figure 7. Two Body Radiation System

All bodies radiate energy at a rate proportional to its absolute temperature. When this radiant energy reaches another surface, it is either absorbed, reflected, or transmitted through the surface. This gives rise to the following surface parameters:

α = absorptivity, the fraction of radiation that is absorbed

ρ = reflectivity, the fraction of radiation that is reflected

τ = transmissivity, the fraction of radiation that is transmitted

A black body emits the maximum amount of radiation possible for a given temperature. Real bodies emit radiation at a lesser rate. The emissivity, ε, is used to account for this and is defined as follows:

$$\varepsilon = \frac{E}{E_b} = \frac{\text{actual emissive power}}{\text{emissive power of a black body}}$$

The absorptivity is related to the emissivity through Kirchhoff's law which states that at thermal equilibrium $\varepsilon = \alpha$.

All surfaces radiate energy, but there will be a net energy exchange only if the bodies are at different temperatures. The shape factor, F_{12} represents the fraction of radiation emitted from surface 1 that is intercepted by surface 2. If surface 1 "sees" only surface 2, the $F_{12} = 1$ and $F_{21} = (A_1/A_2)$.

For black body radiation between surface 1 and surface 2, the net heat transfer from 1 $\rightarrow$ 2 is expressed as:

$$q_{12} = A_1F_{12}(E_{b1} - E_{b2}) = \sigma\, A_1F_{12}(T_1{}^4 - T_2{}^4)$$

To account for reflections that occur in real surfaces, the surfaces are assumed to be "gray." The net heat transfer between two bodies that see each other and nothing else (for example, two infinite parallel planes or a small object surrounded completely by a large object) is given by:

$$q_{12} = \frac{\sigma\left(T_1^4 - T_2^4\right)}{\dfrac{1-\varepsilon_1}{\varepsilon_1 A_1} + \dfrac{1}{A_2 F_{12}} + \dfrac{1-\varepsilon_2}{\varepsilon_2 A_2}}$$

SHAPE FACTOR RELATIONS

Reciprocity relations:

$$A_i F_{ij} = A_j F_{ji}$$

Summation rule:

$$\sum_{j=1}^{N} F_{ij} = 1$$

RADIATION SHIELDS

Radiation shields as depicted in Figure 8 have one-dimensional geometry with a low-emissivity shield inserted between two parallel plates. The relationship is expressed as:

$$q_{12} = \frac{\sigma\left(T_1^4 - T_2^4\right)}{\dfrac{1-\varepsilon_1}{\varepsilon_1 A_1} + \dfrac{1}{A_2 F_{13}} + \dfrac{1-\varepsilon_{3,1}}{\varepsilon_{3,1} A_3} + \dfrac{1-\varepsilon_{3,2}}{\varepsilon_{3,2} A_3} + \dfrac{1}{A_3 F_{32}} + \dfrac{1-\varepsilon_2}{\varepsilon_2 A_2}}$$

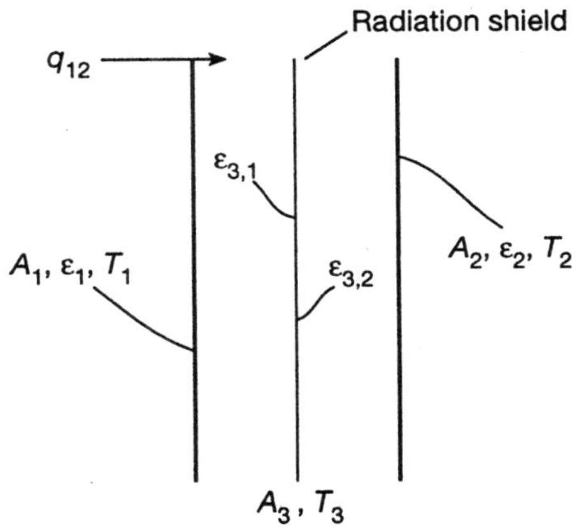

Figure 8. Radiation Shield

Reradiating Surface

Reradiating surfaces are considered to be insulated or adiabatic and are depicted in Figure 9.

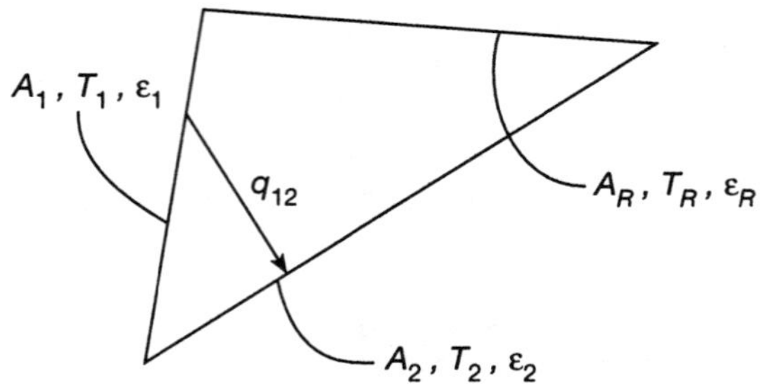

Figure 9. Reradiating Surface

This produces the following relationship:

$$q_{12} = \frac{\sigma\left(T_1^4 - T_2^4\right)}{\dfrac{1-\varepsilon_1}{\varepsilon_1 A_1} + \dfrac{1}{A_1 F_{12} + \left[\left(\dfrac{1}{A_1 F_{1R}} + \dfrac{1}{A_2 F_{2R}}\right)\right]^{-1}} + \dfrac{1-\varepsilon_2}{\varepsilon_2 A_2}}$$

PROBLEM 8:

A 1.5 m by 1.5 m vertical plate maintained at 278 K is exposed to saturated water vapor at 286 K resulting in condensation on the plate. Find:

1) The Nu for this process,

2) The heat transfer coefficient, h_L, and

3) The rate of condensation on the plate

SOLUTION:

1) The Nu is given by the following expression:

$$Nu_l = \frac{h_L L}{k} = 0.943\left[\frac{\rho_l g(\rho_l - \rho_v)\lambda L^3}{\mu_l k_l (T_{sat} - T_s)}\right]^{1/4}$$

λ, the heat of vaporization for water, is evaluated at 286 K.

$$\lambda = 2{,}471 \text{ kJ/kg} = 2.471 \text{ x } 10^6 \text{ J/kg.}$$

The remaining properties are evaluated at $T_f = (278 + 286)/2 = 282$ K

$\rho = 1{,}000$ kg/m³ $\qquad k = 0.585$ W/m-K $\qquad T_{sat} = 286$ K

$\mu = 1.34 \text{ x } 10^{-3}$ $\qquad L = 1.5$ m $\qquad g = 9.8$ m/s²

$$Nu_l = 0.943\left[\frac{(1{,}000)^2(9.8))(2.47\times10^6)(1.5^3)}{(1.34\times10^{-3})(0.585)(286-278)}\right]^{1/4} = 10{,}100$$

2) The heat transfer coefficient, h_L, is given by:

$$h_L = Nu_L \times (k/L) = (10{,}100) \times (0.585) \,/\, (1.5) = 3{,}930 \text{ W/m}^2\text{-K}$$

3) The rate of condensation on the plate is given by the following expression:

$$\dot{m} = \frac{q}{h_{fg}};\; q = hA(T_{sat} - T_s)$$

$$= (3{,}930) \times (1.5 \times 1.5) \times (286 - 278) = 70{,}740 \text{ W}$$

$$\dot{m} = \frac{70{,}740}{2.471\times10^6} = 0.0286 \text{ kg/s} = 103 \text{ kg/hr}$$

PROBLEM 9:

Air flows at 5 kg/h through an electrically heated 0.5 cm diameter tube, 0.5 m long, entering at 100°C. The electric power dissipation is 200 W. Determine:

1) The outlet temperature of the air,

2) The *Re* of the air flowing in the tube,

3) The connection coefficient, *h,* for this process, and

4) The maximum tube temperature

SOLUTION:

1) Using a first law energy balance on the pipe and assuming the air behaves as an ideal gas:

$$q = \dot{m}(h_2 - h_1) = \dot{m}c_p(T_2 - T_1)$$

$$T_2 = T_1 + \frac{q}{\dot{m}c_p} = 100 + \frac{(200)\times(3,600)}{(5)\times(1.004)\times(10^3)} = 243°\text{C}$$

2) The *Re* of the air flowing in the tube is given by the following:

$$Re = \frac{VD\rho}{\mu} = \frac{\dot{m}D}{A\mu} = \frac{4\dot{m}}{\pi\mu D}$$

The properties are to be evaluated at the average temperature of 170°C:

$\mu = 2.48 \times 10^{-5}$ N-s/m^2

$k = 0.0368$ W/m-K

$Pr = 0.69$

$$Re = \frac{4\times 5}{3,600\times\pi\times 2.48\times 10^{-5}\times 0.005} = 14,260$$

3) The convective heat transfer coefficient for this process is given by the following:

$$h = Nu\ (k/D)$$

Since the flow is definitely turbulent and this is a heating process, then:

$Nu = 0.023\ Re^{0.8}\ Pr^{0.4} = 0.023\ (14{,}260)^{0.8}\ (0.69)^{0.4} = 41.7$

$h = 41.7 \times 0.0368/0.005 = 307$ W/m^2—°C

4) The maximum tube wall temperature is found from the following:

$q = hA(T_s - T_m)$ 200 W

$(T_s - T_m) = 200/(hA) = 200/(h)(\pi DL)$

$= 200/(307)(\pi)(0.005)(0.5) = 82.9°\text{C}$

The maximum surface temperature becomes 243 + 83 = 326°C.

PROBLEM 10:

Two parallel plates 1 m by 2 m are maintained at 1,000 K and 500 K, respectively. The configuration factor between the plates, F_{12}, equals 0.5.

If the plates are both assumed to be black bodies, find the net heat transfer between them.

SOLUTION:

$$q = A_1 F_{12} \sigma (T_1{}^4 - T_2{}^4)$$
$$= (2 \times 1)(0.5)(5.67 \times 10^{-8})((1{,}000)^4 - (500)^4) = 53{,}000 \text{ W}$$

PROBLEM 11:

If the plates are now assumed to be gray, each surface having an emissivity of 0.5, find the net heat transfer between them.

SOLUTION:

$$q_{12} = \frac{\sigma\left(T_1^4 - T_2^4\right)}{\dfrac{1-\varepsilon_1}{\varepsilon_1 A_1} + \dfrac{1}{A_2 F_{12}} + \dfrac{1-\varepsilon_2}{\varepsilon_2 A_2}}$$

$$= \frac{5.67 \times 10^{-8}\left(1000^4 - 500^4\right)}{\dfrac{1-0.5}{(0.5)\times(2)} + \dfrac{1}{(2)\times(0.5)} + \dfrac{1-0.5}{(0.5)\times(2)}} = 26{,}600 \text{ W}$$

PROBLEM 12:

In an effort to reduce the radiation exchange between the two plates, a polished shield is now placed between the original plates. Assume that this shield has an emissivity of 0.1 on both sides. The configuration factor between all the plates and the shield, and the shield and each plate both now equal 0.7. Find net heat transfer between the plates.

SOLUTION:

$$q_{12} = \frac{\sigma\left(T_1^4 - T_2^4\right)}{\dfrac{1-\varepsilon_1}{\varepsilon_1 A_1} + \dfrac{1}{A_2 F_{13}} + \dfrac{1-\varepsilon_{3,1}}{\varepsilon_{3,1} A_3} + \dfrac{1-\varepsilon_{3,2}}{\varepsilon_{3,2} A_3} + \dfrac{1}{A_3 F_{32}} + \dfrac{1-\varepsilon_2}{\varepsilon_2 A_2}}$$

$$= \frac{5.67 \times 10^{-8}\left(1{,}000^4 - 500^4\right)}{\dfrac{1-0.5}{(0.5)\times(2)} + \dfrac{1}{(2)\times(0.7)} + \dfrac{1-0.1}{(0.1)\times(2)} + \dfrac{1-0.1}{(0.1)\times(2)} + \dfrac{1}{(2)\times(.7)} + \dfrac{1-0.5}{(0.5)\times(2)}}$$

$$= 4{,}651 \text{ W}$$

FE/EIT

FE: PM Mechanical Engineering Exam

CHAPTER 9

Material Behavior

CHAPTER 9

MATERIAL BEHAVIOR

Mechanical engineers must choose the materials of which components are to be made. One must consider the material's ability to provide an adequate service life for the application. Some of these considerations include the strength required, resistance to corrosion, electrical/magnetic properties, weight considerations, resistance to fatigue loading, and suitability for high or low temperatures. Aesthetic considerations are also frequently required, including color, texture, shape, etc. The ease of fabrication and maintainability must also be considered. Finally, one must consider the cost of producing the component. Seldom, if ever, is there a material that maximized all of the considerations and, therefore, material selection is a trade-off. This section deals with potential engineering materials and their fabrication.

ENGINEERING MATERIALS—METALLIC

Iron-Based Alloys

Of the metallic materials, Iron-based alloys are certainly the most commonly used. Central to understanding Iron-based alloys is the combination of Iron with Carbon. Figure 1 is the equilibrium phase diagram for Iron and Carbon.

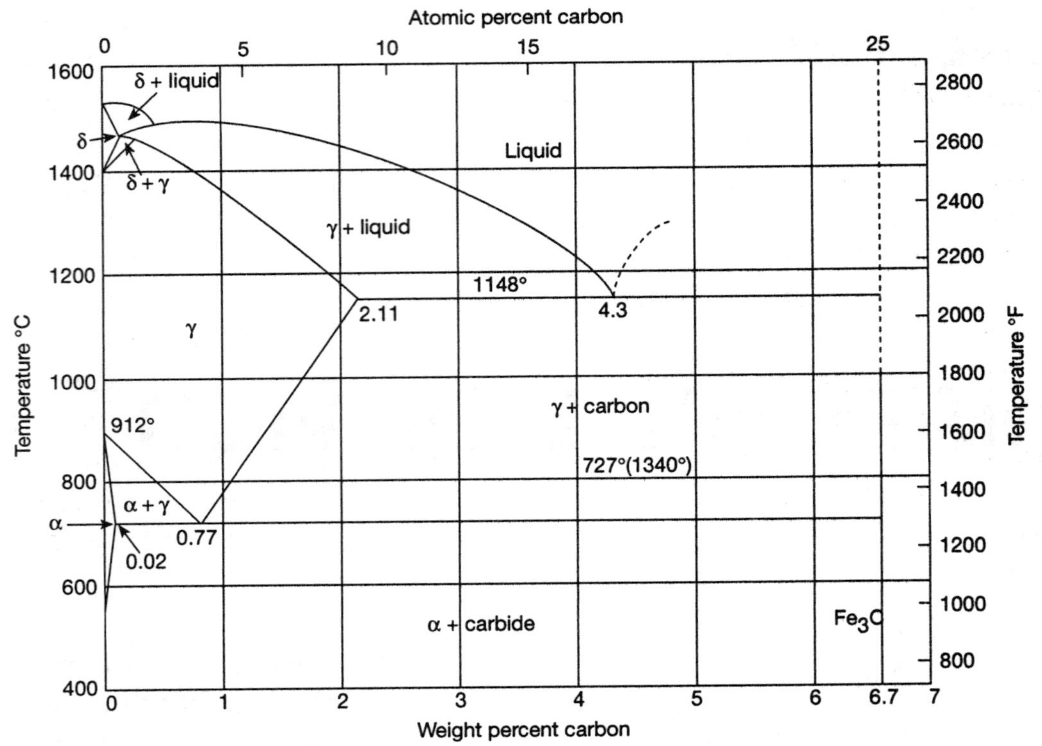

Figure 1. Iron-Carbon Equilibrium Phase Diagram

Steels

Steels are Iron-based alloys with less than 2.0 weight percent Carbon. In Figure 1, note the eutectoid at 0.77 weight percent Carbon and 727°C. Plain Carbon steels with less than 0.77 weight percent Carbon are called hypoeutectoid steels, those with greater than 0.77 weight percent Carbon are called hypereutectoid steels, and steel with 0.77 weight percent Carbon is eutectoid steel. The following are the most important equilibrium constituents for the Iron-Carbon system.

Austenite (sometimes called gamma Iron) is a Face Center Cubic (FCC) matrix of Iron in which the Carbon is in solution. This is an equilibrium phase at temperatures above the eutectoid temperature (727°C for Iron alloyed with only Carbon). When eutectoid Austenite is slowly cooled just below 727°C, the equilibrium constituents are Ferrite (sometimes called alpha Iron), consisting of a saturated Carbon solution in Body Center Cubic (BCC) Iron and Cementite (Fe_3C), an intermetallic compound of Iron and Carbon. In the transformation from eutectoid Austenite to Ferrite and Cementite, the Carbon segregates to form layers of Cementite, and between the layers of Cementite, Ferrite forms. This lamellar structure is called Pearlite.

Now let's examine a hypoeutectoid steel with 0.5 weight percent Carbon, as shown in Figure 2 (a). As the Austenite cools below approximately 780°C, Ferrite begins to form along with the Austenite. This is called proeutectoid or primary Ferrite. Because the Ferrite only has approximately 0.01 weight percent Carbon, the remaining Austenite becomes richer in Carbon. As the steel is slowly cooled to just above the eutectoid temperature, the remaining Austenite is at the eutectoid composition (0.77 weight percent Carbon). As the steel is cooled to just below the eutectoid temperature, the remaining Austenite turns to Pearlite, as discussed above. Similarly, a hypoeutectoid steel when cooled will form proeutectoid or primary Cementite then, below the eutectoid temperature, Pearlite will form from the remaining Austenite. The equilibrium composition and fraction of the steel that is in various phases can be calculated by using the lever rule. The following problem will demonstrate the use of the lever rule.

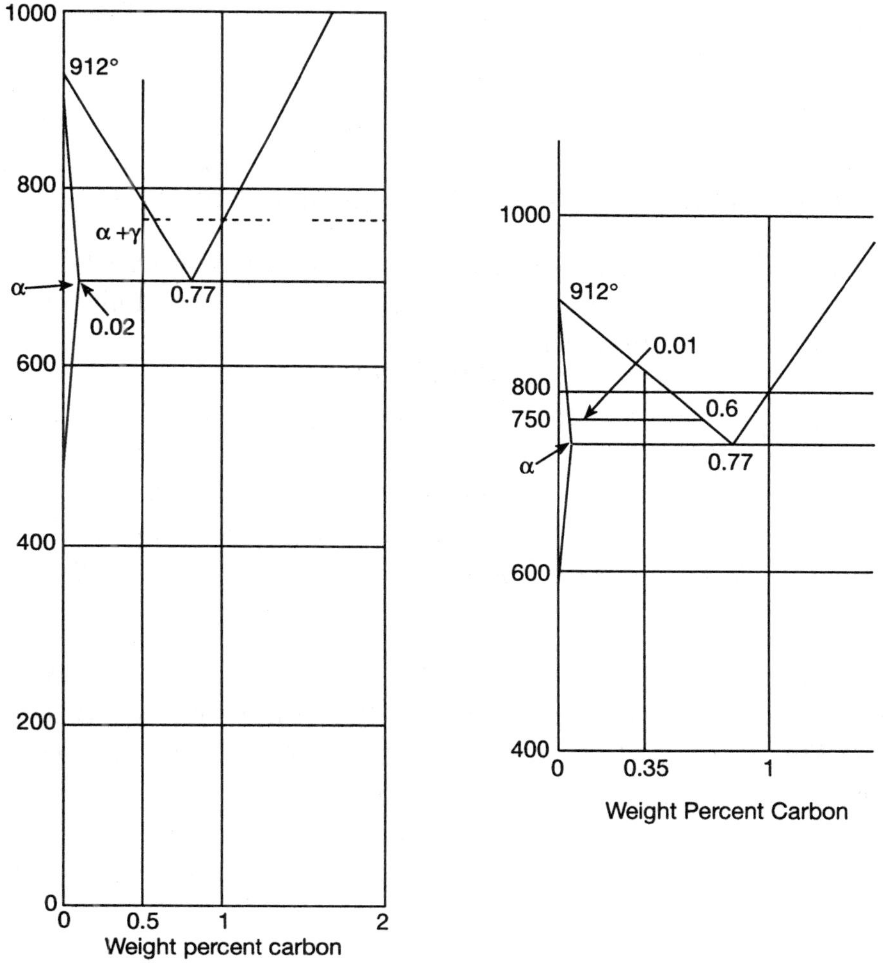

(a) 0.5 weight % Carbon **(b) 0.35 weight % Carbon**

Figure 2. Iron-Carbon Equilibrium Phase Diagram

PROBLEM 1:

A plain carbon steel with 0.35 weight percent Carbon is Austenitized (heated above the eutectoid temperature to ensure that it completely changes to Austenite). It is slowly cooled, such that it can be considered to be in equilibrium.

(a) As the steel is cooled, what is the primary phase that forms and at what temperature will it start to appear?

(b) At 750°C, what is the composition of the primary phase and the Austenite? What weight percent of the steel is in the primary phase and what weight percent is Austenite?

(c) At room temperature, what weight percent of the steel is the primary phase and what weight percent is the Pearlite? What weight percent is Ferrite and what weight percent is Cementite?

Refer to Figure 2 (b).

SOLUTION:

(a) The primary phase is Ferrite, since this is a hypoeutectoid steel. The primary Ferrite will start to form at approximately 810°C.

(b) At 750°C, the equilibrium composition of the primary Ferrite is approximately 0.01 weight percent Carbon and of the Austenite is approximately 0.6. Using the lever rule:

Ferrite: $$\text{wt \% } \alpha = \frac{x_\gamma - x}{x_\gamma - x_\alpha}(100) = \frac{0.6 - 0.35}{0.6 - 0.01}(100) = 42.4\%$$

Austenite: $$\text{wt \% } \gamma = \frac{x - x_\alpha}{x_\gamma - x_\alpha}(100) = \frac{0.35 - 0.01}{0.6 - 0.01}(100) = 57.6\%$$

(c) The formation of all of the primary Ferrite occurs just as the steel reaches the eutectoid temperature. The Austenite that remains becomes Pearlite. Therefore we have to look at the compositions and weight percentages at the eutectoid temperature, as follows:

Primary Ferrite: $$\text{wt \% } \alpha_\rho = \frac{x_\gamma - x}{x_\gamma - x_\alpha}(100) = \frac{0.77 - 0.35}{0.77 - 0.02}(100) = 56\%$$

Pearlite: $$\text{wt \% Pearlite} = \frac{x - x_\alpha}{x_\gamma - x_\alpha}(100) = \frac{0.35 - 0.02}{0.77 - 0.02}(100) = 44\%$$

Total Ferrite: $$\text{wt \% } \alpha = \frac{x_{Fe_3C} - x}{x_{Fe_3C} - x_\alpha}(100) = \frac{6.7 - 0.35}{6.7 - 0.02}(100) = 95\%$$

Total Carmentite: $$\text{wt \% } Fe_3C = \frac{x - x_\alpha}{x_{Fe_3C} - x_\alpha}(100) = \frac{0.35 - 0.02}{6.7 - 0.02}(100) = 5\%$$

Nonequilibrium Structures

If the steel alloy is cooled slowly, proeutectoid Ferrite (hypoeutectoid steel) or Cementite (hypereutectoid steel) forms. If the steel is cooled faster, the proeutectoid constituent grains are smaller and more numerous and the lamellae in the Pearlite are smaller and smaller. At even faster cooling rates, the proeutectoid constituent and Pearlite lamellar structure no longer appear and the structure is called coarse (upper) Bainite. If cooled even faster, a diffusionless transformation occurs by a shear between atom planes to a Body Center Tetragonal form of steel called Martensite. The transformations to Pearlite and Bainite are dependent on both temperature and time. The transformation to Martensite, however, is only dependent on temperature. The Martensite begins to form at the Martensite start temperature (M_s) and is fully complete at the Martensite finish

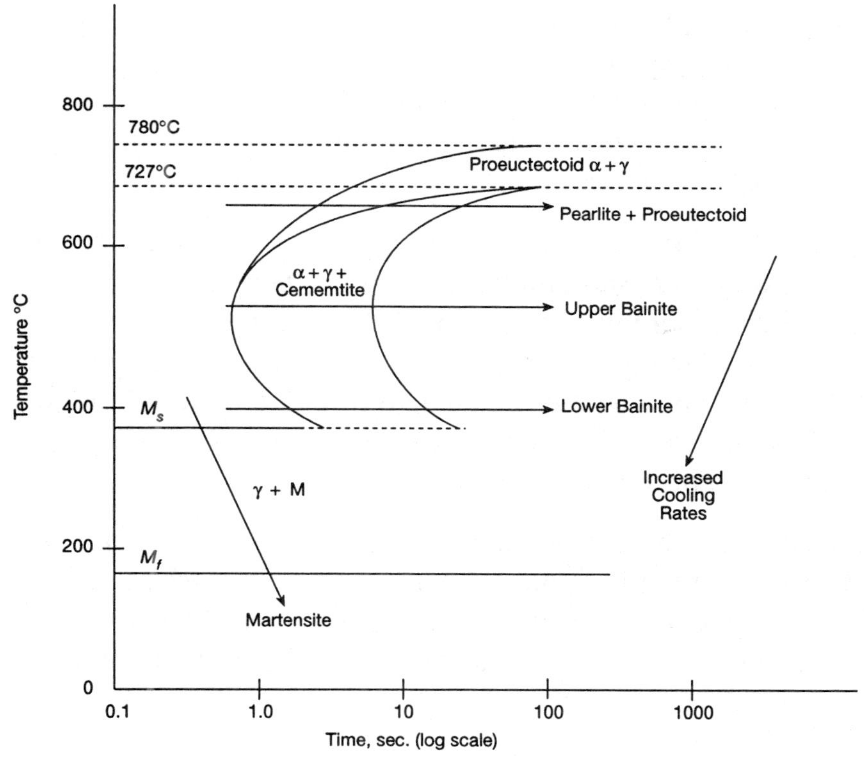

Figure 3. Time-Temperature-Transformation Curve

temperature (M_f). Figure 3 shows a representative Time-Temperature-Transition curve for a plain Carbon steel. It is important to note that in some steels, the transformation to the proeutectoid constituent and Pearlite is so rapid that the Martensitic transformation can only be produced in very thin sections (too thin for practical applications) that are quenched extremely rapidly. For practical applications, Martensite will not be produced in these alloys.

Heat Treatments of Steel

A common heat treatment to produce a very strong, hard steel is to quench the steel to form Martensite. Martensite is very strong and hard, but quite brittle. If the temperature of Martensite is raised to allow limited diffusion, to relieve some of the distortion of the crystalline matrix, the brittleness is rapidly reduced, with a modest reduction in strength and hardness. This process is called Tempering and the resultant steel is called tempered Martensite. Cracking (quench cracks) may occur when larger sections are quenched because the surface material shrinks and forms Martensite while the interior is still hot. This can be avoided by Martempering (Marquenching). The steel is quenched to just above the Martensite start temperature (M_s) and held to allow the interior to cool but not so long as to form lower Bainite. Then the alloy is quenched to below the Martensite finish temperature (M_f) to complete the Martensite transformation. The Martensite is then tempered. Alternatively, the steel can be Austempered by holding the temperature just above the M_s temperature to allow lower Bainite to form. This produces a hard material similar to tempered Martensite in a single heat treatment.

When steel and most metals are cold worked, the grains are distorted causing Strain or Work Hardening (the material becomes harder and more brittle). If sufficient cold work is performed, cracking can occur. To eliminate the strain hardening, the material is annealed. This is performed by heating the alloy to just below the eutectoid temperature, allowing nucleation and grain growth of undistorted grains. The alloy is thus softened and cold work can continue.

Some steels can be Maraged. The steel has a low carbon content and is alloyed to allow for precipitation hardening. When Maraged steels are quenched, a relatively soft Martensitic structure forms. The alloy can be formed to the desired shape and heat treated to allow precipitation hardening.

Surface Hardening

Surface hardening is used for applications in which the surface must be very hard and resistant to wear, but the bulk of the material must be very tough. Journal bearing surfaces, gear teeth and bearing surfaces on tools are examples where surface hardening can be utilized. Carburizing is accomplished by placing the part to be case hardened in a Carbon rich atmosphere (typically Carbon Monoxide). It is performed at high temperature (up to approximately 925°C) to promote diffusion. The Carbon rich surface provides the hardness and wear resistance desired. Nitriding is accomplished by exposing the material to be hardened to an ammonia-rich atmosphere at approximately 510 to 540°C for periods of 50 to 90 hours. The nitrogen diffuses, forming nitrides with alloying elements, such as Aluminum and Chromium, placed in the steel for this purpose. The nitrides are small and well dispersed. Carbonitriding (Cyaniding) allows diffusion of both Carbon and Nitrogen by immersing the steel into a Cyanide bath.

Chromizing is a diffusion process to increase the Chromium content on the surface of the steel for corrosion resistance. The steel can be packed in a mixture of Chromium and Alumina powder and heated above 1,260°C for several (typically three to four) hours.

The processes are dependent on diffusion. Recall, the diffusion coefficient (D) is related to the activation energy (Q) and the absolute temperature (T) by the Arrhenius equation. The constants are the universal gas constant (R) [1.987 cal/(g mol-K)] and the Diffusion proportionality constant (D_0). D_0 is dependent on both the base material and the diffusing element and are listed in materials handbooks. The following formula is used to calculate the diffusion coefficient:

$$D = D_0 e^{-\frac{Q}{RT}}$$

The rate of diffusion is dependent on the diffusion coefficient and the concentration gradient. The steeper the concentration gradient, the faster the diffusion. This is expressed mathematically in Fick's first law:

$$J = -D\frac{dc}{dx}$$

The flux (J) is the net rate of diffusion across a plane, with the units of atoms/cm^2 and (c) is the concentration (atoms/cm^3).

Frequently, the concentration and gradient are constantly changing as diffusion continues. Such is the case with Carburizing, Nitriding, and Chrominizing as noted. These cases are governed by the following differential equation, known as Fick's second law:

$$\frac{dc}{dt} = D\frac{d^2c}{dx^2}$$

The solution of this equation is dependent upon the boundary conditions, including the geometry of the part. Let's consider one solution of particular importance. A semi-infinite slab (large dimensions compared to the diffusion distance) of initial composition c_0, the concentration at the surface of the slab for $t > 0$ is maintained at a value c_1. The solution is as follows:

$$\frac{c(x,t) - c_0}{c_1 - c_0} = 1 - erf\left(\frac{x}{2\sqrt{Dt}}\right)$$

The *erf* is defined as the Gaussian error function. It is tabulated in mathematical tables. The *erf* can be approximated by:

$$erf\,(y) = y \text{ for } y \le 0.5$$

Generally, when one case hardens or Chrominizes a part, one is interested in the time it takes for the diffusing element to diffuse to a specified depth at a specified concentration. Frequently, the specified concentration is such that:

$$\frac{c(x,t) - c_0}{c_1 - c_0} = 1 - erf\left(\frac{x}{2\sqrt{Dt}}\right) = 0.5$$

This implies that:

$$\frac{x}{2\sqrt{Dt}} = 0.5 \text{ or } x = \sqrt{Dt}$$

In general, the size of parts is large compared to the diffusion distance, and the solution to Fick's second law is sufficient to produce the desired results.

PROBLEM 2:

A steel gear made of 1030 steel (containing 0.30 weight percent Carbon) is to be carborized in a furnace rich in Carbon at 1000°C. The surface of the gear reaches 0.9 weight percent Carbon very rapidly and remains at that concentration. A Carbon concentration of 0.6 weight percent Carbon is desired at a depth of 0.3 mm. How long should the gear be left in the carburizing oven? Given: D_0 = 0.25 cm²/sec, Q = 34,500 cal/mole and R = 1.987 cal/mole-K.

SOLUTION:

$$D = D_0 e^{-\frac{Q}{RT}} = 0.25\, e^{-\frac{34{,}500}{(1.987)\,(1{,}273)}} = 2.98 \times 10^{-7}\,\text{cm}^2/\text{sec}$$

$$\frac{c(x,t) - c_0}{c_1 - c_0} = \frac{0.6 - 0.3}{0.9 - 0.3} = 0.5$$

Therefore:

$$x = \sqrt{Dt} \ \text{ and } \ t = \frac{x^2}{D} = \frac{0.03^2}{2.98 \times 10^{-7}} = 3{,}020 \text{ sec} \approx 50 \text{ min}$$

Clad Steels

Steels are frequently clad to add desirable surface qualities. The clad can be applied in several ways including welding, powder metal spraying, explosive bonding, and rolling at high temperatures.

Steel-alloying Elements

Now that we have discussed the Iron-Carbon diagram, it is important to know that steel normally has several alloying elements besides Carbon. Some of the most common are included in Table 1 shown on the following page.

These alloying constituents, as well as others, are used to produce desired material properties, such as corrosion resistance (like stainless steels), increase strength, increase hardness, increase toughness, retard the transformation to Pearlite/Bainite so that Martensite can be produced at slower cooling rates, etc.

Hardenability

Hardenability is a measure of the ease of the transformation of the steel to Martensite. If a steel is very hardenable, it can be cooled slowly without the proeutectoid, Pearlite, or Bainite transformations occurring.

Alloying element	Crystal structure @ 20°C	Typical weight %	Effects on steel
Sulfur		0.04 max	Trace amounts are left in the steel-smelting process
or			
Phosphorus		0.05 max	Reduces strength, increases brittleness, and reduces toughness Sometimes added to make "free cutting" steels
Manganese	Cubic (complex)	0.25–1.90	Increases hardenability Lowers eutectoid temperature Forms intermetallic compounds with Sulfur and Phosphorus
Chromium	BCC	0.40–1.20	Increases hardenability if dissolved in Austenite Reduces Austenite grain size if undissolved compounds remain in Austenite Increases high temperature strength
Nickel	FCC	0.20–2.00	Austenite stabilizer Increases hardenability if dissolved in Austenite
Silicon	Diamond cubic	0.2–1.2	Increases hardenability if dissolved in Austenite
Molybdenum	BCC	0.08–0.40	Increases hardenability and toughness
Vanadium	BCC	0.1–0.15	Increases hardenability
Tungsten	HCP	0.1–0.50	Increases high temperature hardness and strength

Table 1. Steel Alloying Elements

In large parts, for example, the maximum cooling rate in the center is highly dependent on conduction of heat energy to the surface of the part. If it is desired that Martensite be formed throughout the material, a steel with high hardenability must be used.

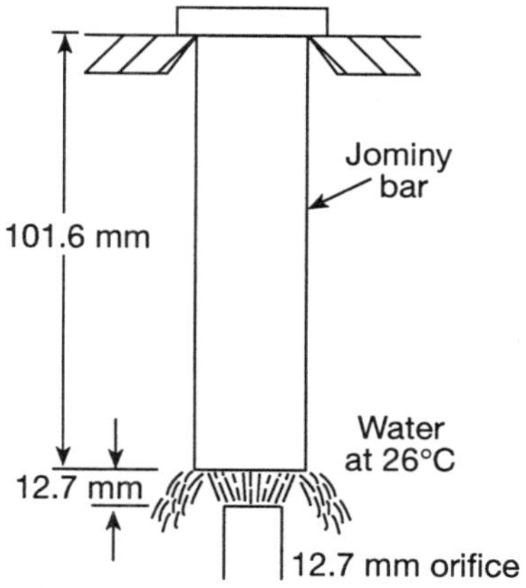

Figure 4a. Quench Test Configuration

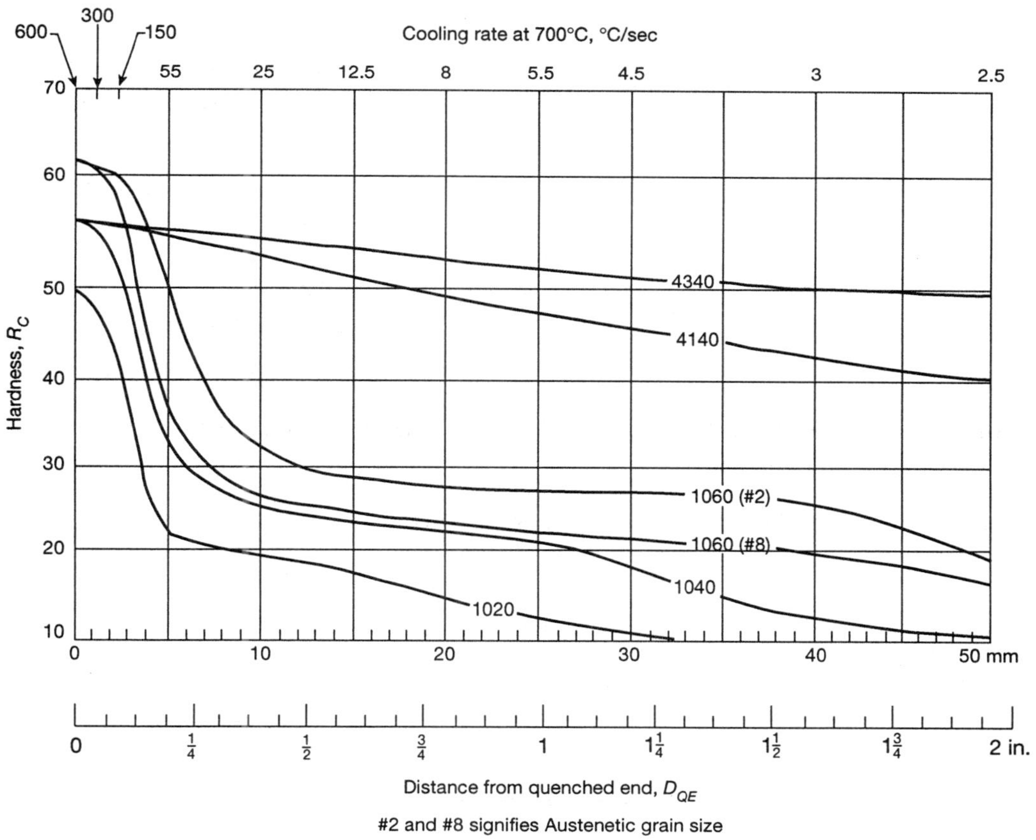

Figure 4b. Hardness v. D_{QE}

Figure 4a and 4b. Jominy Hardenability Test

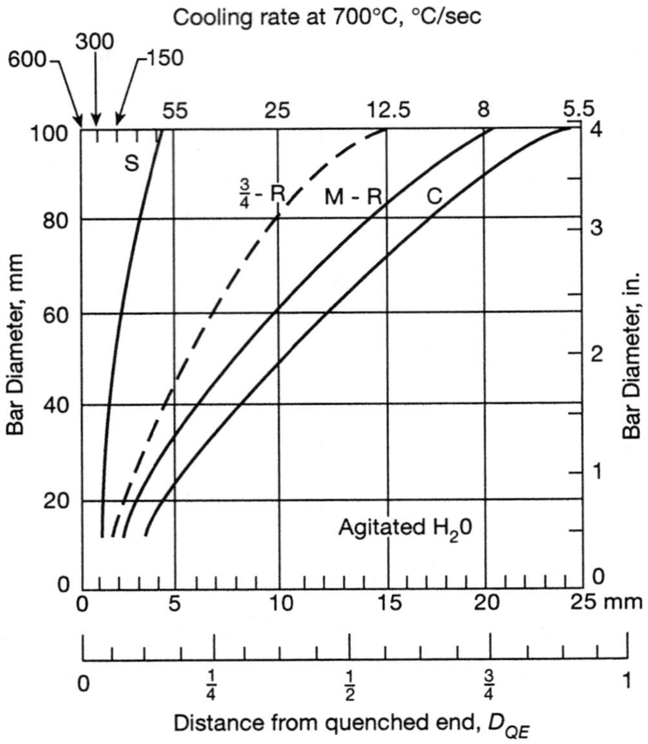

Figure 5a. Agitated Water Quench

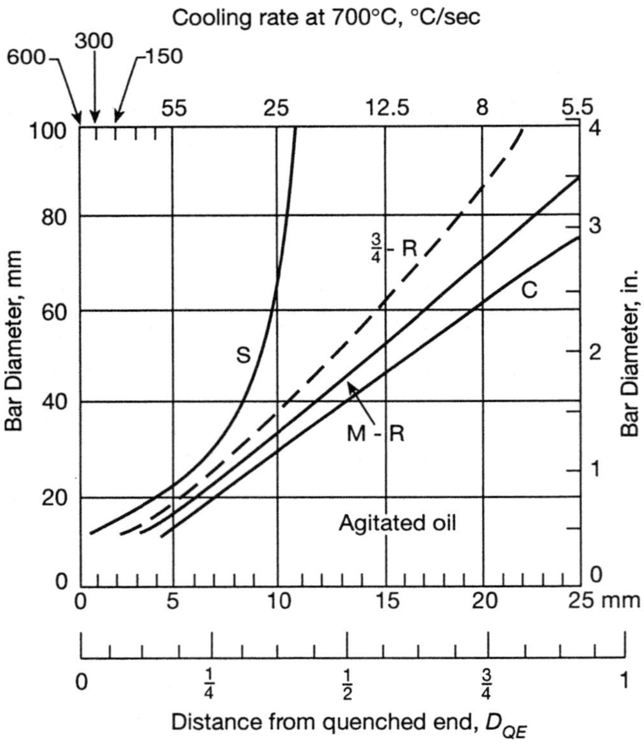

Figure 5b. Agitated Oil Quench

Figure 5a and 5b. Cooling Rates of Round Bars

A material is tested for hardenability using the Jominy hardenability test. A standard specimen is Austenitized. Then one end of the specimen is cooled in a controlled stream of water as shown in Figure 4 (a). The cooling rate at each distance from the quenched end (D_{QE}) is shown in Figure 4 (b). After the specimen is cooled, a flat is ground along the side, using water-cooled grinding equipment to prevent microstructure changes due to the grinding. Then the hardness is tested at various distances from the quenched end and plotted. Figure 4 (b) shows the hardness on the Rockwell "C" (R_c) scale of various materials as a function of D_{QE}. The cooling rates of a designed component can be calculated and the microstructure at any point can be predicted with the hardenability data derived from the Jominy hardenability tests. Figure 5 shows the cooling rate of round stock (or the round stock equivalent to our part) at the center (C), mid-radius (MR), 3/4 radius (3/4R), and surface (S). Figure 5 (a) is for a quench in agitated water and Figure 5 (b) is for agitated oil.

Three main factors affect the hardenability of steel: the Austenite composition; Austenite grain size; and amount, nature, and distribution of undissolved particles in the Austenite. Of particular importance is the Austenite grain size. If the steel is Austenitized at a "low" temperature just above the transformation temperature, the grains will be smaller. If the steel is Austenitized near the melting temperature, the grains will be larger. Similarly, if undissolved particles are small, numerous, and well distributed, the Austenite grains will be small. Small Austenite grain size has a greater density of grain boundaries, which provides the nucleation sites for proeutectoid Ferrite (Cementite) and Pearlite. Therefore, small Austenite grains cause a faster transformation and reduce the hardenability of the material.

Austenite grain size can be measured by quenching a specimen, then polishing, etching, and examining under a metallurgical microscope. The number of grains exposed on the surface is counted. The ASTM index number (n), can then be calculated. First, the number of grains per 0.0645 mm is calculated using a ratio of the counted grains divided by the area over which they were counted as follows:

$$\frac{N_{actual}}{\text{Actual Area}} = \frac{N_{0.0645\ \text{mm}^2}}{0.0645\ \text{mm}^2}$$

and

$$N_{0.0645\ \text{mm}^2} = 2^{(n-1)}$$

For steels, normally there will be 1 to 128 grains/0.0645 mm^2 for an

ASTM index of 1 to 8. Note, n is rounded to the nearest integer of 1 or greater.

PROBLEM 3:

A part, with an equivalent bar diameter of 40 mm is to be made of 1060 steel. Under microscopic examination, 1850 grains are observed in a 1 mm^2 area. What is the expected harness at 3/4 radius if the part is quenched in agitated oil? If quenched in agitated water?

SOLUTION:

First, calculate the grain index number:

$$N_{0.0645\ \mathrm{mm}^2} = \frac{N_{actual}}{A_{actual}}(0.0645) = \frac{1850}{1}(0.0645) = 119$$

$$N_{0.0645\ \mathrm{mm}^2} = 2^{(n-1)}$$

$$n = 1 + \frac{\ln(119)}{\ln(2)} = 7.9 \text{ (now round to 8)}$$

From Figure 5, D_{QE} for $^3/_4R$ in agitated oil is approximately 9 mm and for agitated water is approximately 4 mm. From Figure 4 (b), the hardness of 1060 steel with #8 grain index at D_{QE} = 9 mm is approximately 28 R_c and for D_{QE} = 4 mm is approximately 40 R_c. Therefore, the predicted harness at 3/4R for agitated oil is 28 R_c and for agitated water is 40 R_c. If this hardness is not satisfactory, select another material, using the same D_{QE}'s, that meets the requirement.

Low-Alloy Steels and Brittle Fracture

Commonly used low-alloy steels can become brittle at low ambient temperatures, especially those encountered during the winter in the Northern states and Canada. Brittle fracture is catastrophic, with no warning. Very little energy is required to cause the fractured surface and the crack propagates at the speed of sound in the material. Early investigations of brittle fracture used an impact test (such as the Charpy impact tester) to measure the energy absorbed in fracturing the specimen. The tester has a weighted pendulum, which is dropped from a known height, swings down, and impacts and fractures the specimen at the bottom of the swing. The remaining kinetic energy of the pendulum will cause it to swing up until all remaining kinetic energy is converted to potential energy. The difference between the potential energy at the start and end of the pendulum swing is, neglecting friction, the energy to cause fracture of the specimen. Specimens of a single material are tested at various temperatures and the fracture energy as a function of temperature is plotted. The temperature at which the steepest part of the curve occurs is considered the Reference Transition Temperature (RTT), the temperature above which brittle fracture will not occur. The RTT increases substantially with Carbon content and decreases with Manganese content. Figure 6 shows typical curves with various Carbon contents.

More recent investigation of brittle fracture has centered around the fracture toughness of the material. The fracture toughness is dependent on the material, temperature, and the geometry of the initiating flaw.

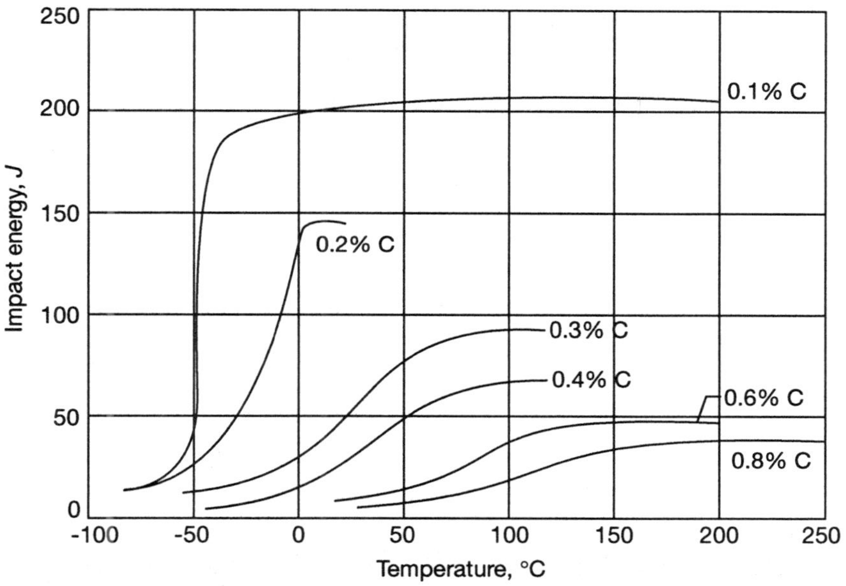

Figure 6. Impact Energy v. Temperature For Several Low-Alloy Steels

Stainless Steels

Stainless steels are a class of steels that have at least 11% Chromium as an alloying element. Generally these materials are selected because of their resistance to general corrosion (rusting). It is very important to note, however, that these alloys are frequently much more susceptible to local corrosion, including stress corrosion cracking and oxygen pitting. Most of the stainless steel produced is Austenitic stainless steel, which generally has approximately eight percent Nickel as an alloying element. The Nickel causes the steel to retain the FCC Austenite structure at room temperature. Normally, Austenitic stainless steels are nonmagnetic but can become magnetic with substantial work hardening. If little or no Nickel is added,

Ferritic stainless steel (has the Ferrite BCC crystalline structure and is generally magnetic) can be produced. Additionally, some stainless steels with no Nickel and high Carbon content are capable of undergoing the Martensitic transformation and are called Martensitic stainless steels. Other stainless steels are capable of being precipitation hardened.

PROBLEM 4:

What classification of stainless steel is most appropriate for making a knife blade?

SOLUTION:

Martensitic stainless steel or precipitation hardening stainless steels can be heat treated to produce the strength and hardness necessary to hold a sharp edge required for a good knife. Austenitic stainless steel is quite ductile and therefore would have to be sharpened often.

Cast Irons

Cast Iron includes a wide range of alloys containing 2.0 to 4.0 percent Carbon and 0.5 to 3.0 percent Silicon. These alloys generally contain other elements, like Manganese, Sulfur, Phosphorus, Chromium, etc. Generally, Cast Irons are classified as Gray Iron, Ductile Iron, Malleable Iron, Compacted Graphite Iron, and White Iron.

Gray Iron generally has 2.5 to 4.0 percent Carbon and 1.0 to 3.0 percent Silicon. Its microstructure is characterized by flakes of graphite that are interconnected. When fractured, the fracture path follows the graphite, giving a gray appearance, hence the name. Gray Iron has excellent machinability, resists galling, and provides excellent wear resistance and excellent damping capacity. It also exhibits essentially no ductility and very low-impact strength.

Ductile Iron is produced by adding Magnesium to the molten metal, under controlled conditions, to cause the graphite to become spherical or nodular. Ductility and strength are improved over Gray Iron, but machinability and thermal conductivity are reduced.

Malleable Iron, when cast, has no free graphite. When annealed for several days, the graphite is present as irregularly shaped nodules called temper Carbon. Because of the lower cost of production, Ductile Iron has mostly replaced Malleable Iron in castings.

Compacted Graphite Iron has a graphite microstructure partially with flakes, like Gray Iron, and partially of spheroid particles, like Ductile Iron. As would be expected, its properties are also a compromise between those of Ductile Iron and Gray Iron, which makes it suitable for applications requiring some ductility, wear resistance, thermal conductivity, etc.

As in cast *White Iron*, virtually all of the Carbon appears in the form of carbides. White Iron is very hard and brittle and fractures with a white surface. It has excellent wear resistance.

Alloying element	Crystal structure @ 20°C	Typical weight %	Effects on aluminum
Copper	FCC	0–11%	For precipitation hardening
Manganese	Cubic	0–1.6%	Contributes to strain hardening
Silicon	Diamond cubic	0–13%	Reduces melting temperature for weld filler and brazing alloys
Magnesium	HCP	0–10%	Contributes to strain hardening
Zinc	HCP	0–8%	Increases strength

Table 2. Alloying Elements for Aluminum

ALUMINUM AND ALUMINUM ALLOYS

Aluminum, as refined, is frequently used as an engineering material. It is light-weight, relatively strong, corrosion resistant, has high electrical and thermal conductivities, and is metallic. As can be predicted by its FCC crystalline structure, it is quite ductile and readily strain hardens with cold work. It is easily cast, drawn, formed, machined, and welded. Addition of alloying elements produces a wide range of properties, including alloys with the strength of many steel alloys. Some major alloying elements are indicated in Table 2, as shown above.

Aluminum can be strengthened by cold work (strain hardening) and precipitation hardening. Precipitation hardening is accomplished by subjecting the material to a solution heat treatment at elevated temperatures, which allows alloying elements to go into solution. The alloy is then cooled and formed while the alloying elements are in a supersaturated solution. For some alloys, the precipitation elements will diffuse sufficiently fast at room temperature to nucleate small, well-dispersed particles, which increases strength and hardness. This process is called natural aging. Other alloys must have the temperature raised to an intermediate temperature to increase the diffusion rate and increase the precipitation process. Specific temperature and aging times are tabulated for various alloys in materials handbooks.

Aluminum alloys are designated by a four-digit number, as specified by ANSI H35.1. This alloy designation is sometimes followed by a letter and number designation. A designation with an "H" indicates the treatment process and extent of cold work and the letter "T" represents the treatment process and extent of precipitation hardening.

COPPER AND COPPER ALLOYS

Copper and its alloys are generally used for their electrical and thermal conductivity, ductility, and corrosion resistance to many harsh environments. They can be cast, formed, drawn, welded and readily strain hardened with cold work. They have relatively low strength. Major Copper alloy classifications include: Coppers, High Copper alloys, Brasses, Bronzes, Copper Nickels, and Nickel Silvers.

Coppers have a minimum copper content of 99.3%. Some alloying elements can be added, such as Silver to improve resistance to softening and fatigue strength; Tellurium, Sulfur, or Lead to improve machinability; and Chromium and Zirconium for high temperature strength. Beryllium produces precipitation hardenable alloys.

High Coppers have Copper content between 96% and 99.3% in wrought form and 94% and 99.3% in cast form.

Brasses are copper alloys that have Zinc (generally 5 to 40 weight percent) as the major alloying element. Minor alloying elements (generally 1 percent or less) include Tin, Lead, Manganese, Silicon, Aluminum, and Iron.

Bronzes are Copper alloys with major alloying elements other than Zinc and Nickel. The major alloying elements are used to name the type bronze. Included in wrought alloys are Copper-Tin Phosphorous alloys (phosphor bronzes), Copper-Tin-Lead-Phosphorous alloys (leaded phosphor bronzes), Copper-Aluminum alloys (aluminum bronzes), and Copper-Silicon alloys (silicon bronzes). Included in cast alloys are Copper-Tin alloys (Tin-Bronzes), Copper-Tin-Lead alloys (leaded tin bronzes), Copper-Tin-Nickel alloys (nickel-tin bronzes), and Copper-Aluminum alloys (aluminum bronzes).

Copper Nickels are generally named for the percentage of each in the alloy (70–30: Copper-Nickel has 70% Copper and 30% Nickel). These alloys are extremely malleable and can be extensively cold worked without annealing. They have excellent corrosion resistance and high temperature characteristics.

Nickel Silvers are Copper-Nickel-Zinc alloys named for their white color. They are tarnish resistant and have been used widely as the base for silver-plated ware.

NICKEL ALLOYS

Nickel alloys in general have superior corrosion resistance and high temperature characteristics. Major alloying elements including Chromium, Iron, and Molybdenum and minor alloying elements including Cobalt, Niobium, and Titanium are used to produce the desired alloy. Some Nickel-Iron alloys have superior ferromagnetic properties.

Of special note are Nickel-Copper alloys, which have excellent corrosion resistance and high toughness. They are capable of performing well at relatively high temperatures. The addition of small amounts of Aluminum and Titanium produces an alloy that is precipitation hardenable.

TITANIUM ALLOYS

Titanium alloys are noted for their light weight, ductility, relatively high strength, and corrosion resistance.

MAGNESIUM ALLOYS

Magnesium alloys are normally cast and are light weight with good strength to weight ratio.

SPECIALTY METALS

Specialized metallic elements/alloys, such as Zirconium, Hafnium, Gadolinium, Beryllium, Cadmium, Plutonium and Uranium (nuclear industry applications), Tungsten (light bulb filaments and high temperature applications), and Tantalum (surgical implants), do not have widespread use.

FABRICATION OF METALS

Casting is the forming of a solid of desired shape from molten metal. Virtually all metals are at least initially cast into ingots and then formed or recast into the desired component parts. Sand casting uses a mixture of sand and clay to form a mold, into which the molten metal is poured. The sand is broken away from the casting and is reused. Permanent mold castings use a mold made of graphite or metal that can be reused. Die casting uses a permanent mold but the molten metal is injected under pressure. It is most frequently used for metal alloys with low melting points. In investment casting, a wax or polymer replica of the desired part is covered in a cement-type material. The mold is heated and the wax/polymer is melted away. The metal is then either poured or forced into the mold by centrifugal force to produce small intricate parts, such as jewelry and dental castings. In continuous castings, the solid (generally a round or square bar) issues from the bottom through a water-cooled die orifice.

Castings inherently have several types of defects that must be considered. Included are voids, inclusions, nonsymmetric grains, and composition variations. The casting will also shrink as it is cooled from molten temperatures to room temperature.

Metal forming includes bulk forming, where the cross-sectional area is reduced, and sheet forming, where the cross section does not change significantly but the sheet is formed into the desired shape (as in car body panels). Bulk forming includes rolling, drawing, forging, and other techniques. Hot rolling is performed above the recrystallization temperature of the metal. The advantages to hot rolling are that less bulky and powerful equipment is required for a given decrease in area and annealing between rolling passes is not required. The disadvantages are a rough surface and that oxidation can occur on the surface. Cold rolling produces a much finer surface finish but produces strain hardening due to cold work and must be annealed at various intermediate stages. Therefore, less reduction in area is acceptable per pass. Forging is a technique using impact to change the shape of hot metal. Horseshoes were made by blacksmiths using this technique. Drop forging normally includes a pair of dies, one of which is dropped on the other, with the hot metal placed between the dies. This is used frequently to produce tools, crankshafts, etc. Sheet metal stamping is used to change the shape of sheet metal to the desired shape (such as automobile body parts).

Metal joining can be accomplished by welding, brazing, or soldering. Welding can be accomplished without filler material when the two pieces of material diffuse together to form a bond. Explosive bonding and spot welding are examples. Metal can also be joined using filler metal. The most common is shielded metal arc welding (stick welding) and inert gas arc welding. Inert gas arc welding is used extensively for metals that in the molten state are very reactive with the atmosphere (such as Aluminum and Titanium). Oxy-acetylene gas welding is also common. Brazing and Soldering use filler material with a temperature less than the melting temperature of the metals to be joined. The melting temperature for the filler material is greater than 538°C for brazing and less than 538°C for soldering.

Cutting of metals is performed by a tool shaving off a portion of the metal. Examples include cutting with a lathe, milling machine, drill, or shaper. A sharp tool edge is moved relative to the material to remove a small slice of the metal. The metal that is removed can chip off (free cutting) or remain as a long-generally spiral shaving. Regardless, cutting causes extensive deformation and produces heat. The temperature at the tool edge can be reduced with lubrication, various artificial cooling methods, and decreasing the depth of the cut with each pass.

Powder Metallurgy is the forming of parts with metal powders. Typically, the metal powders are selected, mixed, then compacted with a pressure on the order 550±150 MPa. The metal powder is raised to an appropriate temperature and held at that temperature to allow time for sintering (a process where the metal particles bond together by diffusion). This method results in very precise parts.

ENGINEERING MATERIALS—PLASTICS

Plastics are a class of materials that are made of long chain-like molecules called *high polymers*. Plastics are normally divided into three groups: Thermo-plastics (thermoplasts), Thermo-setting plastics (thermosets), and elastomers. Thermo-plastics, when heated, become a highly viscous liquid which can be recast/reformed, making them readily recyclable. Depending on the structure that forms from the molten state, the resultant solid can be elastic and tough or brittle. Thermo-setting plastics (such as epoxies) cure by a chain-linking chemical reaction. Cure temperatures are typically 120 to 180°C, although curing at room temperature can be achieved with the use of catalysts.

Additives for plastics include:

Fillers—Wood flour, mica, silica, clay, and various fibers are added to reduce weight, the amount of resin required, and improve strength.

Blowing or Foaming Agents—Used to evolve gas to form foam products

Mold-Release Compound—minimizes sticking of molded plastics to the mold surface

Lubricants—Added to ease fabrication and impart lubricating properties to the plastic.

Antistatic Components—reduces the static charge which can accumulate on the surface of plastic components.

Colorants—Dyes or pigments that impart the desired color to the plastic.

Flame Retardants—raise the ignition temperature and retards the combustion rate in plastic for safety reasons.

Heat Stabilizers—raise the temperature at which the plastic will become unstable.

Impact Modifiers—increase impact strength and toughness

Plasticizers—increase softness and flexibility of the plastic.

Ultraviolet Stabilizers—retard or prevent degradation of strength properties and color which occur with exposure to sunlight.

FIBER COMPOSITE MATERIAL

Fiber composite materials are composed of two distinct constituents. Fibers, which provide strength, are embedded into a matrix. Commonly used fibers have high tensile strength in the range of 1.8 to 4.5 GPa and can be arranged along the axis of greatest force. The matrix is commonly thermo-setting resins. Fiber composite materials can be produced to have very high strength and stiffness to weight ratios. Common fibers currently used include glass fiber, Aramid (trade name Kevlar) Fiber, Carbon fiber, and Boron fiber.

ELASTOMERS

Elastomers are rubber and rubber-like material. In the raw state, elastomers are soft and sticky when hot, hard and brittle when cold. Vulcanization (combining with sulfur/sulfur compounds and heated to cause cross linking between polymer chains) improves mechanical properties, extends the temperatures at which they are elastic and flexible, and makes them insoluble in virtually all known solvents. Various alloying agents are added to improve specific properties, including strength, hardness, stiffness, toughness, color, etc. Major types of elastomers include:

Natural Rubber is prepared by coagulating the latex of the *hevea brasilenis* tree.

GR-S is made from butadiene and styrene, which are produced from petroleum. It is by far the most commonly produced synthetic elastomer and is used in tires.

Neoprene is made from acetylene, which through intermediate steps is converted to Chloroprene. Chloroprene is polymerized to Neoprene.

Nitrile rubbers are made from butadiene and acrylonitrile.

Butyl is made by copolymerization of isobutylene and a small portion of Butadiene or isoprene.

Fabrication of Plastics

Plastics can be formed by numerous methods, most of which involve the use of molds. Thermosets are frequently cast by mixing the components, pouring them into a mold, and allowing the polymerization reaction to take place. Compression molding is a technique in which powdered or pelletized plastic (normally thermosets) are placed in a mold at a high temperature and pressure to form the part. Thermoplasts are generally heated and forced into a mold. In the injected mold process, the thermoplast is heated then forced by a ram into a mold. Bottles are formed using blow molding, a technique where an extruded tube is placed into a mold and the tube is blown against the mold walls. Calendering is a rolling process where the softened plastic is rolled between hot rollers to form sheet material or to bond a plastic coating to another material. Thermoforming is used to form plastic sheet into the desired shape using air pressure, vacuum, or mechanical force.

ENGINEERING MATERIALS—CERAMICS

Ceramics is a classification of materials that includes various minerals, intermetallic compounds, glasses, clay products, whiteware, refractories, and cements. Ceramics are in general very hard, can have excellent high temperature characteristics, but are very brittle. The compressive strength is many times greater than the tensile strength. A summary of common ceramic materials is included in Table 3.

GLASS

When melted glass is cooled, it becomes more and more viscous, turns to a soft pliable solid, then a hard brittle amorphic solid. There is no specific melting or freezing point. A plot of the specific volume of glass as a function of temperature shows a change in slope (see Figure 7). The temperature at which this occurs is called the Glass Transition Temperature (T_g). The behavior of glass is related to the structure of silica (SiO_2), generally the major constituent in glass. Silica forms chains, with each silicon atom surrounded by four Oxygen atoms in the form of a tetrahedron (known as a silicon tetrahedron). At higher temperatures, these chains can slide past each other due to thermal vibrations. At lower temperatures, below the glass transition temperature (Tg), the chains are pinned and only elastic strain can occur. If the force exceeds the elastic limit, brittle fracture occurs. This glassy amorphous state is called *vitreous silica*. Note silica is polymorphic and is found in nature in its various crystalline forms such as the mineral quartz.

Various oxides are added to silica to form glasses suitable for the application. These constituent groups include:

1. Other glass formers, such as Boron Oxide, B_2O_3. The valence of the metal ion is usually 3 or greater and the ion is small.

2. Modifiers, such as oxides of Sodium and Potassium or other low-valance elements. Tend to break up the continuity of the chains, lowers melting temperature, and simplifies processing. Can only be added in small amounts.

3. Intermediate oxides, or intermediates. These oxides do not form glasses by themselves but add to the chain of silica maintaining the glass state. An example is Lead Oxide (PbO). It is added in large quantities to produce glass of great brilliance (for lead crystal/cut glass stemware, decanters, etc.) and to attenuate radiation (X-rays and Gamma rays).

Material	Composition	Structure
Silica		
cristobalite	SiO_2	Cubic
tridymite	SiO_2	Complex
quartz	SiO_2	Complex
vitreous silica	SiO_2	Amorphous (glass)
Silicates		
Li, Al glass	$SiO_2+MgO+Al_2O_3+TiO_2+Li_2O$	Glass and crystalline
Ceramics	$SiO_2+Al_2O_3$	Hexagonal
Mullite	Complex aluminum silicate +	
Kalinite & other clays	other elements	
Spinel	$A^{2+}B_2{}^{3}+0_4$ example $MgAl_2O_4$	Cubic
Alumina	Al_2O_3	Hexagonal
Magnesia	MgO	Cubic
Domomite	$MgO+CaO$	Complex
Olivine	$MgO+SiO_2$	Complex
Cordierite	$MgO+Al_2O_3+SiO_2$	Complex
Zircon	ZrO_2• SiO_2	Cubic
Zirconia	ZrO_2 (CaO stabilized)	Cubic
Carbon		
Graphite		Hexagonal
Pyrolytic		Hexagonal
Diamond		Diamond cubic
Carbides Metallic	Ti, V, Cr, Y, Zr, Mo, Hf, Ta, W	Cubic
Carbides Nonmetallic	Si, B	Cubic
Nitrides Metallic	Al, Ti, V, Cr, Y, Zr, Mo	Cubic or hexagonal
Nitrides Nonmetallic	B, Si	Other
Silicides	Mo, Transition elements	Tetrahedral
Borides	Ta, V, Cr, Zr, Mo, Ti, C	Cubic + other

Table 3. Ceramic Materials

Common types of glass include (typical compositions are included):

Fused silica (99.5+% SiO_2): Used for high temperature applications up to 1,000°C. Hard to melt and fabricate.

Soda lime: Plate glass (71–73% SiO_2; 12–14% Na_2O; 10–12% CaO; 1–4%MgO; and 0.5–1.5% Al_2O_3). Easily fabricated. Used extensively for windows, containers, and electric bulbs.

Borosilicate: Low expansion (Pyrex) (80.5% SiO_2; 3.8% Na_2O; 0.4% K_2O; 12.9% B_2O_3; and 2.2% Al_2O_3). Low thermal expansion and good thermal shock resistance.

Lead silicate (35% SiO_2; 7.2% K_2O; 58% PbO): Easily melted and fabricated, good electrical properties, high refractive index.

Aluminoborosilicate: Low alkali (E-glass) (54.5% SiO_2; 0.5% Na_2O; 22% CaO; 8.5% B_2O_3; and 14.5% Al_2O_3). Widely used for fibers in glass resin composites.

Tempered glass is much more resistant to breaking in bending. The outer surfaces are cooled usually with air, while the glass is soft, causing the outside layer to contract. The inside layer flows to relieve the compressive stresses. As the inside layer cools tension is produced on the inside and compression is produced on the surface. Bending stresses must overcome the residual compression before tensile failure occurs. When a crack enters the center of the glass, which is in tension, it propagates rapidly,

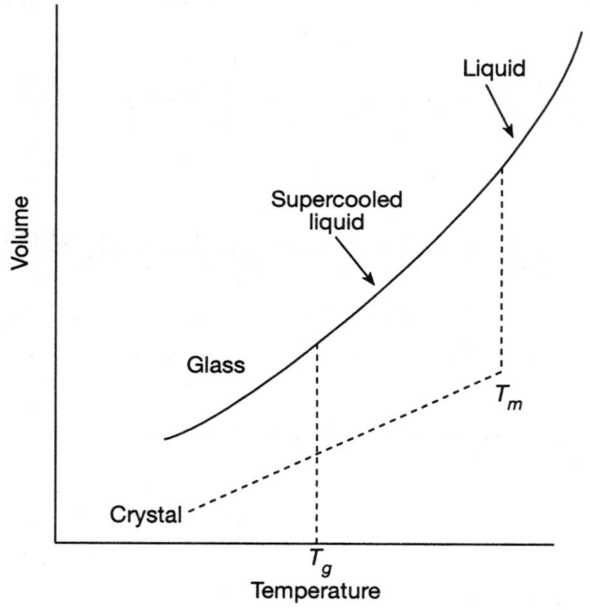

Figure 7. Glass/Crystal Transition

producing small square shaped pieces as opposed to the shards produced when ordinary glass is broken.

Glass Ceramics are glasses that are devitrified (changed into a crystalline state). The advantage is that in the glass state the material is easily formed, cut, and ground to shape. The temperature is then raised to allow nucleation and grain growth of the crystalline structure.

BRICK AND TILE

Made of low-cost, easily fused clay. This clay contains high silica, high alkali, high-FeO. The clay is pressed into molds then fired at relatively low temperatures.

EARTHENWARE, STONEWARE, CHINA, OVENWARE AND PORCELAIN

Earthenware is made of clay. It is fired at relatively low temperatures and therefore has very little fusion (Sintering). It has low strength and is quite porous. In better grades of earthenware, clay Silica, Feldspar mixtures are used. Higher firing temperatures are used, causing some formation of glass. This improves strength and reduces porosity.

Stoneware is fired at a higher temperature than earthenware and therefore is even less porous (< 5%). Generally, the composition is more closely controlled. Stoneware is an excellent material for ovenware.

China is obtained by firing to a high temperature causing a large portion of its mixture—Clay, Silicon, and Feldspar—to be converted to clear glass. This gives fine china its translucent properties. English bone china contains approximately 45% bone ash, 25% clay, and 30% Feldspar and Quartz. $CaPO_4$ from the bone ash lowers the melting temperature and some of the phosphate substitutes for silica as a glass former.

Porcelain is fired at higher temperatures than the rest of the group. It usually contains no flux and is much harder than the rest of the group.

ABRASIVES/CUTTING TOOLS

Hard ceramic material is used to make abrasive wheels and papers. The abrasive material is held in a matrix until it is worn, when it spalls off to expose fresh sharp ceramic particles. Most of the grinding wheels are manufactured from synthetic Alumina and the rest Silicon Carbide. Silicon Carbide is much harder but more fragile.

Similarly, saw blade tips, with diamond (10 on Mohs hardness and hardest material known) particles in the matrix are used to cut very hard materials, such as carbides. Carbides, Nitrides, and Borides (9+ Mohs hardness) are very hard and are suitable for tips of various cutting tools.

CEMENTS (PORTLAND, HIGH-ALUMINA, AND OTHERS)

There are numerous grades of cement, based on their constituents and additives. Portland cement (used in concrete and mortar, etc.) contains several minerals: Tricalcium Silicate ($3CaO{\bullet}SiO$), Dicalcium Silicate ($2CaO{\bullet}SiO$), Tricalcium Aluminate ($3CaO{\bullet}Al_2O_3$), Tetracalcium Aluminum Ferrite ($4CaO{\bullet}Al_2O_3{\bullet}FeO$). Typically, the component percentages are as follows: CaO (60–65%), SiO_2 (19–25%), Al_2O_3 (5–9%), and FeO (2–4%), but the compositions can vary greatly based on the application. When water is added to Portland cement, first a solution is formed, then recrystallization and precipitation into a sheet structure of silicate groups, with Oxygen and Calcium in the interstitial sites. Water molecules separate the sheets, providing hydraulic bonding. This is an exothermic reaction, the heat of which can damage large structures. Therefore, low-heat cements are made by reducing the amount of Tricalcium Aluminate. Approximately 2% Gypsum ($CaSO_4{\bullet}H_2O$), which reacts with the Tricalcium Aluminate, is frequently added to retard setting and reduce shrinkage. In general, the less water used in the cement, the greater the strength. High Alumina cement has approximately 38–48% Al_2O_3, 35–42% CaO, 3–11% SiO_2, and 2–15% FeO. Setting is produced by producing Aluminum hydrate crystals from the Tricalcium Aluminate. It is quick setting, obtaining in 24 hours the same strength that takes 30 days for Portland cement. Silica Cement is made from Sodium Silicate (Na_2SiO_2) and fine quartz. The reaction results in a silica gel that results from the breakdown of sodium silicate. This is used in the foundry for molds. Mineral Gypsum is heated to convert the double hydrate ($CaSO_4{\bullet}2H_2O$) to a hemihydrate ($CaSO_4{\bullet}1/2H_2O$). The hemihydrate added to water will form the solid double hydrate. It is extensively used for wallboard in the construction industry.

COMPOSITE MATERIALS

Composite materials combine desirable properties of two or more materials to produce a material superior (including economically superior) in the application to either one alone. Although one generally thinks of fiber-reinforced plastic when one mentions composite materials, there are many others in general use and in development. One can divide the composites by matrix and second material. Examples are shown in Table 4. Additionally, there are many hybrids that contain more than one second material and/or have an interlocking matrix so that both materials are continuous.

Some intrinsic properties of composite materials are approximately related to the volume fraction of the components. Included are the following:

$$\rho_c = \sum f_i \rho_i$$

$$C_c = \sum f_i C_i$$

$$E_c = \sum f_i E_i$$

where:

ρ_c = density of composite

C_c = heat capacity of composite per unit volume

	Matrix		
Second phase	**Ceramic**	**Metal**	**Polymer**
Ceramic	Cer-Cer Al_2O_3-SiC	Met-Cer Tungsten Carbide in Cobalt	Poly-Cer Glass/Graphite/Boron Fiber-reinforced plastics
Metal	Cer-Met Al-Al_2O_3 Steel-reinforced concrete	Met-Met Powdered Metallurgy Bonded metal	Poly-Met Steel-belted radial tires
Polymer			Poly-Poly Kevlar Fiber-reinforced plastic

Table 4. Composite Material Matrix

E_c = Young's modulus of composite

f_i = volume fraction of individual material

C_i = heat capacity of individual material per unit volume

E_i = Young's modulus of individual material

Frequently, composites are made with two or more materials, which extend uninterrupted from one end to the other. An example is steel-reinforced concrete. When these supports are placed in tension or compression and/or the temperature of the support is changed, both materials still have to retain equal length. In other words, for both materials to maintain the same positions with respect to each other on each end, the length and change in length of both materials must be the same. Therefore, the strain must be the same as follows:

$$\left(\frac{\Delta L}{L}\right)_1 = \left(\frac{\Delta L}{L}\right)_2$$

The strain is composed of two parts. The first is the elongation per unit length due to a temperature change ($\alpha \Delta T$ term) and strain due to force (e term).

$$(\alpha \Delta T + e)_1 = (\alpha \Delta T + e)_2$$

For a bar in tension and compression, the strain due to force is equal to σ/E, in accordance with Hooke's law, and $\sigma = F/A$. The equation becomes:

$$\left(\alpha \Delta T + \frac{(F/A)}{E}\right)_1 = \left(\alpha \Delta T + \frac{(F/A)}{E}\right)_2$$

where:

ΔL = change in length of a material

L = original length of the material

α = coefficient of thermal expansion for a material

ΔT = change in temperature for the material

e = engineering strain

F = force in a material

A = cross-sectional area of the material

E = Young's modulus for the material

PROBLEM 5:

A bottom of a frying pan is made of three plies, each 1.0 mm in thickness. The top and bottom plies are made of 304 stainless steel and the center ply is made of copper. There are no internal forces on the three plies at 25°C. The pan is heated to 170°C. Calculate the density and heat capacity of the composite. Determine the stress generated in the Stainless Steel and the Copper due to the thermal expansion. The characteristics of the materials are as follows:

Stainless Steel:	$\rho = 7.8\ \text{Mg/m}^3$	Copper	$\rho = 9.0\ \text{Mg/m}^3$
	$\alpha = 13.9 \times 10^{-6}\ \text{K}^{-1}$		$\alpha = 17.6 \times 10^{-6}\ \text{K}^{-1}$
	$c = 0.46\ \text{J/(g-K)}$		$c = 0.38\ \text{J/(g-K)}$
	$E = 205\ \text{GPa}$		$E = 104\ \text{GPa}$

SOLUTION:

Consider a one-centimeter square of the material. The volume fraction of each ply is 1/3. Therefore, the volume fraction of the copper is 1/3 and the volume fraction of the stainless steel is 2/3.

$$\rho_c = \sum f_i \rho_i = \left(\frac{2}{3}\right)7.8 + \left(\frac{1}{3}\right)9.0 = 8.2\ \text{Mg/m}^3$$

$$C_c = \sum f_i C_i = \left(\frac{2}{3}\right)0.46 + \left(\frac{1}{3}\right)0.38 = 0.43\ \text{J/(g-K)}$$

To determine the forces, use the following formula:

$$\left(\alpha \Delta T + \frac{(F/A)}{E}\right)_1 = \left(\alpha \Delta T + \frac{(F/A)}{E}\right)_2$$

or

$$\left(\alpha \Delta T + \frac{\sigma}{E}\right)_1 = \left(\alpha \Delta T + \frac{\sigma}{E}\right)_2$$

For static equilibrium the force in the stainless steel must be equal and opposite to the force in the Copper. Let the force in the stainless steel equal F and the force in the Copper equal ($-F$). Also, the area of the stainless steel is twice the area of the Copper. Therefore, let the area of the Copper equal A and the area of the stainless steel equal $2A$ and therefore:

$$\sigma_c = \frac{F}{A_c}$$

and
$$\sigma_{ss} = \frac{F}{A_{ss}}$$

and
$$A_c = \frac{A_{ss}}{2}$$

Therefore,
$$\sigma_c = -2\,\sigma_{ss}$$

Substituting,

$$\left(\alpha\Delta T + \frac{\sigma}{E}\right)_1 = \left(\alpha\Delta T + \frac{\sigma}{E}\right)_2$$

$$\left(13.9\times10^{-6}(145) + \frac{\sigma_{ss}}{205\times10^9}\right) = \left(17.6\times10^{-6}(145) - \frac{2\sigma_{ss}}{104\times10^9}\right)$$

$$\frac{\sigma_{ss}}{205\times10^9} + \frac{2\sigma_{ss}}{104\times10^9} = (17.6\times10^{-6} - 13.9\times10^{-6})\,145$$

$$\sigma_{ss} = 22.3 \text{ MPa (tension)}$$

and
$$\sigma_c = 44.6 \text{ MPa (compression)}$$

FE/EIT

FE: PM Mechanical Engineering Exam

CHAPTER 10

Measurement and Instrumentation

CHAPTER 10

MEASUREMENT AND INSTRUMENTATION

MEASUREMENT FUNDAMENTALS

An *instrument* is a device for determining the value of a quantity or variable. The variables of interest depend on the particular system being investigated. In an industrial process, measurement and control of temperature, pressure, flow rates, etc., are required to determine the quality and efficiency of production. Measurements may be *direct*, such as using a ruler or micrometer to measure a dimension, or *indirect*, such as using a thermistor to determine temperature change.

Practical measurements always have some error. The accuracy of an instrument is the proximity of the measured value to the actual value being measured. *Precision* deals with the reproducibility of the measurements, for example, how successive readings of the same variable differ from each other. *Sensitivity* is the ratio of output signal of the instrument to the change of input value of the variable. *Resolution* refers to the smallest change in the measured value to which the instrument will respond.

Error, which is the difference between actual and measured value, may be classified as *systematic* or *random*. An example of a systematic error, for example, would be the application of excessive pressure to the spindle of a micrometer. A systematic error can also be introduced when room temperatures being monitored by a thermometer change faster than the response time of the thermometer. In this case, a device with a much shorter response time should be employed. Random errors are those that cannot be directly determined.

INSTRUMENTS

There are three essential parts to any instrument: the *sensing element*, the *transmitting element*, and the *indicating element*. The *sensing element* is selected to respond directly to the measured quantity producing a motion, pressure, or electrical signal. This signal is transmitted via linkage, tubing, or wire to a pointer or other device to indicate the value of the measurement. The measuring system may be quite simple as in the case of a thermometer where the volume change of the contained fluid is used to measure temperature change. It can also be fairly complex as in the case of a strain gage where minute changes in electrical wire resistance is translated into strain and ultimately into stress of the system.

An important consideration regarding any temperature measuring device is that the device will produce an output based on its temperature and not the temperature of the medium under investigation. As an example, consider a thermometer that is exposed to sunlight. Due to high solar flux, the temperature of the thermometer fluid will be considerably higher than the temperature of the surrounding air, which it may be attempting to measure. Consequently, the thermometer will record a reading that is much higher than the surrounding air. The thermocouple will give a similar error if proper radiation shields are not employed.

Counting is used to measure the number of events that occur. Counting is the only measurement that can ever be exact. All other measurements are subject to a degree of inaccuracy.

Mass is the measure of the amount of matter contained in an object. The terms of mass and weight are often used interchangeably even though they are completely different. The most accurate means of determining mass is comparison to a known quantity on a beam balance. This is generally limited to small masses. For large masses, unequal arms are used to multiply the reference mass to the unknown mass. Since deflection of a spring (within the elastic limits) is directly proportional to applied force, the spring scale and torsional balance are simple and inexpensive weighing devices. A spring scale measures force and not mass directly. These two quantities are linked by the gravitational acceleration, g, as follows:

$$W = m\frac{g}{g_c}$$

where:

$W =$ weight,

$m =$ mass

$g =$ gravitational acceleration

and $g_c =$ conversion factor between mass and weight.

Displacement measuring devices are used to measure dimensions, distance between points, and such derived quantities as area, velocity, etc. These devices fall into two categories: those based on comparison to a known reference and those based on some fixed physical relationship. *Comparative devices* include a machinist's scale, folding rule, tape measure, vernier caliper, and micrometer. Many of these devices depend on the operator's care and skill.

The vernier and micrometer are two methods of increasing the sensitivity and precision of displacement measurement. The vernier is an auxiliary scale which slides along the main scale. It is uniformly divided so that ten divisions on the vernier scale correspond to exactly nine divisions on the main scale. If a measurement falls between two scale subdivisions, the fraction of this interval is determined by the number of the vernier divisions that coincide with the main scale reading. (The micrometer will be discussed further in the section on error analysis.) *Fixed physical relationship devices* use measured quantities to find linear displacements. For example, displacement can be measured electrically through its effect on the resistance, inductance, or capacitance of an appropriate sensing device. *Ultrasonic* is yet another way of measuring distance. Here, the time required for a sound wave to complete a round trip is translated into distance of separation.

FLUID VELOCITY MEASUREMENTS

Fluid velocity measurements can be obtained with a Pitot Tube. The straight horizontal tube shown in Figure 1 will measure the total pressure of a moving stream if it is connected to a manometer as shown. The static pressure tap in the same figure will likewise measure the static pressure of the flow if it is connected to a manometer. As configured in the diagram, the total pressure is applied to one leg of the manometer while the static pressure is applied to the other leg. The change of fluid heights in the manometer, Δh, represents the difference between the total and static pressure heads and is known as the velocity head. Application of Bernoulli's equation results in the following equation for velocity:

$$V = \sqrt{2g\Delta h}$$

where:

g = gravitational acceleration.

The equation assumes that the flow and manometer fluids are the same. If they are different, the velocity is given by the following equation:

$$V = C\sqrt{2g\Delta h_m\left(\frac{S_0}{S_1} - 1\right)}$$

where:

Δh_m = change in height of the manometer fluid

S_0 = specific gravity of the manometer fluid

S_1 = specific gravity of the flowing fluid

C = correction coefficient

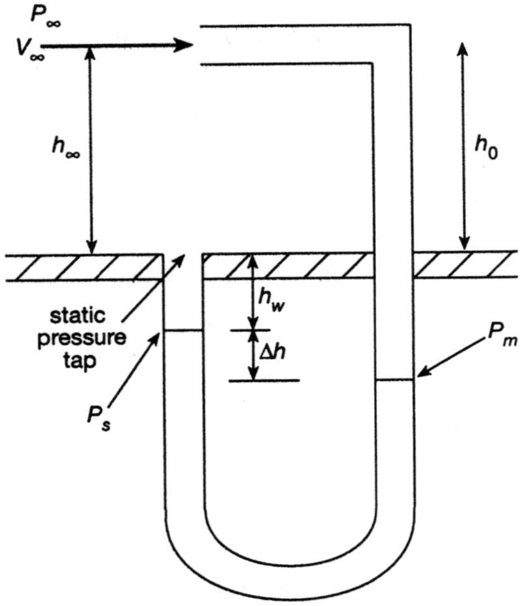

Figure 1. Pitot Tube

PROBLEM 1:

The velocity of gasoline (S.G. = 0.68) is to be measured with a mercury (S.G. = 13.5) manometer. The deflection of the mercury is 25 cm. Determine the resulting velocity of the gasoline.

SOLUTION:

The velocity can be determined using the following equation with $\Delta h_m = 0.25$ m. Assuming $C = 1$.

$$V = C\sqrt{2g\Delta h_m\left(\frac{S_0}{S_1} - 1\right)}$$

$$V = \sqrt{2 \times 9.81\ \text{m/s}^2 (0.25\ \text{m})\left(\frac{13.5}{0.68} - 1\right)} = 9.62\ \text{m/s}$$

The static tube and the Pitot Tube may be combined into one instrument called a Pitot-Static Tube. For total head tubes that have reasonably long horizontal legs relative to tube diameter, the tube correction coefficient is commonly unity. However, Pitot Tubes with damaged tips, short tips, tube burrs, and short tips need calibration checks to determine correct tube coefficients that may deviate considerably from unity.

FLUID FLOW MEASUREMENT

Most devices measure flow indirectly. Flow measuring devices are commonly classified into those that sense or measure velocity and those that measure pressure or head. Some of the water measuring devices that use measurement of head, h, or pressure, p, to determine discharge, Q, are orifices and Venturi meters.

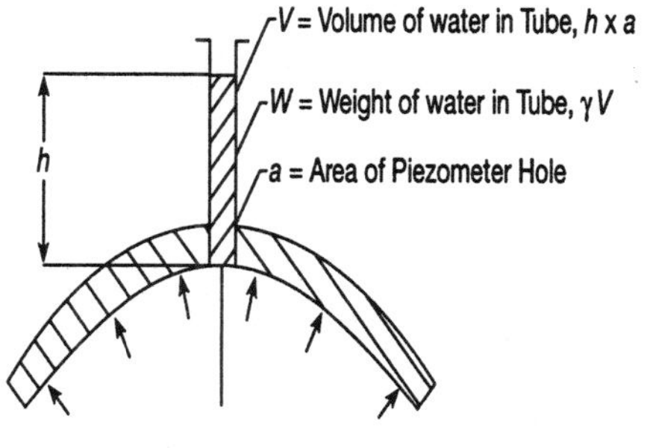

Figure 2. Pressure Head

Pressure, p, is the force per unit area (as shown in Figure 2) that acts in every direction normal to object boundaries. If an open vertical tube is inserted through and flush with the wall of a pipe under pressure, water will rise to a height, h, until the weight, W, of water in the tube balances the pressure force, F_p, on the wall opening area, a, at the wall connection. These tubes are called piezometers. The volume of water in the piezometer tube is $h \times a$. The volume times the unit weight of water, $h \times a$, is the weight, W. The pressure force, F_p, on the tap connection area is designated $p \times a$. The weight and pressure force are equal, and dividing both by the area, a, gives the unit pressure on the wall of the pipe in terms of head, h, written as:

$$p = \gamma h$$

Thus, head is pressure, p, divided by unit weight of water, γ.

Head has units of length (m or ft). Pressure is often expressed in Pa (N/m^2) or psi (pounds per square inch (lb/in^2)).

DIFFERENTIAL HEAD FLOWMETERS

This class of flowmeters includes Venturi, nozzle, and orifice meters. These meters have no moving parts but use the principle of accelerating flow by some form of constriction. Heads are measured upstream where the meter is the size of the approach pipe and downstream where the area is reduced to a minimum. The velocity at one of these locations is solved for in terms of the difference of heads between the two locations. Using the product of the upstream velocity and area results in discharge expressed as:

$$Q_a = CA_1A_2\sqrt{\frac{2g(h_1 - h_2)}{A_1^2 - A_2^2}}$$

where:

Q_a = discharge

A_1 = upstream approach area

A_2 = area of the throat or orifice opening

h_1 = upstream head

h_2 = downstream head

g = gravity constant

C = coefficient determined experimentally

The term $h_1 - h_2$, often written in shorter form as Δh, is the differential head that gives the name to this class of meters. If the change of head is determined with a manometer using a different fluid than that which is flowing, the equation becomes:

$$Q_a = CA_1A_2\sqrt{\frac{2g\Delta h_m(S_0 / S_1 - 1)}{A_1^2 - A_2^2}}$$

where:

Δh_m = change of head of the manometer fluid

S_0 = specific gravity of manometer fluid

S_1 = specific gravity of flowing fluid.

This equation is valid for the Venturi, nozzle, and orifice meters using proper respective effective coefficients. Each kind of flow meter has a different value of effective discharge coefficient.

VENTURI METERS

Venturi meters, shown in Figure 3, are one of the most accurate types of flow measuring devices that can be used to measure water flow. They contain no moving parts, require very little maintenance, and cause very little head loss. Tables or diagrams of the head difference versus rate of flow can be prepared, and flow indicators or flow recorders can be used to display the differential or rate of flow. Venturi meters are often used in the laboratory to calibrate other closed conduit flow measuring devices.

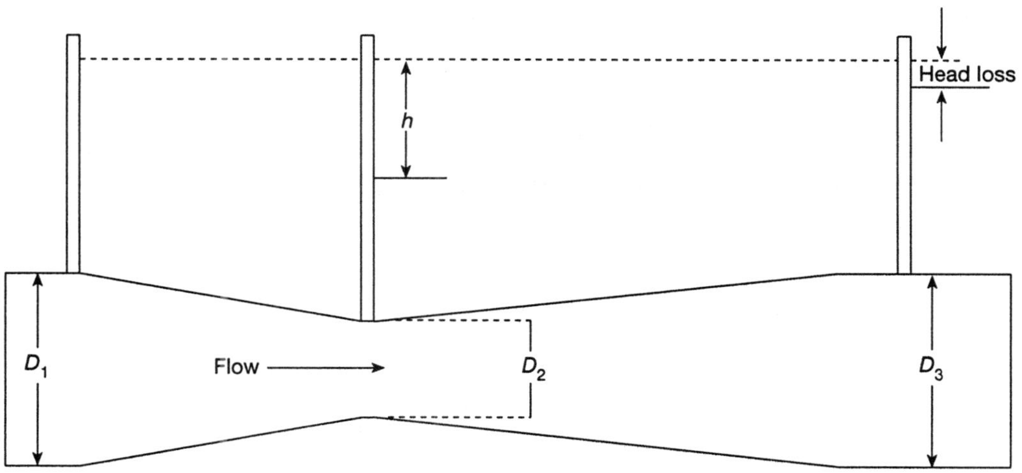

Figure 3. Venturi Meter

PROBLEM 2:

Benzene at 20°C flows through a Venturi tube whose inlet diameter is 3.15 cm and whose throat diameter is 1.38 cm. The differential pressure is measured by a U-tube manometer containing mercury. The differential level of the mercury is 1.57 cm. Find the volumetric flow rate assuming the discharge coefficient to be 0.995.

SOLUTION:

At the stated temperature, the S.G. of benzene is 0.880 and the S.G. of mercury is 13.5. The volume flow rate is given by the following:

$$Q_a = CA_1A_2\sqrt{\frac{2g\Delta h_m(S_0 / S_1 - 1)}{A_1^2 - A_2^2}}$$

The length dimensions must be converted to meters. Substituting:

$$Q_a = (0.995)\left(\tfrac{\pi}{4}\right)(0.0315\text{ m})^2\left(\tfrac{\pi}{4}\right)(0.0138\text{ m})^2 \times$$

$$\sqrt{\frac{2(9.81\text{ m/s}^2)(0.0157\text{ m})(13.5/0.88 - 1)}{\left(\tfrac{\pi}{4}\right)^2(0.0315\text{ m})^4 - \left(\tfrac{\pi}{4}\right)^2(0.0138\text{ m})^4}}$$

$$= 3.19 \times 10^{-4}\text{ m}^3/\text{s}$$

NOZZLE METERS

In effect, the flow nozzle is a Venturi meter that has been simplified and shortened by eliminating the gradual downstream expansion (Figure 4). The streamlined entrance of the nozzle causes a straight jet without contraction, so its effective discharge coefficient is nearly the same as the Venturi meter. Flow nozzles allow the jet to expand of its own accord. This feature causes a greater amount of turbulent expansion head loss than the loss that occurs in Venturi meters, which suppress exit turbulence with a gradually expanding tube boundary.

The effective coefficient of discharge for flow nozzles in pipelines varies from 0.96 to 1.2 for turbulent flow and increases as the throat-to-pipe-diameter ratio increases.

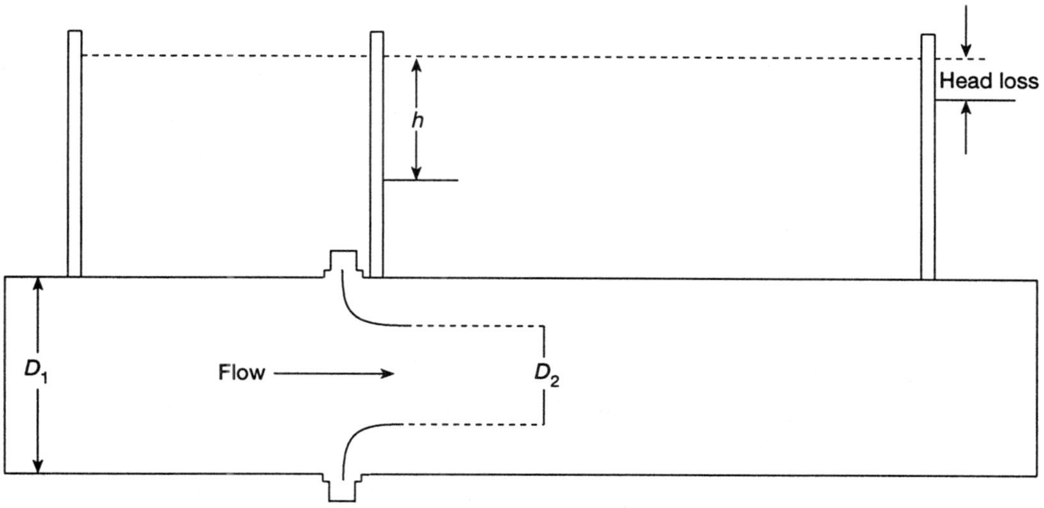

Figure 4. Sectional View of Nozzle Meter

ORIFICE METERS

The most common differential-pressure type flowmeter used in pipelines is the sharp-edged orifice plate (Figure 5).

The equation for the orifice plate meter can be as follows:

$$Q_a = CA_2\sqrt{2g\Delta h_m\left(\frac{S_0}{S_1 - 1}\right)}$$

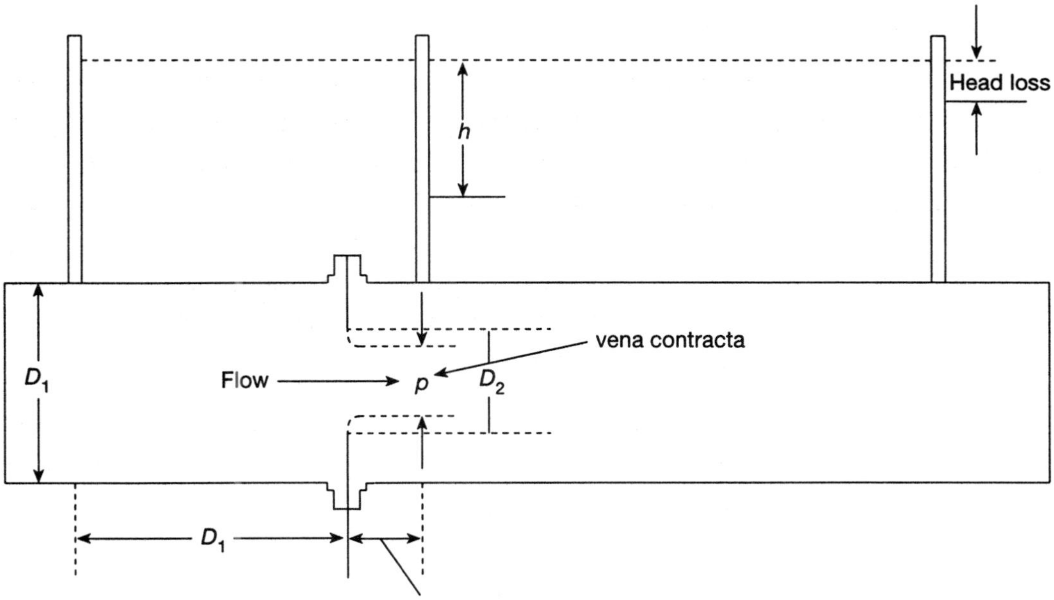

Figure 5. Sectional View of Orifice Meter

where C is given as follows:

$$C = \frac{C_v C_c}{\sqrt{1 - C_c^2 (A_2 / A_1)^2}}$$

where:

C_v = velocity coefficient

C_c = contraction coefficient

While elaborate correlations exist to determine precise values of C, the typical values of C for general cases is 0.61 for a sharp-edged orifice.

Advantages of the orifice plate are its simplicity and the ability to select a proper calibration on the basis of geometry. Disadvantages of the orifice plate include the long, straight pipe length requirements and the limited practical discharge range ratio of about one to three for a single orifice hole size.

PROBLEM 3:

An orifice plate must be selected to meet the following criterion: 400 gpm of water at 20°C is to flow inside a 6-in Sch 40 steel pipe. The pressure taps on either side of the orifice plate are connected to a U-shaped manometer containing mercury. The deflection of the mercury is not to exceed 15 cm. Determine the required orifice diameter. Assume the typical value of $C = 0.61$ for a sharp-edged orifice.

SOLUTION:

400 gpm = 0.0252 m^3/s. The inside pipe diameter, D_1, is 14.63 cm. Assume the S.G. of water to be 1.00 and the S.G. of mercury to be 13.5. The following equation is required:

$$Q_a = C A_2 \sqrt{2 g \Delta h_m \left(\frac{S_0}{S_1 - 1} \right)}$$

Substituting values:

$$0.0252 \text{ m}^3/\text{s} = (0.61) \left(\tfrac{\pi}{4}\right) (D_2^2) \sqrt{2 \, (9.81 \text{ m/s}^2) \, (0.15 \text{ m}) \left(\frac{13.5}{1-1} \right)}$$

$$D_2 = 0.093 \text{ m} = 9.3 \text{ cm}$$

STRAIN GAGES

A strain gage attached to the desired mechanical element measures the strain in the element. The most common type of strain gage used today for stress analysis is the bonded resistance strain gage (Figure 6).

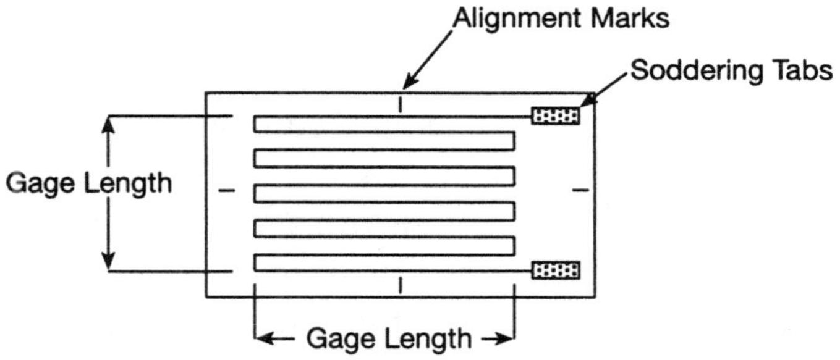

Figure 6. Strain Gage

These gages use a grid of fine wire or a constantan metal foil grid encapsulated in a thin resin backing. The gage is glued to the carefully prepared test specimen by a thin layer of epoxy. The epoxy acts as the carrier matrix to transfer the strain in the specimen to the strain gage. As the gage changes in length, the tiny wires either contract or elongate depending upon a tensile or compressive state of stress in the specimen. The cross-sectional area will increase for compression and decrease in tension. Because the wire has an electrical resistance that is proportional to the inverse of the cross-sectional area,

$$R \propto \frac{1}{A},$$

a measure of the change in resistance can indicate the strain in the material.

Bonded resistance strain gages are produced in a variety of sizes, patterns, and resistances. Many factors must be considered in choosing the right gage for a particular application. Operating temperature, state of strain, and stability of installation all influence gage selection. Bonded resistance strain gages are well suited for making accurate and practical strain measurements because of their high sensitivity of strains, low costs, and simple operation.

The measure of change in electrical resistance when the strain gage is strained is known as the *gage factor*. The gage factor is defined as the fractional change in resistance divided by the fractional change in length along the axis of the gage. Common gage factors are in the range of 1.5–2 for most resistive strain gages.

$$GF = \frac{\frac{\Delta R}{R}}{\frac{\Delta L}{L}} = \frac{\frac{\Delta R}{R}}{\in}$$

In order to reduce the gage length, common strain gages utilize a grid pattern as opposed to a straight length of wire. This grid pattern causes the gage to be sensitive to deformations transverse to the gage length. Therefore, corrections for transverse strains need to be computed and applied to the strain data. The transverse sensitivity factor, K_t, is defined as the transverse gage factor divided by the longitudinal gage factor. These sensitivity values are expressed as a percentage and vary from 0–10%.

$$K_t = \frac{GF_{transverse}}{GF_{longitudinal}}$$

A final consideration for maintaining accurate strain measurement is temperature compensation. The resistance of the gage and the gage factor will change due to the variation of resistivity and the strain sensitivity with temperature. Strain gages are produced with different temperature expansion coefficients. It is only necessary to match the expansion coefficients of the strain gage with the specimen material.

The change in resistance of bonded resistance strain gages for most strain measurements is very small. From a simple calculation, for a strain of 1 μ or 10^{-6} with a 120Ω gage and a gage factor of 2, the change in resistance produced by the gage is:

$$\Delta R = 1 \times 10^{-6} \times 120 \times 2 = 240 \times 10^{-6}\,\Omega$$

Furthermore, it is the fractional change in resistance that is important and the number to be measured will be in the range of μΩ (micro ohms). For large strains a simple multimeter may suffice, but in order to acquire sensitive measurements in the μΩ range, a Wheatstone bridge circuit is necessary.

WHEATSTONE BRIDGE CIRCUIT

Due to their outstanding sensitivity, Wheatstone bridge circuits are very advantageous for the measurement of resistance, inductance, and capacitance. Wheatstone bridges are widely used for strain measurements. A Wheatstone bridge is shown in Figure 7.

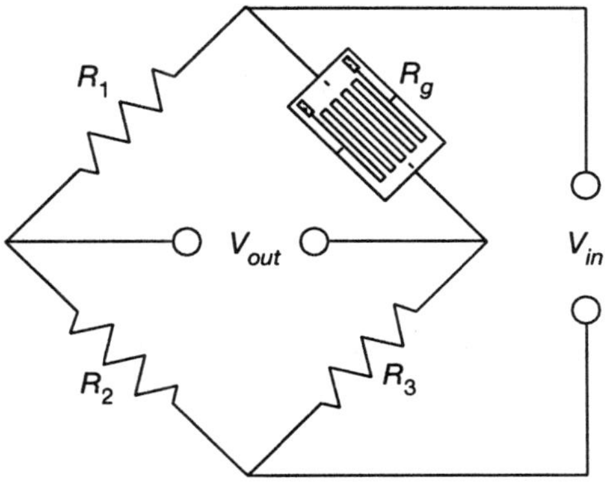

Figure 7. Wheatstone Bridge Circuit

The Wheatstone bridge circuit consists of four resistors arranged in a diamond orientation. An input DC voltage, or excitation voltage, is applied between the top and bottom of the diamond and the output voltage is measured across the middle. When the output voltage is zero, the bridge is said to be balanced. One or more of the legs of the bridge may be a resistive transducer, such as a strain gage. The other legs of the bridge are simply completion resistors with resistance equal to that of the strain gage(s). As the resistance of one of the legs changes, by a change in strain from a resistive strain gage for example, the previously balanced bridge is now unbalanced. This unbalance causes a voltage to appear across the middle of the bridge. This induced voltage may be measured with a voltmeter or the resistor in the opposite leg may be adjusted to rebalance the bridge. In either case the change in resistance that caused the induced voltage may be measured and converted to obtain the engineering units of strain.

If it is inconvenient to balance the Wheatstone bridge circuit, the strain can be determined by measuring the output voltage, V_{out}. Assuming that $R_1 = R_2$, and that R_3 is selected to have the nominal resistance of the strain gage, R_g, the resistance in the strained state is given by the following:

$$R_g = R_3 \left(\frac{V_{in} - 2V_{out}}{V_{in} + 2V_{out}} \right)$$

The difference between the nominal resistance of the strain gage and R_g yields ΔR. From the value of ΔR, the strain can be determined. The following problem illustrates the procedure.

PROBLEM 4:

Consider the previous case with a 120Ω strain gage. Determine the strain that would correspond to the following conditions:

$$V_{in} = 6\text{ V},\ V_{out} = 0.001\text{ V (1 millivolt)},$$

$$R_1 = R_2 \text{ and } R_3 = 120\Omega$$

SOLUTION:

First the value of R_g must be determined. Using the previous equation to find R_g:

$$R_g = 120\Omega\left(\frac{6\text{ V} - 2\times 0.001\text{ V}}{6\text{ V} + 2\times 0.001\text{ V}}\right) = 119.92\Omega$$

$$\Delta R = R_g - 120 = 119.92 = -0.08\Omega$$

From the gage factor equation the strain can be determined as follows:

$$GF = \frac{\frac{\Delta R}{R}}{\frac{\Delta L}{L}} = \frac{\frac{\Delta R}{R}}{\in}$$

Solving for the strain:

$$\in = \frac{\frac{\Delta R}{R}}{GF} = \frac{\frac{-0.08\Omega}{120}}{2} = -3.33\times 10^{-4}$$

TEMPERATURE MEASUREMENT

Temperature can be measured with several types of sensors. All of them infer temperature by sensing some change in a physical characteristic. Six types with which the engineer is likely to come into contact are: *thermocouples*, *resistive temperature devices* (RTDs and thermistors), *infrared radiators*, *bimetallic devices*, *liquid expansion devices*, and *change-of-state devices*. A brief review of each follows.

Thermocouples consist essentially of two strips or wires made of different metals and joined at one end. Changes in the temperature at that juncture induce a change in electromotive force (emf) between the other ends. As temperature goes up, output voltage of the thermocouple rises, though not necessarily linearly.

Resistive temperature devices capitalize on the fact that the electrical resistance of a material changes as its temperature changes. Two key types are the metallic devices (commonly referred to as RTDs), and thermistors. As their name indicates, RTDs rely on resistance change in a metal, with the resistance rising more or less linearly with temperature. Thermistors are based on resistance change in a ceramic semiconductor; the resistance drops nonlinearly with temperature rise.

Infrared sensors are noncontact devices. They infer temperature by measuring the thermal radiation emitted by a material.

Bimetallic devices take advantage of the difference in rate of thermal expansion between different metals. Strips of two metals are bonded together. When heated, one side will expand more than the other, and the resulting bending is translated into a temperature reading by mechanical linkage to a pointer. These devices are portable and they do not require a power supply, but they are usually not as accurate as thermocouples or RTDs and they do not readily lend themselves to temperature recording.

PROBLEM 5:

Consider a bimetallic strip made of two pieces of metal with different coefficients of expansion bonded together. Heating or cooling from the bonding temperature will cause the strip to bend in an arc. The radius of curvature of the resulting arc is given as follows:

$$r = t \times \frac{\{3(1+m)^2 + (1+mn)\,[m^2 + (1/mn)]\}}{6(\alpha_2 - \alpha_1)\,(T - T_0)\,(1+m)^2}$$

where:

t = combined thickness of strip

m = ratio of thickness of low-to-high expansion materials

n = ratio of modulus of elasticity of low-to-high materials

α_1 = lower coefficient of expansion

α_2 = higher coefficient of expansion

T = temperature

T_0 = initial bonding temperature

A bimetallic strip is constructed of strips of yellow brass ($\alpha = 2.20 \times 10^{-5}$ °C^{-1}, E = 96.5 GN/m^2) and Invar ($\alpha = 1.7 \times 10^{-6}$ °C^{-1}, E = 147 GN/m^2) bonded together at 30°C. Each has a thickness of 0.3 mm. Calculate the radius of curvature when the strip is subject to a temperature of 100°C.

SOLUTION:

The radius of curvature is given by the following equation:

$$r = t \times \frac{\{3(1+m)^2 + (1+mn)\,[m^2 + (1/mn)]\}}{6(\alpha_2 - \alpha_1)\,(T - T_0)\,(1+m)^2}$$

$(T - T_0) = 70°\text{C}$ $\qquad m = 1$

$n = \dfrac{147}{96.5} = 1.52$ $\qquad t = 2 \times (0.3 \times 10^{-3}\ \text{m}) = 0.6 \times 10^{-3}\ \text{m}$

Substituting:

$$r = \frac{(0.6 \times 10^{-3})\,[3(1+1)^2 + (1+1.52)\,(1+1/1.52)]}{6(2.02 - 0.17)\,10^{-5}\,(70)\,(1+1)^2} = 0.312\ \text{m}$$

Liquid-expansion devices, exemplified by the household thermometer, generally come in two main arrangements: the mercury type and the organic-liquid type. Versions employing gas instead of liquid are also available. Liquid-expansion sensors require no electric power and are stable even after repeated cycling. On the other hand, they do not generate data that are easily recorded or transmitted, and they cannot make spot measurements.

Change-of-state temperature sensors consist of labels, crayons, lacquers, or liquid crystals whose appearance changes once a certain temperature is reached. They are used to monitor the temperature of thermal devices. For example, when a thermal device exceeds a certain temperature, a white dot on a sensor label attached to the thermal device will turn black. Slow response time on the order of minutes prohibit their use for transient temperature changes, and accuracy is lower than with other types of sensors. With the exception of the case of liquid-crystal displays, the change in state is irreversible.

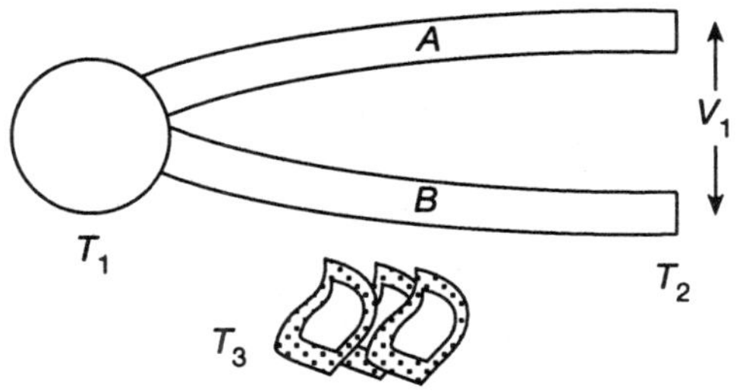

Figure 8. Thermocouple

In the chemical process industries, the most commonly used temperature sensors are: *thermocouples*, *resistive devices*, and *infrared thermometers*.

Thermocouples: A thermocouple consists of two alloys joined together at one end and opened at the other. The voltage at the output end (the open end; V_1 in Figure 8) is a function of the temperature T_1 at the closed end. As the temperature rises, the voltage goes up. Thermocouples are often located inside a metal or glass shield that protects them from harsh environments.

The open-end voltage is a function of not only the closed-end temperature (i.e., the temperature at the point of measurement) but also the temperature at the open end (T_2 in Figure 8). Only by holding T_2 at a standard temperature can the measured voltage be considered a direct function of the change in T_1.

The accepted standard for T_2 is 0°C; therefore, most tables and charts make the assumption that T_2 is at that level. Table 1 (shown on the following page) gives representative millivolt output for various thermocouples with a cold reference junction (T_2) of 0°C. In commercial equipment, the difference between the actual temperature at T_2 and 0°C is usually corrected for electronically, within the device. This voltage adjustment is referred to as the cold-junction correction.

PROBLEM 6:

You are to design an experiment that requires the accurate measurement of a temperature in the range of 200–300°C. Which of the thermocouples would provide the greatest precision?

SOLUTION:

The greatest precision would be provided by the thermocouple that had the greatest Seebeck coefficient (*SC*), which is the change in voltage

	Temperature-Millivolt Relation for Thermocouples			
T (°C)	**J-Type Iron + Constantan –**	**T-type Copper + Constantan –**	**K-type Chromel + Alumel –**	**S-Type Platinum + 10% Rhodium –**
50	2.585	2.036	2.023	0.299
100	5.269	4.279	4.096	0.646
150	8.01	6.704	6.138	1.029
200	10.779	9.288	8.138	1.441
250	13.555	12.013	10.153	1.874
300	16.327	14.862	12.209	2.323
350	19.09	17.819	14.293	2.786
400	21.848		16.397	3.259
450	24.61		18.516	3.742
500	27.393		20.644	4.233
600	33.102		24.905	5.239
700	39.132		29.129	6.275
800	45.494		33.275	7.345
900	51.877		37.326	8.449
1,000	57.953		41.276	9.587

Table 1. Voltage Output of Select Thermocouples

divided by the corresponding change in resistance. For a temperature of 250°C, it is found as follows for Type-J thermocouple:

$$SC_J = \frac{\Delta V}{\Delta T} = \frac{V_{300} - V_{200}}{300 - 200}$$

$$= \frac{16.327 - 10.779}{100}$$

$$= 5.55 \times 10^{-2} \text{ MV/°C}$$

The results for the other thermocouples are as follows:

$$SC_T = 5.57 \times 10^{-2}\ \text{MV/°C}$$

$$SC_K = 4.08 \times 10^{-2}\ \text{MV/°C}$$

$$SC_S = 8.82 \times 10^{-3}\ \text{MV/°C}$$

Since the T-type thermocouple has the greatest value of the Seebeck coefficient at the given temperature of 250°C, it would be the best choice. The J-type might also be considered since its value is only slightly different from the T-type.

Temperature changes in the wiring between the input and output ends do not affect the output voltage, provided that the wiring is of the same material as the thermocouple. For example, if a thermocouple is measuring the temperature in a furnace and the instrument that shows the reading is some distance away, the wiring between the two could pass near another furnace and not be affected by its temperature, unless it becomes hot enough to melt the wire.

The composition of the junction itself does not affect the thermocouple action in any way, so long as the temperature, T_1, is kept constant throughout the junction and the junction material is electrically conductive. Similarly, the reading is not affected by insertion of non-thermocouple alloys in either or both leads, provided that the temperature at the ends of the insertion material is the same (Figure 9).

This ability of the thermocouple to work with an inserted metal in the transmission path enables the use of a number of specialized devices, such as thermocouple switches. Whereas the transmission wiring itself is normally the thermoelectric equivalent of the thermocouple alloy, properly operating thermocouple switches must be made of gold-plated or silver-plated copper alloy elements with appropriate steel springs to ensure good

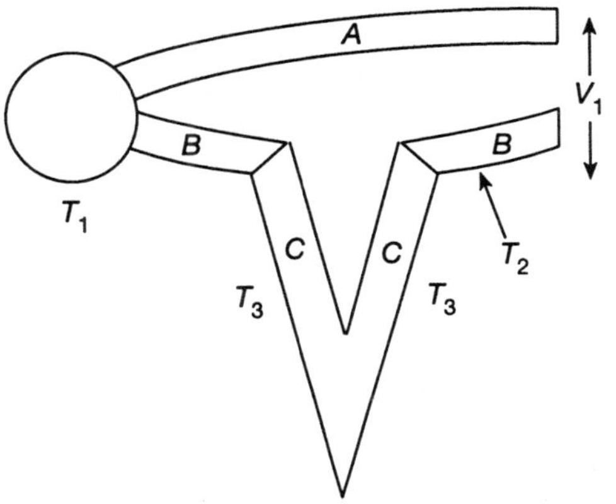

Figure 9. Thermocouple Junctions

contact. As long as the temperature at the input and output junctions of the switch are equal, this change in composition makes no difference.

Resistance Temperature Devices: A typical RTD consists of a fine platinum wire wrapped around a core and covered with a protective coating. Usually, the core and coating are glass. The size of the wire used will determine the electrical resistance of the wire. For platinum at 0°C, the resistance can vary from 10 ohms to several thousand ohms for very thin film RTD. The most common value is 100 ohms at 0°C. A temperature coefficient for platinum of 0.00385Ω/Ω/°C, results in a resistance change of 0.385 ohms for each °C at 0°C.

Although the resistance temperature curve is relatively linear, accurately converting measured resistance to temperature requires curve fitting. The Callendar-Van Dusen equation is commonly used to approximate the RTD curve:

$$R_t = R_0\,[1 + At + Bt^2 + C(t - 100)^3]$$

where:

R_t is the resistance of the RTD at temperature = t,

R_0 is the resistance of the RTD at 0°C,

A, B, and C are the Callendar-Van Dusen coefficients shown in Table 2,

t is the temperature in degrees Celsius.

For temperatures above 0°C, the C coefficient equals zero. Therefore, for temperatures above 0°C, this equation reduces to a quadratic.

Callendar-Van Dusen Coefficients

α	0.00385
A	3.9080×10^{-3} Ω/°C
B	-5.8019×10^{-7} Ω/°C^2
C	-4.2735×10^{-12} Ω/°C^3

Table 2. Callendar-Van Dusen Coefficients for Given α

PROBLEM 7:

An RTD made of platinum has a temperature coefficient $\alpha = 0.00385$ and a resistance of 100Ω at 0°C. Determine the resistance of the RTD at 100°C.

SOLUTION:

The resistance can be found by using the Callendar-Van Dusen equation:

$$R_t = R_0\,[1 + At + Bt^2 + C(t - 100)^3]$$

where:

$R_0 = 100\Omega$,

$t = 100°\text{C}$,

A and B given in Table 2,

and $C = 0$ in this problem since $T > 0°\text{C}$.

Substituting:

$$R_t = 100\Omega[1 + (3.908 \times 10^{-3}\ \Omega/°\text{C})(100°\text{C}) + (-5.8019 \times 10^{-7}\ \Omega/°\text{C}^2)\,(100°\text{C})^2]$$

$$R_t = 138.49981 = 138.5\Omega \text{ (4 significant figures)}$$

The most commonly used standard slope, pertaining to platinum of a particular purity and composition, has a value of 0.00385 (assuming that the resistance is measured in ohms and the temperature in degrees Celsius). A resistance versus temperature curve drawn with this slope is a so-called European curve, because RTDs of this composition were first used extensively on that continent.

Thermistors: The resistance-temperature relationship of a thermistor is negative and highly nonlinear. This nonlinearity poses a serious problem for designers who must develop their own circuitry. This difficulty can be eased by using thermistors in matched pairs, in such a way that the nonlinearities offset each other.

Thermistors are usually designated in accordance with their resistance at 25°C. The most common of these ratings is 2,252 ohms; among the others are 5,000 and 10,000 ohms. If not specified to the contrary, most instruments will accept the 2,252 type of thermistor. Of the major temperature sensing devices, the thermistor exhibits by far the largest parametric change with temperature.

Infrared sensors: These measure the amount of radiation emitted by a surface. Electromagnetic energy radiates from all matter regardless of its temperature. In many process situations, the energy is in the infrared region. As the temperature goes up, the amount of infrared radiation and its average frequency go up.

Different materials radiate at different levels of efficiency. This efficiency is quantified as emissivity, a decimal number or percentage ranging between 0 and 1 or 0% and 100%. Most organic materials, including skin, are very efficient, frequently exhibiting emissivities of 0.95. Most polished metals, on the other hand, tend to be inefficient radiators at room temperature, with emissivity or efficiency often 20% or less.

To function properly, an infrared measurement device must take into account the emissivity of the surface being measured. This can often be looked up in a reference table.

THERMOCOUPLE REFERENCE JUNCTION PRINCIPLES

When accurate thermocouple measurements are required, it is common practice to reference both legs to copper lead wire at the ice point so that copper leads may be connected to the voltage readout instrument. This procedure avoids the generation of thermal voltages at the terminals of the readout instrument.

Changes in reference junction temperature influence the output signal, and practical instruments must be provided with a means to cancel this potential source of error. The voltage generated is dependent on a difference in temperature, so in order to make a measurement the reference must be known. This is shown schematically in Figure 10 and can be accomplished by placing the reference junction in an ice water bath at a constant 0°C (32°F). In this case, the output voltage corresponds to the values in Table 1.

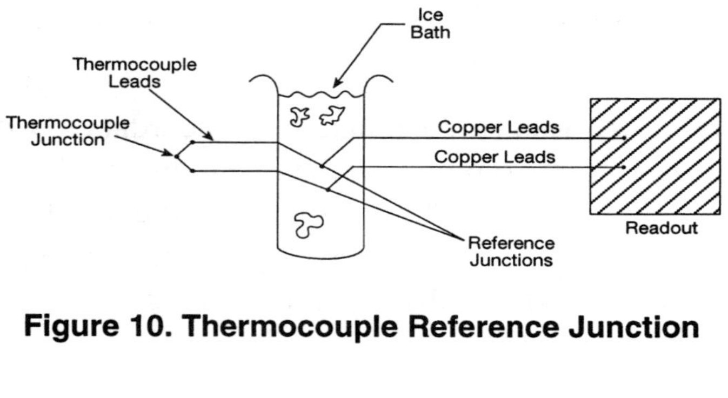

Figure 10. Thermocouple Reference Junction

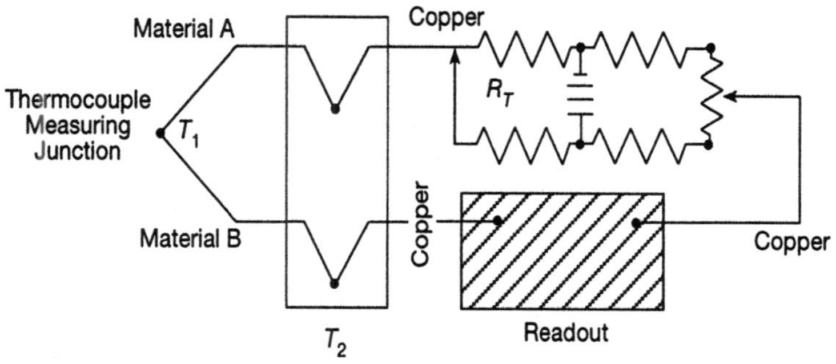

Figure 11. Thermocouple Bridge Circuit

Another method is the electrical bridge method which usually employs a self-compensating electrical bridge network as shown in Figure 11. This system incorporates a temperature sensitive resistance element (R_T), which is in one leg of the bridge network and thermally integrated with the cold junction T_2. The bridge is usually energized from a mercury battery or stable DC power source. The output voltage is proportional to the unbalance created between the preset equivalent reference temperature at T_2 and the hot junction T_1. In this system, the reference temperature of 0°C or 32°F may be chosen.

As the ambient temperature surrounding the cold junction T_2 varies, a thermally generated voltage appears and produces an error in the output. However, an automatic equal and opposite voltage is introduced in series with the thermal error. This cancels the error and maintains the equivalent reference junction temperature over a wide ambient temperature range with a high degree of accuracy. By integrating copper leads with the cold junction, the thermocouple material itself is not connected to the output terminal of the measurement device, thereby eliminating secondary errors.

UNCERTAINTY ANALYSIS

No measured quantity is exact. There is always uncertainty associated with a measured quantity. When more than one quantity is used to obtain a result, the uncertainties of the individual quantities must be combined to obtain the uncertainty of the final results.

Simple methods can be used to determine uncertainty. A simple method is to assume that the error in the results equals the maximum error in any parameter used to calculate the results.

PROBLEM 8:

It is desired to calculate electrical power, P, resulting from a voltage, $E = 100\ \text{V} \pm 2\ \text{V}$, and a current, $I = 10\ \text{A} \pm 0.2\ \text{A}$.

SOLUTION:

Since $P = EI$, the nominal value of power is:

$$(100\ \text{V}) \times (10\ \text{A}) = 1{,}000\ \text{W}$$

Since the error for both E and A is 2%, then the error in the result is assumed to be 2% or:

$$P = 1{,}000\ \text{W} \pm 20\ \text{W}$$

PROBLEM 9:

Using the data from Problem 8, assume that the errors occur in the most detrimental manner. Determine the error in this case.

SOLUTION:

$$P_{max} = (100 + 2\ \text{V}) \times (10 + 0.2\ \text{A}) = 1{,}040.4\ \text{W}$$

$$P_{min} = (100 - 2\ \text{V}) \times (10 - 0.2\ \text{A}) = 960.4\ \text{W}$$

The uncertainty in power is +4.04% and –3.96%

Next, consider an improvement to the above methods by evaluating another example, as shown in Figure 12. Figure 12 is a cylinder of diameter, D, length, L, T_{amb} = Ambient Temperature, T = Body temperature p, c, are properties of the cylinder. Suppose an experiment is conducted to

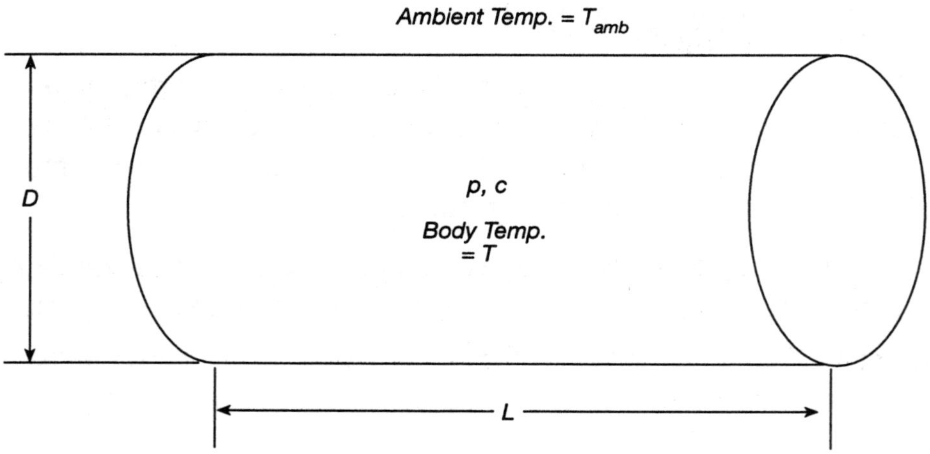

Figure 12. Internal Energy of a Body

determine the internal energy, U, of a body that is at a temperature ΔT above the room temperature.

The equation for this is:

$$U = \rho V c \Delta T \tag{1}$$

where:

$$DT = T - T_{amb} \tag{2}$$

and:

$$V = \frac{\pi D^2}{4} L \tag{3}$$

In the above equations, assume that the *measured* quantities are the body temperature T, the ambient temperature T_{amb}, the diameter D of the body, and its length L. Assume that the body density ρ and specific heat c are *reference* data—that is, they are obtained from a textbook or handbook. Suppose the data is as follows:

Measured:

$$T = 200°\text{F}$$

$$T_{amb} = 70.0°\text{F}$$

$$D = 3.00 \text{ in.}$$

$$L = 6.00 \text{ in.}$$

Reference:

$$\rho = 5.00 \text{ slugs/ft}^3$$

$$c = 2.00 \text{ Btu/slug-°F}$$

Using this data in equations (1), (2), and (3) yields:

$$V = 0.0245 \text{ ft}^3$$

$$\Delta T = 130°\text{F}$$

$$U = 31.9 \text{ Btu}$$

The measured values of the two temperatures, the diameter, and length will all have uncertainties. This will result in uncertainty in the computed value of U. While reference values can at times have uncertainty, in this example, assume no uncertainty is assigned to the reference data.

Now, in general the uncertainty in any measurement X can be given the symbol dX, so instead of a measurement X the experimenter should always measure $X \pm dX$. The measurement X is called the *nominal* value, and dX is called the *estimate of the uncertainty*. This uncertainty depends on the accuracy of the measurement device. By convention, the uncertainty dX is usually taken to be half of the smallest scale reading of the instrument. For example, in Figure 13, the instrument used to measure the diameter and length of the body was a micrometer.

The smallest division of the instrument shown is 0.005". This means that dD and dL should both be taken as 0.0025". The reasoning behind this is that, with this particular micrometer, an experimenter could distinguish readings that were an exact increment of 0.005", and perhaps even that a reading was midway between an increment, but no better than that. In other words, an instrument with 0.005" increments can distinguish lengths to ±0.0025".

Using this concept, and assuming that the temperatures are measured using thermocouples with an uncertainty of dT = 0.5°F, all the measurements can be written in the following form:

$$T \pm dT = 200 \pm 0.5°\text{F}$$

$$T_{amb} \pm dT_{amb} = 70.0 \pm 0.5°\text{F}$$

$$D \pm dD = 3.00 \pm 0.0025''$$

$$L \pm dL = 6.00 \pm 0.0025''$$

Then, with this data, the question is, what is dU in $U \pm dU$?

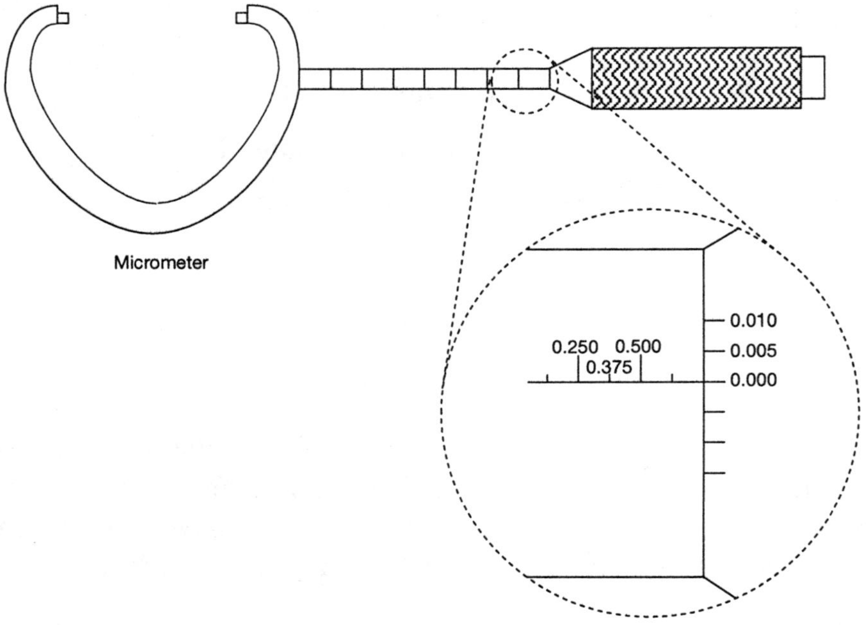

Figure 13. Micrometer

Given a calculated result R, which is a function of independent variables, x_i,

$$R = R\,(x_1, x_2, x_3, \ldots, x_n,) \tag{4}$$

where:

$$W_R = \text{uncertainty in the results}$$

$$w_1, w_2, \ldots, w_n = \text{uncertainty in the independent variables}$$

If the uncertainty in the independent variables are all given with the same odds, then the uncertainty in the results having these odds is:

$$W_R = \left[\left(\frac{\partial R}{\partial x_1} w_1 \right)^2 + \left(\frac{\partial R}{\partial x_2} w_2 \right)^2 + \ldots + \left(\frac{\partial R}{\partial x_n} w_n \right)^2 \right]^{0.5} \tag{5}$$

For the problem stated above:

$$R = U$$

$$x_1 = \mathrm{D}$$

$$x_2 = \mathrm{L}$$

$$x_3 = \mathrm{T}$$

$$x_4 = T_{amb}$$

Use the formulas to derive the nominal results. $U = 31.9$ Btu in the above example.

$$\begin{aligned}
\frac{\partial U}{\partial x_1} &= \frac{\partial U}{\partial D} = \rho c \left(\frac{\pi D}{2} \right) L\,(T - T_{amb}) = \frac{2}{D} U \\
\frac{\partial U}{\partial x_2} &= \frac{\partial U}{\partial L} = \rho c \left(\frac{\pi D^2}{4} \right) (T - T_{amb}) = \frac{U}{L} \\
\frac{\partial U}{\partial x_3} &= \frac{\partial U}{\partial T} = \rho c \left(\frac{\pi D^2}{4} \right) L = \frac{U}{(T - T_{amb})} \\
\frac{\partial U}{\partial x_4} &= \frac{\partial U}{\partial T_{amb}} = -\rho c \left(\frac{\pi D^2}{4} \right) L = -\frac{U}{(T - T_{amb})}
\end{aligned} \tag{6}$$

With the partial derivatives and uncertainties determined, these values are substituted into Equation 5. Note how the resulting partial derivatives can be expressed in terms of U. This will simplify the evaluation of the uncertainty. Substituting into Equation 5:

$$W_R = \left[\left(\frac{\partial R}{\partial x_1}w_1\right)^2 + \left(\frac{\partial R}{\partial x_2}w_2\right)^2 + \ldots + \left(\frac{\partial R}{\partial x_n}w_n\right)^2\right]^{0.5}$$

$$W = \sqrt{\begin{array}{l}\left(\frac{2U}{D}\right)^2 (.0025)^2 + \left(\frac{U}{L}\right)^2 (.0025)^2 \\ + \left(\frac{U}{(T - T_{amb})}\right)^2 (.5)^2 + \left(\frac{U}{(T - T_{amb})}\right)^2 (.5)^2\end{array}}$$

$$\frac{W}{U} = \sqrt{\left(\frac{2 \times .0025}{3.00}\right)^2 + \left(\frac{.0025}{6.00}\right)^2 + \frac{(0.5^2 + 0.5^2)}{(130^2)}}$$

$$\frac{W}{U} = 0.00570$$

$$U = 31.9 \pm 0.182 \text{ Btu}$$

Uncertainty analysis is important in determining the accuracy of the experimental results. In addition, it is very useful in indicating where an experiment can be improved. In the above example, the error in measuring T or T_{amb} was five times the error in measuring D. This suggests that an improved temperature measuring device would be more effective in reducing the overall error of the experiment than, say, trying to measure the diameter of the cylinder more precisely.

PROBLEM 10:

Determine the uncertainty in the power P of Problem 8 using the uncertainty analysis theory previously described.

SOLUTION:

Using the analysis described:

$$W_R = \left[\left(\frac{\partial R}{\partial x_1} w_1\right)^2 + \left(\frac{\partial R}{\partial x_2} w_2\right)^2 + \ldots + \left(\frac{\partial R}{\partial x_n} w_n\right)^2\right]^{0.5}$$

where $R = P = EI = 1{,}000$ W.

$x_1 = I$ $\qquad\qquad$ $w_1 = 0.2$ A

$x_2 = E$ $\qquad\qquad$ $w_2 = 2$ V

$$\frac{\partial R}{\partial E} = I = 10 \text{ A};$$

$$\frac{\partial R}{\partial I} = E = 100 \text{ V}$$

Substituting into the equation for W_R:

$$W_R = [(100 \text{ V} \times 0.2 \text{ A})^2 + (10 \text{ A} \times \text{V})^2]^{0.5}$$

$$= 28.3 \text{ W}$$

The percent error is 28.3/1000 or 2.83%.

FE/EIT

FE: PM Mechanical Engineering Exam

CHAPTER 11

Mechanical Design

CHAPTER 11

MECHANICAL DESIGN

MECHANICAL DESIGN

Mechanical design and machine component analysis rely on the forces on the component (using statics and dynamics fundamentals), the development of stress in the material (using mechanics of materials fundamentals), and the ability of the material to withstand the stress (using materials science fundamentals). Machine components can be analyzed as a bar (axial tension and compression), a shaft (torsion), a beam (bending moment and shear force), or a combination of the three. In the elastic (linear) region of material response, the stress (normalized force) and strain (normalized displacement) are related as follows:

$$\sigma = E\varepsilon$$

$$\tau = G\gamma$$

and

$$G = \frac{E}{2(1+\nu)}$$

From Mechanics of Materials, the maximum normal and shear stress, and displacement can be related to the load (bending moment, shear force) as follows:

Tension and compression of a bar:

$$\sigma = \frac{P}{A} \text{ and } \delta = \frac{PL}{AE}$$

Shear stress in a shaft:

$$\tau = \frac{Tr}{J} \text{ and } \theta = \frac{TL}{JG}$$

Normal and shear stress in a beam:

$$\sigma = \frac{My}{I} \text{ and } \tau = \frac{VQ}{Ib}$$

where: σ = normal stress

δ = deflection

ε = strain

ν = Poisson's ratio

P = load

M = bending moment

r = radius from centroid

y = distance from neutral axis

V = shear force

G = Shear Modulus (Modulus of Rigidity)

Q = first moment of area above the shear plane with respect to the neutral axis

τ = shear stress

θ = angular displacement

γ = shear strain

J = polar moment of inertia

T = torque

A = cross-sectional area

I = moment of inertia

b = width at shear plane

E = Modulus of elasticity (Young's modulus)

L = length

The stress on a point is the total of stress due to the various loadings. The stresses can be summed because the response of the material is linear (principle of superposition) below the yield stress. This stress is represented as a tensor with three normal stresses and six shear stresses as shown in Figure 1.

$$\begin{vmatrix} \sigma_x & \tau_{xy} & \tau_{xz} \\ \tau_{yx} & \sigma_y & \tau_{yz} \\ \tau_{zx} & \tau_{zy} & \sigma_z \end{vmatrix}$$

Stress Tensor

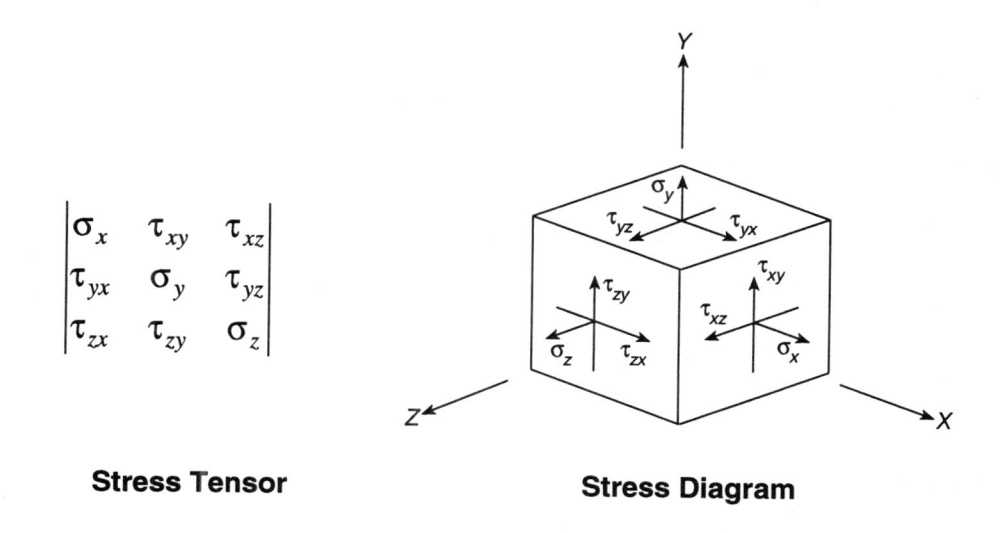

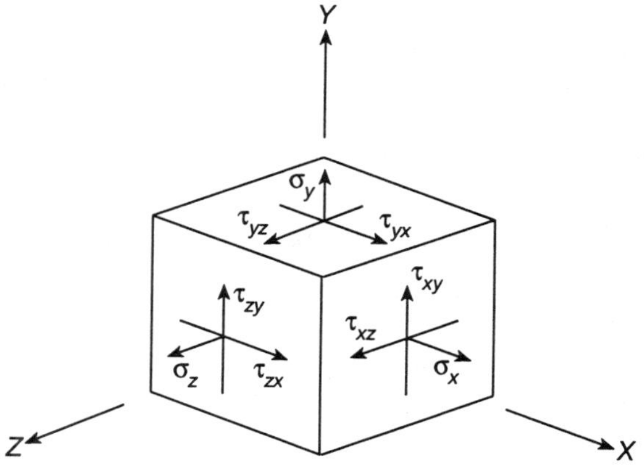

Stress Diagram

Figure 1. Three–Dimensional Stress Tensor

Analyzing shear stresses on a differential element subjected to plane stress (i.e., no normal stress in the z direction and no shear stresses on the z face plane), we sum the moments generated by the shear stresses about the origin. See Figure 2.

$$\sum M_0 = \tau_{xy} A_x \quad dx - \tau_{xy} A_y \quad dy = 0$$

$$\sum M_0 = \tau_{xy} dy \quad dz \quad dx - \tau_{ys} dx \quad dz \quad dy = 0$$

Therefore:

$$\tau_{xy} = \tau_{ys}$$

Similarly:

$$\tau_{xz} = \tau_{zs}$$

$$\tau_{zy} = \tau_{yz}$$

The stress tensor is therefore symmetrical about the main diagonal, and the stress state on a three-dimensional element can be represented by three normal and three shear stresses. If one were to rotate the orientation of the axis for this element, the magnitude of the normal and shear stresses will change. There exists a set of axes that the shear stresses will reduce to zero and only normal stresses will exists. These are called principal normal stresses. This is the Eigenvalue transformation studied in linear algebra to produce a diagonal matrix. This has been solved for our specific case.

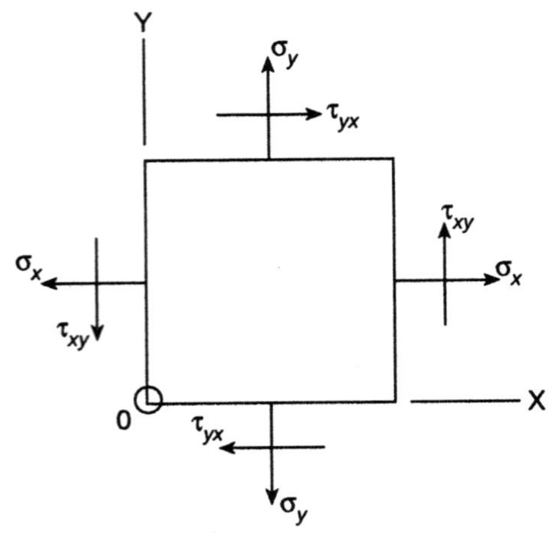

Figure 2. Plane Stress

Principle Stresses and Mohr's Circle

Given the stress state of an element, we need to discover how that stress state relates to the stress state of an element in an tensile test specimen. After all, our tabulated data includes yield and ultimate tensile strength of tensile test specimens. To better understand equivalent stress states, let's sum shear and normal forces on an oblique line at angle ϕ from the y-axis of a plane stress differential element as shown in Figure 3:

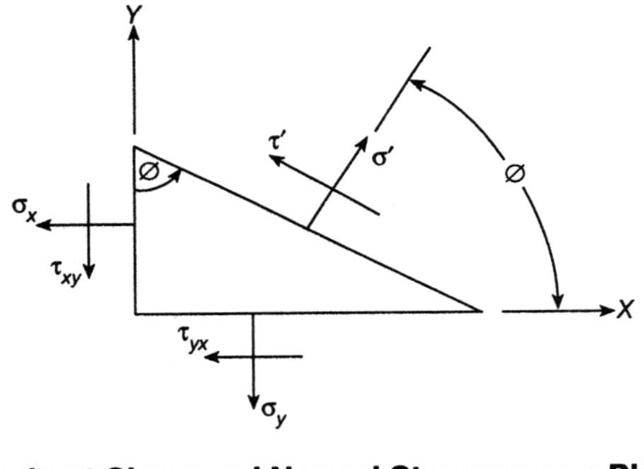

Figure 3. Resultant Shear and Normal Stresses on a Plane at Angle ϕ

The normal and shear stresses on the oblique line (after some manipulation and substitution of double angle formulas) are:

$$\sigma' = \frac{\sigma_x + \sigma_y}{2} + \frac{\sigma_x - \sigma_y}{2}\cos 2\phi + \tau_{xy}\sin 2\phi$$

$$\tau' = \frac{\sigma_x - \sigma_y}{2}\sin 2\phi + \tau_{xy}\cos 2\phi$$

The resultant normal and shear forces as a function of the angle 2ϕ forms a circle centered on the σ axis at:

$$\sigma = \frac{\sigma_x + \sigma_y}{2}$$

with a radius of:

$$R = \sqrt{\left(\frac{\sigma_x - \sigma_y}{2}\right)^2 + \tau_{xy}^{\ 2}}$$

and principle stresses of:

$$\sigma_1,\ \sigma_2 = \frac{\sigma_x + \sigma_y}{2} \pm \sqrt{\left(\frac{\sigma_x - \sigma_y}{2}\right)^2 + \tau_{xy}^{\ 2}}$$

The diagrammatic representation of the stress state of a point is called Mohr's circle, as shown in Figure 4.

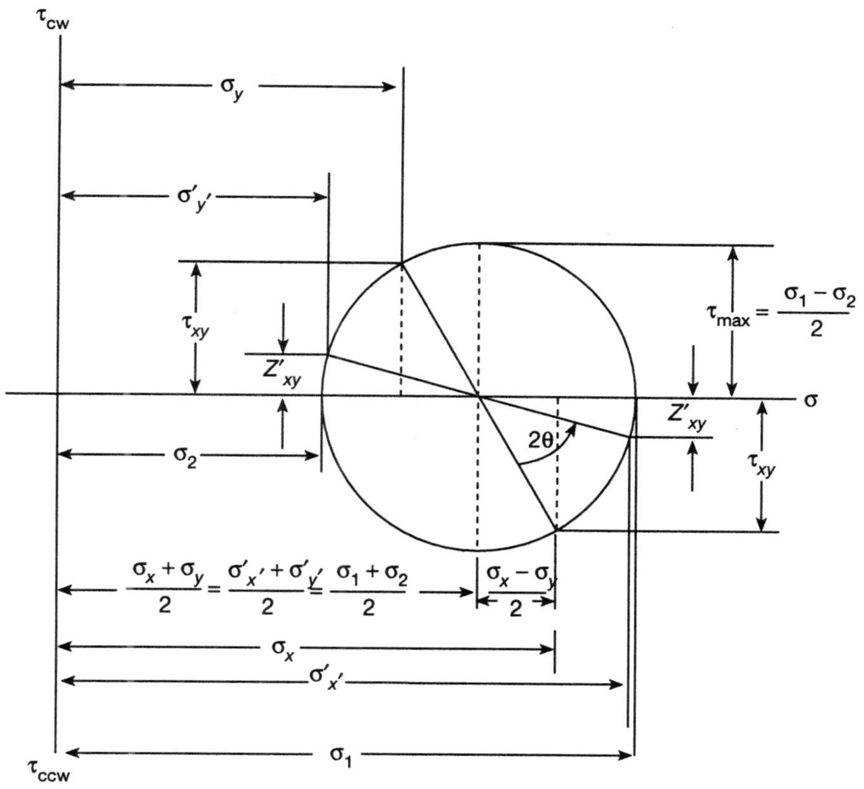

Figure 4. Mohr's Circle

Now let's set the shear stress equal to zero. Two angles of ϕ satisfy the following equation, and the resultant normal stresses at these angles are the principal normal stresses.

$$\tan 2\phi = \frac{\sigma_x - \sigma_y}{2\tau_{xy}}$$

Knowing the principal stresses, we can calculate the strain induced at the point of interest. It should be noted that many of our analyses involve components where the principal directions and stresses are known and the following can be used directly.

Loading	Principal stresses	Principal strains
Uniaxial	$\sigma_1 = E\varepsilon_1$ $\sigma_2 = 0$ $\sigma_3 = 0$	$\varepsilon_1 = \frac{\sigma_1}{E}$ $\varepsilon_2 = -\nu\varepsilon_1$ $\varepsilon_3 = -\nu\varepsilon_1$
Biaxial	$\sigma_1 = \frac{E(\varepsilon_1 + \nu\sigma_2)}{1+\nu^2}$ $\sigma_2 = \frac{E(\varepsilon_2 + \nu\sigma_1)}{1+\nu^2}$ $\sigma_3 = 0$	$\varepsilon_1 = \frac{\sigma_1}{E} - \frac{\nu\sigma_2}{E}$ $\varepsilon_2 = \frac{\sigma_2}{E} - \frac{\nu\sigma_1}{E}$ $\varepsilon_3 = -\frac{\nu\sigma_1}{E} - \frac{\nu\sigma_2}{E}$
Triaxial	$\sigma_1 = \frac{E\varepsilon_1(1-\nu) + \nu E(\varepsilon_2 + \varepsilon_3)}{1-\nu-2\nu^2}$ $\sigma_2 = \frac{E\varepsilon_2(1-\nu) + \nu E(\varepsilon_1 + \varepsilon_3)}{1-\nu-2\nu^2}$ $\sigma_3 = \frac{E\varepsilon_3(1-\nu) + \nu E(\varepsilon_1 + \varepsilon_2)}{1-\nu-2\nu^2}$	$\varepsilon_1 = \frac{\sigma_1}{E} - \frac{\nu\sigma_2}{E} - \frac{\nu\sigma_3}{E}$ $\varepsilon_2 = \frac{\sigma_2}{E} - \frac{\nu\sigma_1}{E} - \frac{\nu\sigma_3}{E}$ $\varepsilon_3 = \frac{\sigma_3}{E} - \frac{\nu\sigma_1}{E} - \frac{\nu\sigma_2}{E}$

Table 1. Stress and Strain Relationships

PROBLEM 1:

A solid 5 cm diameter steel shaft is subjected to the load shown on Figure 5. Find the state of stress at point *A*. Show the stress state on a properly oriented cube. Calculate the principal stresses for this location. Draw Mohr's circle.

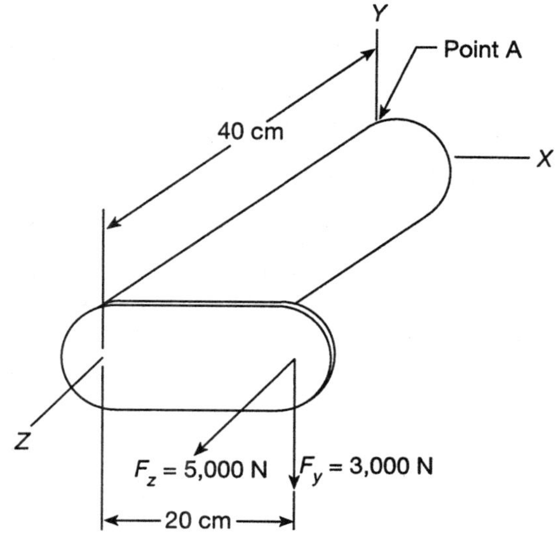

Figure 5

SOLUTION:

Draw a free-body diagram as shown in Figure 5a.

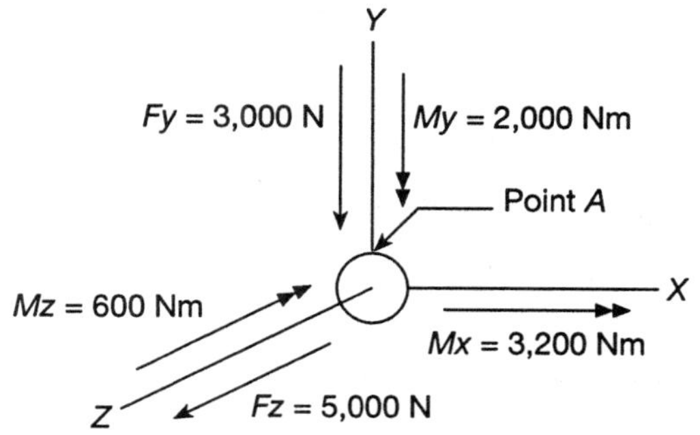

Figure 5a. Free-Body Diagram

Calculate stresses (using superposition). Take each force/moment, analyze separately, then add together.

F_z is axial force (loaded as a bar).

$$\sigma_2 = \frac{P}{A} = \frac{4(5{,}000\text{N})}{\pi(0.05\text{m})^2} = 2.25 \text{ MPa}$$

F_y causes a shear force on the z plane in the y direction:

$$\tau_{zy} = \frac{VQ}{Ib}$$

but $$Q|_a = 0$$

therefore $$\tau_{sy} = 0$$

M_y is a bending moment about the y axis.

Since point A is on the neutral axis,

$$\sigma_x = 0$$

M_x is a bending moment about the x axis, causing a normal stress at point A:

$$\sigma_z = \frac{M_y}{I} = \frac{M\frac{d}{2}}{\frac{\pi d^4}{64}} = \frac{32M}{\pi d^3} = \frac{32(3,200\text{Nm})}{\pi(0.05)^3} = 260.76 \text{ MPa}$$

M_z causes a torque through the shaft and results in a shear stress on the z plane in the x direction at point A:

$$\tau_{zx} = \frac{Tc}{J} = \frac{T\frac{d}{2}}{\frac{\pi d^4}{32}} = \frac{16M_2}{\pi d^3} = \frac{16(600\text{Nm})}{\pi(0.05m)^3} = 24.45 \text{ MPa}$$

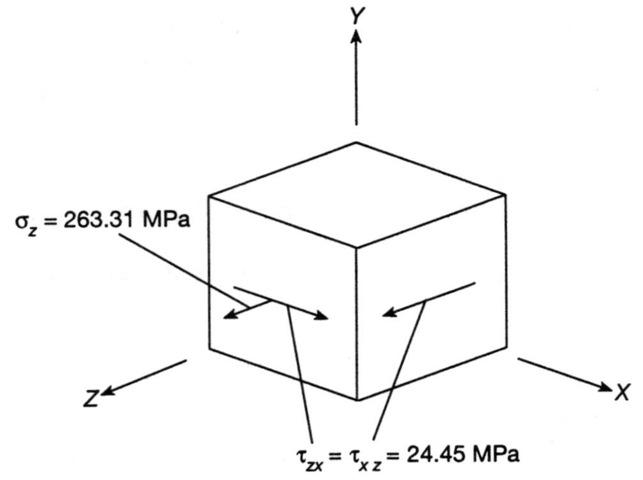

Figure 5b. Stresses on a Properly Oriented Cube

Algebraic summation of stresses:

$$\sigma_z = 2.55 \text{ MPa} + 260.76 \text{ MPa} = 263.31 \text{ MPa}$$

$$\tau_{zx} = 24.45 \text{ MPa}$$

Principal stresses are:

$$\sigma_1, \sigma_2 = \frac{0 + 263.31}{2} \pm \sqrt{\left(\frac{0 - 263.31}{2}\right)^2 + 24.45^2}$$

$$\sigma_1 = 131.655 + 133.905 = 265.56 \text{ MPa}$$

$$\sigma_2 = 131.655 - 133.905 = -2.25 \text{ MPa}$$

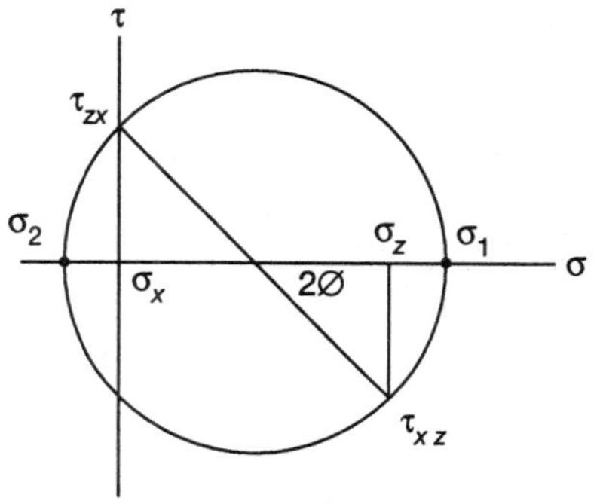

Figure 5c. Mohr's Circle

STATIC FAILURE THEORIES

The object now is to relate the stress state of the point that is limiting in our design to the stress state of a tensile specimen used to generate tabulated data. In sum, to find out how close our complex part comes to reaching failure based on the simple loading of a tensile specimen. Several theories have developed, based on maximum shear stress, maximum normal stress, or energy considerations. There have been proposed modifications as well based on empirical data.

Brittle Materials

Materials such as cast irons, ceramics, and concrete have little yield (less than 0.5%) before fracture and are classified as brittle. Brittle materials generally have a much higher compressive strength than tensile strength.

Maximum normal stress theory: In tensile specimens, the stress state is one-dimensional normal stress with no shear. This theory states that when the maximum principal normal stress (at a point) reaches the ultimate tensile strength (S_t) or ultimate compressive strength (S_c), failure occurs. Note that in two dimensions, the safe envelope would look like Figure 6.

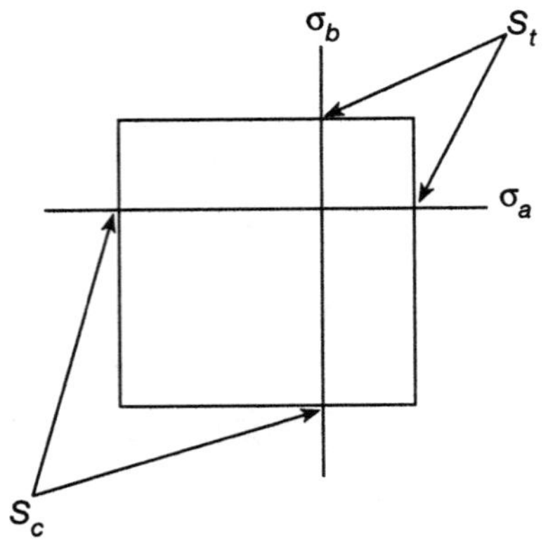

Figure 6. Maximum Normal Stress Theory

This theory works well in the first and third quadrants, but not as well in the second and fourth quadrants. Let's look at a three-dimensional plane strain element and the associated Mohr's Circle shown in Figure 7.

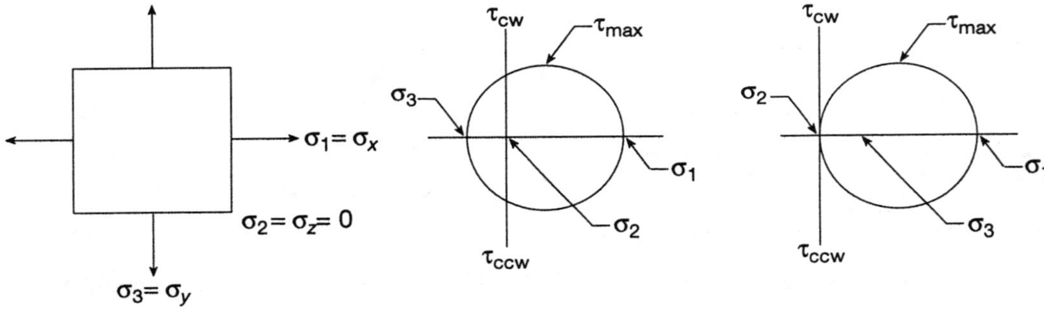

Figure 7. Plain Stress Element and Associated Mohr's Circles

Note that the maximum shear stress is at a 45-degree angle to both the x and y axis instead of at a 45-degree angle to both the x and z axis. Expressed in another way, when the x and y stress (σ_1 and σ_3) are both tension or compression, the maximum shear depends only on the stress of greatest magnitude and the zero stress on the z axis (σ_2). (See Figure 7.) When one of the stresses is tensile and one compressive, the maximum shear is dependent on both. In the Coulomb–Mohr Theory (also known as Internal-friction theory), the second and fourth quadrants have one principal tensile stress and one principal compressive stress. Here, stress failure is represented by straight lines connecting the ultimate tensile stress (S_t) and ultimate compressive stress (S_c) intersections with the axes, as shown in Figure 8.

The Modified Mohr Theory expands the failure envelope based on empirical data, primarily from tests on cast iron. The diagram for both the Coulomb–Mohr Theory and the Modified Mohr Theory is shown in Figure 8.

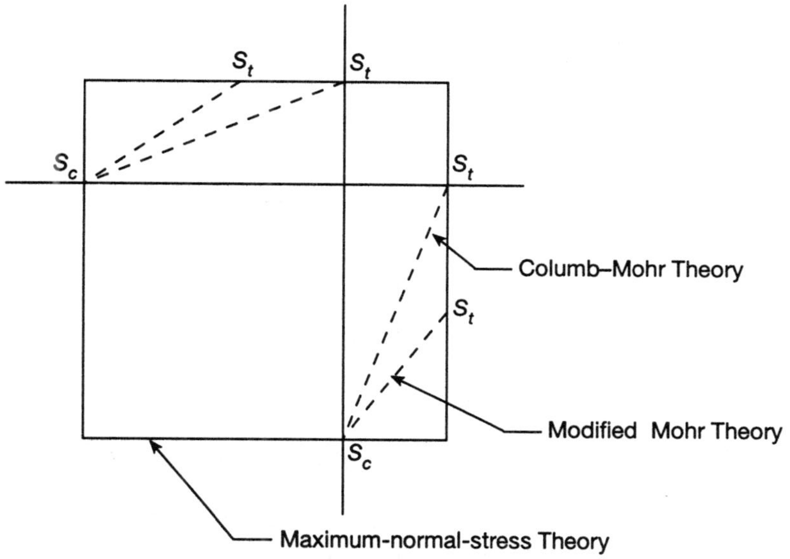

Figure 8. Diagram of Coulomb-Mohr and Modified Mohr Theories

Ductile Materials

In ductile materials, the tabulated data of concern is the yield strength (S_y) instead of the ultimate tensile/compressive strengths (S_t and S_c) used for brittle materials. Two theories are presented, maximum shear stress theory and distortion energy theory.

Maximum shear stress theory: For ductile materials, we assume that the magnitude of tensile and compressive yield stresses are equal. With this assumption, the maximum shear stress is equal to:

$$\tau_{max} = \frac{\sigma_1 + \sigma_3}{2}$$

where: $s_1 > s_2 > s_3$

From this it follows that the shear at yield (τ_y) is related to the yield strength (S_y) as follows:

$$\tau_y = 0.5\, S_y$$

In the first and third quadrant, the limiting shear stress is when σ_1 or σ_2 is equal to the yield stress. In the second and fourth quadrants, the limiting shear stress is where the magnitude of the difference between the two principal stresses equal the magnitude of the yield stress. Diagrammatically, the maximum shear stress theory looks like Figure 9.

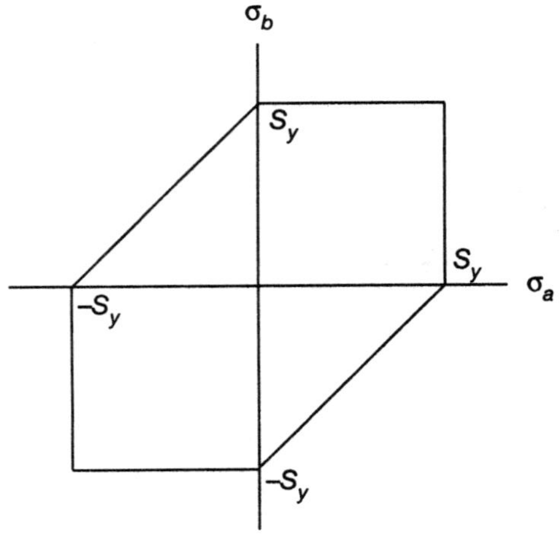

Figure 9. Diagram of Maximum Shear Stress Theory

Although similar to the Coulomb–Mohr Theory, there are significant differences. First, yield strength, rather than ultimate tensile/compressive strength, is used. Second, the compressive yield strength is assumed to be equal to the tensile yield strength.

Distortion Energy theory: (also called *Shear Energy Theory*, *Von Mises-Henkcy Theory*, and *Octahedral-Shear-Stress Theory*). This theory assumes failure occurs when the octahedral shear stress equals or exceeds the octahedral stress for the simple tensile test at failure.

$$\sigma' = \left[\frac{(\sigma_1 - \sigma_2)^2 + (\sigma_2 - \sigma_3)^2 + (\sigma_1 - \sigma_3)^2}{2}\right]^{\frac{1}{2}}$$

Diagrammatically, the distortion energy looks like Figure 10.

Items are worthy of note:

1. The maximum-shear-stress theory is more conservative.
2. The distortion-energy theory passes closer to the center of empirical data and may be better at predicting failure.
3. The relationship between the yield strength and the shear yield strength is:

$\tau_y = 0.5\, S_{yr}$–from maximum shear stress theory

and $\tau_y = 0.577\, S_y$–from distortion energy theory

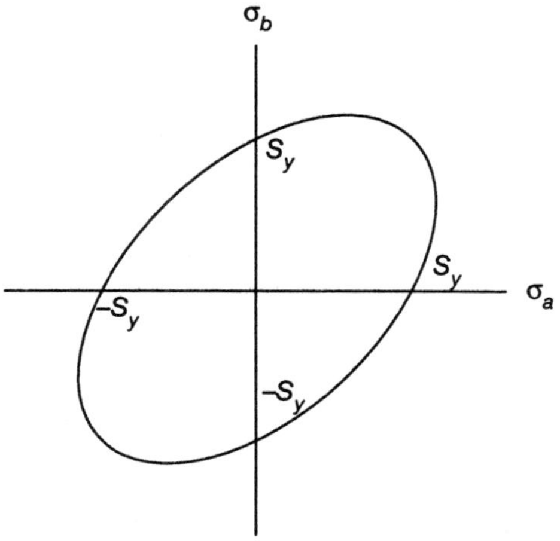

Figure 10. Diagram of Distortion Energy Theory

PROBLEM 2:

Strain gages have been placed in a machine support in an orientation to measure principal stresses. The stresses measured are 75 MPa, 20 MPa, and – 65 MPa. The material used on the support is cast aluminum, with S_y =165 MPa. Calculate the factor of safety using (a) the maximum shear stress theory and (b) the distortion energy theory.

SOLUTION:

a) For maximum shear stress:

$$\tau_{\max} = \frac{\sigma_1 - \sigma_3}{2} = \frac{75-(-65)}{2} = 70 \text{ MPa}$$

$$\tau_y = \frac{S_y}{2} = \frac{165}{2} = 82.5 \text{ MPa}$$

$$S.F. = \frac{\tau_y}{\tau_{\max}} = \frac{82.5}{70} = 1.18$$

(b) For distortion energy:

$$\sigma' = \left[\frac{(\sigma_1-\sigma_2)^2+(\sigma_2-\sigma_3)^2+(\sigma_1-\sigma_3)^2}{2}\right]^{\frac{1}{2}}$$

$$= \left[\frac{(75-20)^2+(20-(-65))^2+(75-(-65))^2}{2}\right]^{\frac{1}{2}}$$

$$\sigma' = 122.2 \text{ MPa}$$

$$S.F. = \frac{S_y}{\sigma'} = \frac{165}{122.2} = 1.35$$

Of note, these safety factors would cause concern about long-term reliability. Typically, a safety factor of two is appropriate, although a reliability approach with periodic testing/inspection may have safety factors similar to those calculated above.

BUCKLING

Columns in compression can fail due to buckling. Buckling is an instability not dependent on the material. The Euler buckling formula gives the critical compressive load, below which buckling should not occur. Since the original Euler equation was developed for columns with rounded ends, the end constraint correction factor (k) is used to correct for the end conditions. The Euler equation and the table of end constraint correction factor k are as follows:

$$P_{cr} = \frac{\pi^2 EI}{(kl)^2}$$

Substituting $I = r^2A$, the Euler equation can be written as:

$$\frac{P_{cr}}{A} = \frac{\pi^2 E}{\left[k\left(\frac{l}{r}\right)\right]^2}$$

where:

P_{cr} = critical load

I = moment of inertia

l = length

r = radius of gyration

k = constant based on the end condition

E = modulus of elasticity (Young's modulus)

A = cross-sectional area

$\frac{l}{r}$ = slenderness ratio for column

Commonly used k values for columns (for use in the Euler equation noted previously)

Column End Conditions	Theoretical Value	Recommended Value
Fixed–free	2.0	2.1
Pinned–pinned	1.0	1.0
Fixed–pinned	0.7	0.8
Fixed–fixed	0.5	0.65

Table 2. *K* Values for Euler's Column Formula

Note: "pinned" means the end can rotate but not translate. "fixed" means the end cannot rotate or translate. "Free" means the end can both translate and rotate without constraint.

Eccentric Loading–Short Columns

When a column is loaded eccentrically, the stress is the sum of the compressive stresses caused by the column acting as a bar (axial load) and the stress caused by the column acting as a beam (bending load). The sum of these two stresses represents the stress in the beam, as follows:

$$\sigma_{max}, \ \sigma_{min} = \frac{F}{A} \pm \frac{Mc}{I}$$

where:

F = load

M = moment caused by eccentric loading

c = distance to outer fiber of column from neutral axis

I = moment of inertia

PROBLEM 3:

A 5 mm diameter steel push rod for a machine is 60 cm long. The ends are free to rotate but not translate. What is the critical load for buckling?

SOLUTION:

$$P_{cr} = \frac{\pi^2 EI}{(kl)^2} = \frac{\pi^2 \left(205 \times 10^9\right)\left(\pi \dfrac{0.0025^4}{4}\right)}{[(1.0)(0.6)]^2} = 173 \text{ N}$$

STRESS CONCENTRATION

In the development of stresses, it was assumed that there were no irregularities that cause a localized increase in stress. In actual parts, there are stress risers such as holes and changes in cross sections, etc., which cause a local increase in stress. The theoretical or geometric stress concentration factors (K_t) have been studied and diagrams have been developed. Figure 11 gives the stress concentration factor for one geometry/loading combination. The K_t is a function of geometry alone. Frequently, the material of which the part is made is ductile and at the stress concentration point the material yields slightly. This relieves some of the stress and thereby reduces the stress concentration at the point from the theoretical

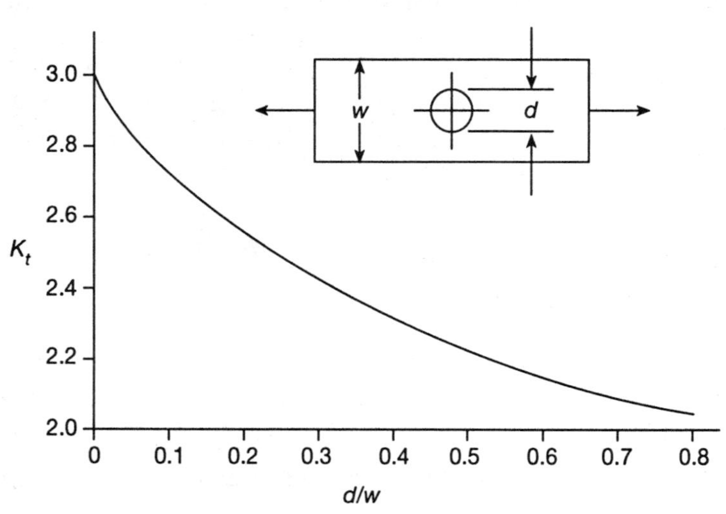

Figure 11. Stress Concentration Factor

amount. Notch sensitivity (q) is a measure of the material's "softening" at the tip of the crack. The following formulas apply:

$$\sigma_{max} = \sigma_o K_f \text{ where } K_f = 1 + q(K_t - 1)$$

Again K_t is only dependent on the geometry, notch sensitivity (q) is dependent on both geometry and the material, and K_f is reduced due to the softening of the ductile material. An example of a notch sensitivity graph is presented in Figure 12. It is always permissible and conservative to use K_t without taking into account the notch sensitivity of the material.

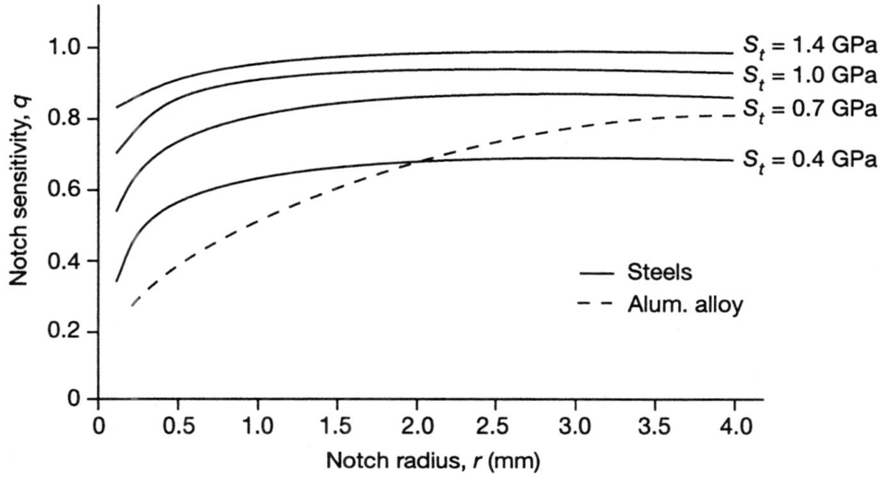

Figure 12. Notch Sensitivity

FATIGUE

Fatigue is a dynamic failure mode, normally with maximum stresses below the yield point. Extensive tests have been conducted to demonstrate the response of materials to cyclic tensile loading. Much of the work has been done using the R. R. Moore rotating-beam machine. A specimen is loaded with a bending moment and rotated to cause a sinusoidal, completely reversed stress at the surface of the specimen.

Load amplitude vs. number of cycles data are then plotted on log-log or semi-log paper. Figure 13 shows graphs that are typical of the fatigue response for steel and aluminum.

Note at the right, there appears to be a load amplitude below which failure does not occur for steel. This load amplitude is called the endurance limit. For aluminum, there is no true endurance limit; however, the fatigue strength (sometimes called endurance limit) of aluminum is generally available for 50×10^7 cycles.

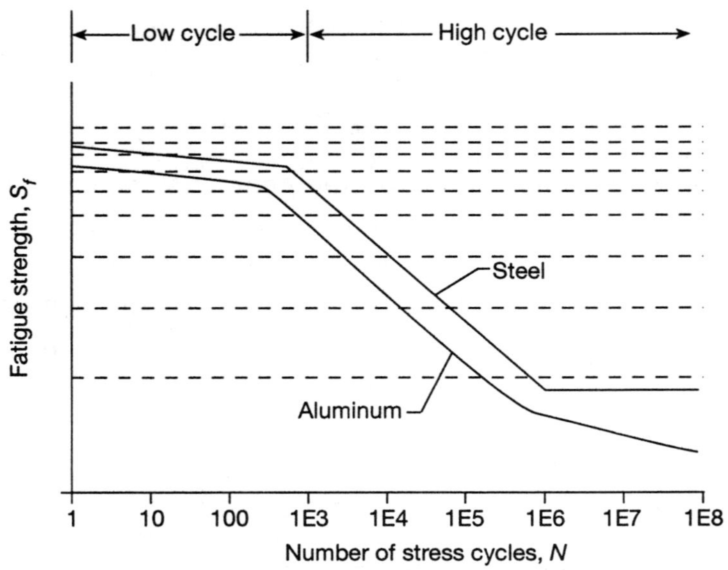

Figure 13. Typical Fatigue Strength Curves

Although it is always best to run fatigue tests on the material used, it is often required to estimate the endurance limit from the tabulated ultimate tensile strength (S_t). The following approximations are suggested for steel:

$$S'_e = 0.504 S_t \quad \text{when} \quad S_t \le 1{,}400 \ \text{MPa}$$

and

$$S'_e = 700 \ \text{MPa} \ \text{when} \ S_t > 1{,}400 \ \text{MPa}$$

Similar relationships have been proposed for other materials; however, the endurance limit varies drastically between alloys. Standard handbook values for the endurance limit on most commonly used alloys are available.

If only the Brinell hardness number (BHN) is known, the ultimate tensile strength for steels can be estimated as follows:

$$S_t = 3.3 \ (\text{BHN}) \ (\text{in MPa})$$

Now, we must somehow relate the fatigue strength endurance limit of our manufactured part (S_e) in our application to the tabulated data from the R. R. Moore specimens (S'_e). This is done using multiplicative factors as follows:

$$S_e = k_a k_b k_c k_d k_e k_f S'_e$$

Surface Factor (k_a): This accounts for the reduced endurance limit of our part because it is not polished as the test specimen is. The factor is tabulated in tables. Table 3 is for steels.

Steel Surface Finish Factors (k_a)									
	S_t (MPa)								
Surface finish	**420**	**560**	**700**	**840**	**980**	**1120**	**1260**	**1400**	**1540**
Machined or cold drawn	0.84	0.78	0.73	0.71	0.69	0.67	0.66	0.64	0.63
Hot rolled	0.70	0.61	0.55	0.50	0.46	0.42	0.39	0.37	0.33
As forged	0.53	0.46	0.40	0.36	0.32	0.30	0.28	0.25	0.22

Table 3. Surface Finish Factor (k_a) for Steel

Size Factor (k_b): This accounts for the effect of the size and potential loading differences between our part and the test specimen. First, find the equivalent diameter (d_e) of our part, to account for the part geometry and loading differences, as follows:

Round stock of diameter D loaded in bending (as opposed to rotating):

$$d_e = 0.370D$$

Rectangular stock with dimensions of $h \times b$ loaded in bending:

$$d_e = 0.808(hb)^{\frac{1}{2}}$$

Then use the following formula to find k_b:

$$k_b = \left(\frac{d_e}{7.62}\right)^{-0.1133}$$

where d_e is in mm and $2.79 \leq d_e \leq 51$ mm.

Load Factor (k_c): This accounts for the loading type, as follows:

$k_c = 0.923$ axial loading $S_t \leq 1{,}520$ MPa

$k_c = 1.00$ axial loading $S_t > 1{,}520$ MPa

$k_c = 1.00$ bending

$k_c = 0.577$ torsion and shear

Temperature Factor (k_d): This accounts for the decrease in strength of materials at high temperatures. Table 4 has suggested values for steel.

°C	k_d	°C	k_d	°C	k_d	°C	k_d	°C	k_d
20	1.000	100	1.002	200	1.020	300	0.975	400	0.922
50	1.010	150	1.025	250	1.000	350	0.927	450	0.840

Table 4. Temperature Factor (k_d)

Stress Concentration Factor (k_e):

where:

$$k_e = \frac{1}{K_f}$$

Note K_f is the same stress concentration factor discussed in the previous section on fatigue. It is applied to the alternating stress only.

Miscellaneous-effects Factor (k_f): Included to emphasize that many things can change endurance strength, including corrosion, electrolytic plating, metal spraying, cyclic frequency, shot peening, desired reliability, etc.

Non-fully Reversing Fatigue

Many times our part does not fully reverse between tension and compression. The stress state can be modeled as a fully reversing stress added to a mean stress. The following relationships are used to equate the non-fully reversing stress to the fully reversing endurance limit:

Soderberg equation:

$$\frac{\sigma_a}{S_e} + \frac{\sigma_m}{S_y} = \frac{1}{n}$$

Goodman relation:

$$\frac{\sigma_a}{S_e} + \frac{\sigma_m}{S_t} = \frac{1}{n}$$

and

Gerber equation:

$$\frac{n\sigma_a}{S_e} + \left(\frac{n\sigma_m}{S_t}\right)^2 = 1$$

where:

σ_a = alternating stress

σ_m = mean stress

S_y = yield strength of the material

S_t = ultimate tensile strength of the material

n = factor of safety

S_e = fatigue endurance limit

Figure 14 shows the relationships in diagram form.

Cumulative Fatigue: If a part is subjected to various stress amplitudes, the damage of each cycle can be thought of as one over the number of cycles to failure for that stress amplitude. When the cumulative damage reaches one, the part fails. Mathematically,

$$\sum \frac{n_i}{N_i} = C \approx 1$$

where:

n_i = number of cycles at the ith stress level

N_i = number of cycles to failure at the ith stress level

A value of one for C is commonly used as suggested above but in reality it is between 0.7 and 2.2, depending on the loading order of the stress amplitudes, etc.

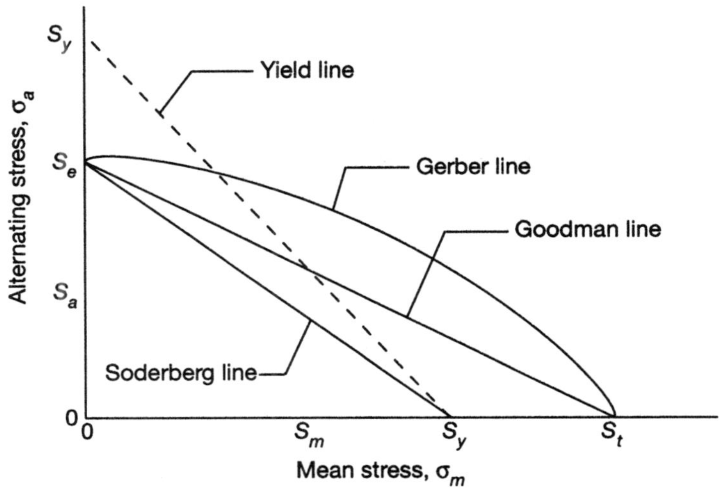

Figure 14. Non-fully Reversing Fatigue Relationships

PROBLEM 4:

A sewing machine part was designed to have a square cross section of 3.5 mm by 3.5 mm. The part is subjected to bending, which fluctuates between 1.5 N-m and 3.0 N-m on each stitch of the sewing machine. The part is to have a machined finish. The steel to be used in manufacturing this part has the following characteristics: S_y = 890 MPa; S_t = 1,560 MPa. From the geometry of the part, a stress concentration is noted and determined to have a theoretical stress concentration factor of K_t = 1.6, and the material has a notch sensitivity of q = 0.8. Determine: (a) the endurance limit for our part (S_e), and (b) the factor of safety using the Goodman relationship.

SOLUTION:

(a) Recall

$$S_e = k_a k_b k_c k_d k_e k_f S'_e$$

The material endurance limit S'_e can be estimated using S_t as follows:

$$S'_e = 700 \text{ MPa}$$

Surface factor (k_a) is taken from Table 3 and determined as k_a = 0.63.

Size factor (k_b) is determined as follows:

$$d_e = 0.808\ (\text{bh})^{\frac{1}{2}} = 0.808\ \{(3.5)(3.5)\}^{\frac{1}{2}} = 2.83 \text{ mm}$$

$$k_b = \left(\frac{d_e}{7.62}\right)^{-0.1133} = \left(\frac{2.83}{7.62}\right)^{-0.1133} = 1.12$$

Load factor (k_c) for bending is k_c = 1.00.

Temperature factor (k_d) — since this machine will be used at room temperature (approximately 20°C), k_d = 1.000.

Stress concentration factor (k_e) — first determine the K_f as follows:

$$K_f = 1 + q(K_t - 1) = 1 + (0.8)(1.6 - 1) = 1.48$$

then:

$$k_e = \frac{1}{K_f} = \frac{1}{1.46} - 0.68$$

Miscellaneous factors (k_f) — since no miscellaneous factors are considered, k_f = 1.00.

$$S_e = k_a k_b k_c k_d k_e k_f S'_e$$
$$= (0.63)(1.12)(1.00)(1.00)(0.68)(1.00)(700)$$
$$= 336 \text{ MPa}$$

(b) Determine the maximum and minimum stress on the part, as follows:

$$\sigma_{\text{min}} = \frac{My}{I} = \frac{M\left(\frac{h}{2}\right)}{\frac{bh^3}{12}} = \frac{6M}{bh^2} = \frac{6(1.5)}{(0.0035)(0.0035)^2} = 210 \text{ MPa}$$

and $$\sigma_{\text{max}} = \frac{My}{I} = \frac{M\left(\frac{h}{2}\right)}{\frac{bh^3}{12}} = \frac{6M}{bh^2} = \frac{6(3.0)}{(0.0035)(0.0035)^2} = 420 \text{ MPa}$$

Mean and alternating stresses are calculated as follows:

$$\sigma_{\text{m}} = \frac{\sigma_{\text{max}} + \sigma_{\text{min}}}{2} = \frac{210 + 420}{2} = 315 \text{ MPa}$$

and $$\sigma'_{\text{a}} = \frac{\sigma_{\text{max}} - \sigma_{\text{min}}}{2} = \frac{420 - 210}{2} = 105 \text{ MPa}$$

Recall that the stress concentration factor applies to the alternating stress as follows:

$$\sigma_a = \sigma'_{\text{a}} \; K_f = 105(1.48) = 155.4 \text{ MPa}$$

Using Goodman relationship:

$$\frac{1}{n} = \frac{\sigma_a}{S_e} + \frac{\sigma_m}{S_{ut}} = \frac{155.4}{336} + \frac{315}{1{,}560} = .664$$

or $$n = \frac{1}{0.664} = 1.51$$

SPRINGS

Stiffness

Stiffness is the slope of the line relates load to displacement. It is important in spring design because it can be calculated for a spring once, then the energy of the spring and displacement for various loads can be quickly calculated. It is sometimes useful to calculate the stiffness (k) for bars, shafts, and beams.

For bars:

$$k = \frac{AE}{l}$$

For shafts:

$$k = \frac{JG}{l}$$

For beams: Stiffness is calculated by dividing the force by the resultant displacement.

Axially Loaded Helical Springs

Stress in a Helical Spring: If we make a cut in a helical spring, we find a shear force and a torque as shown in Figure 15.

The shear stress due to the shear force and due to the torque are additive on the inside portion of the spring. The shear stress on the inside of the coil is then:

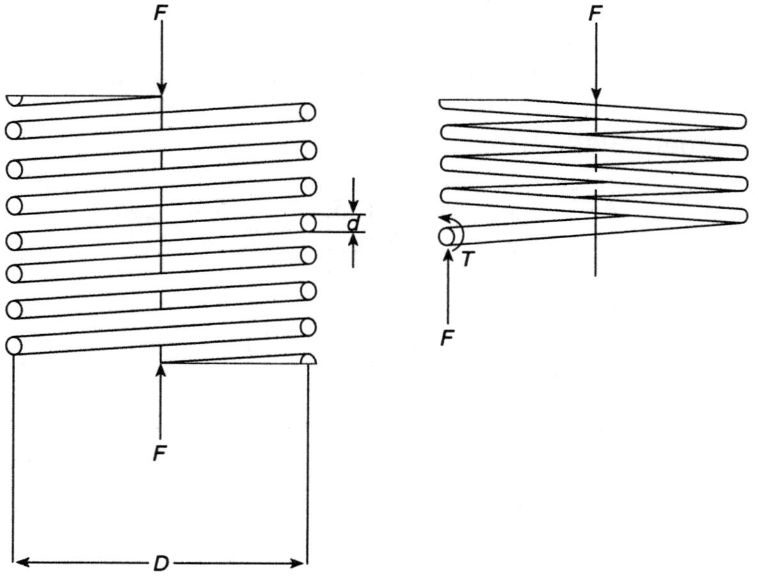

Figure 15. Helical Spring

$$\tau = K_s \frac{8FD}{\pi d^3}$$

where:

F = force on the spring

D = diameter of the spring coil

d = diameter of the spring bar stock

$$C = \frac{D}{d}$$

$$K_s = \frac{2C+1}{2C}$$

The curvature causes a stress concentration on the inside of the coil, and this stress concentration factor is:

$$K_c = \frac{2C(4C+2)}{(4C-3)\,(2C+1)}$$

Recall in fatigue, the stress concentration factor is only applied to the alternating stress, and the same is true on the spring.

The spring constant of a helical coil is:

$$k = \frac{d^4 G}{8D^3 N_a}$$

where:

N_a = number of active coils

G = shear modulus (Modulus of Rigidity)

D = diameter of spring coil

d = diameter of bar stock

Extension Springs: Extension springs are normally "close wound," which means that a torque is placed on the wire as it is being wound on the mandrel to preload the coils, holding them together.

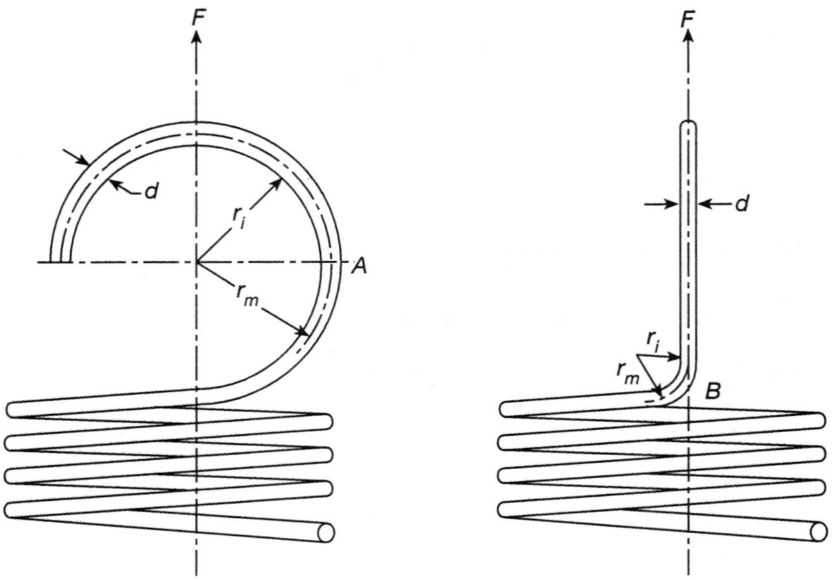

Figure 16. Extension Spring End

Extension springs must also have an attachment point. The cheapest and most common attachment is shown in Figure 16.

The disadvantage of this type of attachment point is the stress concentrating effect of the bend, which is approximated by:

$$K_c = \frac{r_m}{r_i}$$

Some of the stress concentrating effect can be mitigated by tapering the last few turns of the coil before the end. This does not change the stress concentration factor but changes the coil diameter and the moment arm on which the force is acting, thereby reducing the stress.

Compression Helical Springs: The last coil on each end of a compressive spring can be left plain, left plain and ground flat, squared, or both squared and ground as shown in Figure 17. The result is the last coil may or may not contribute to the stiffness of the spring. Table 5 lists the corrections for end effects.

Stability: Compression springs can become unstable if their length-to-diameter ratio is too large. This is analogous to buckling on a compression bar. Absolute stability occurs when:

$$L_0 < \frac{\pi D}{\alpha}\left[\frac{2(E-G)}{2G+E}\right]^{\frac{1}{2}}$$

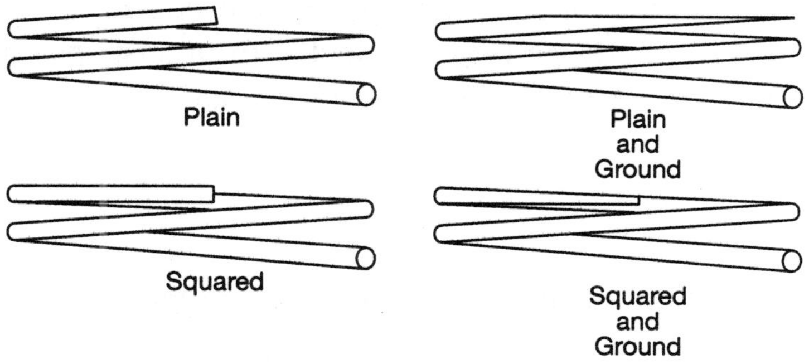

Figure 17. Common Spring Ends for Compression Helical Springs

where:

$\alpha = 0.5$ for springs supported between flat parallel surfaces (fixed ends)

$\alpha = 0.707$ for springs with one fixed end and the other end pivoted (hinged)

$\alpha = 1.0$ for springs with both ends pivoted (hinged)

$\alpha = 2.0$ for springs with one end clamped and the other end free

	Type of Spring Ends			
Term	**Plain**	**Plain & Ground**	**Squared**	**Squared & Ground**
End coil – N_e	0	1	2	2
Total coils – N_t	N_a	N_a+1	N_a+2	N_a+2
Free length – L_0	pN_a+d	$p(N_a+1)$	pN_a+3d	pN_a+2d
Solid length – L_s	$d(N_t+1)$	dN_t	$d(N_t+1)$	dN_t
Pitch – p	$(L_0-d)/N_a$	$L_0/(N_a+1)$	$(L_0-3d)/N_a$	$(L_0-2d)/N_a$

Table 5. Compression Helical Spring Formulas

PROBLEM 5:

A pickup truck has two coil springs supporting the frame at the rear axle. The springs have a stock diameter of 18 mm, coil diameter of 75 mm, 7 coils, with plain and ground ends, and a free length of 325 mm. For steel, G = 83 GPa. The truck weighs 12,000 N, with 4,000 N on the rear wheels. A 22,00 N box is dropped at a height of 500 mm from the bed.

Assume that both springs are equally loaded and neglect the effect of all other suspension and body parts (i.e., shocks, front suspension, etc.). Determine how far the bed will travel downward and if the springs bottom out.

SOLUTION:

First determine the spring constant for the springs.

For plain and ground ends (from Table 5):

$$N_a = N - 1 = 7 - 1 = 6$$

$$k = \frac{d^4 G}{8D^3 N_a} = \frac{(0.018)^4 (83 \times 10^9)}{8(0.075)^3 (6)} = 430 \times 10^3 \text{ N/m}$$

Note: This is for each of the two springs.

Assume the truck bed drops "$\delta_{dynamic}$" meters due to the load being dropped. The increase in potential energy of the springs will equal the change in potential energy of the load, as follows:

$$2{,}200(0.500 + \delta_{dynamic}) = 2\left(\frac{1}{2}\right)(430{,}000)(\delta_{dynamic}{}^2)$$

$$430{,}000\ \delta_{dynamic}{}^2 - 2{,}200\ \delta_{dynamic} - 1{,}100 = 0$$

Use the quadratic formula:

$$\delta_{dynamic} = \frac{-b \pm \sqrt{b^2 - 4ac}}{2a}$$

$$= \frac{-(-2{,}200) \pm \sqrt{(-2{,}200)^2 - 4(430{,}000)(-1{,}100)}}{2(430{,}000)}$$

$$= 0.0026 \pm 0.0506$$

$$\delta_{dynamic} = 0.0532 \text{ m}$$

(Only the positive root has significance to this problem.)

The compression of the spring due to the weight of the truck at the rear wheels is:

$$\delta_{static} = \frac{F}{k} = \frac{4{,}000}{415{,}000} = 0.0096 \text{ m}$$

The total deflection is:

$$\delta_{\text{total}} = \delta_{\text{dynamic}} + \delta_{\text{static}} = 0.0532 + 0.0093 = 0.063 \text{ m}$$

The solid length of the spring (fully compressed) is:

$$L_s = dN_t = (0.018)7 = 0.126$$

Amount spring can compress is:

$$\delta_{\text{maximum}} = L_0 - L_x = 0.325 - 0.126 = 0.199 \text{ m}$$

Since the actual defection is less than the maximum possible deflection, the springs will not bottom out.

Helical Torsion Springs

Helical torsion springs provide their force by bending of the spring. The stress on the spring is calculated using curved beam theory, and the result is similar to a straight beam, with the addition of a correction factor *K*, to account for the curvature. The equation is:

$$\sigma = K\frac{Mc}{I}$$

For round spring stock,

$$I = \frac{\pi d^4}{64}$$

and

$$c = \frac{d}{2}.$$

Further, define radius *r* as $M = Fr$. When the substitutions are made, the maximum stress at the outer and inner fiber of the spring stock can be written as:

$$\delta = K\frac{32Fr}{\pi d^3}$$

The correction factor *K* is different for the inner and outer fibers as follows:

$$K_i = \frac{4C^2 - C - 1}{4C(C-1)}$$

and $$K_o = \frac{4C^2 + C - 1}{4C(C+1)}$$

where: $$C = \frac{D}{d}$$

The limiting fiber is the inner fiber and, therefore, this should be used for design considerations.

Next, the spring constant must be determined. For a helical torsion spring made of round stock, the following applies:

$$k = \frac{Fr}{\theta} = \frac{d^4 E}{64 DN}$$

Spring Materials

It is always best to test spring materials to determine yield and ultimate tensile strength. However, these values can be approximated using the following formula:

$$S_t = \frac{A}{d^m}$$

where S_t is the ultimate tensile strength (MPa) and A and m are empirical constants, as tabulated in Table 6.

Material	ASTM	m	A
Music wire	A228	0.163	2,060
Oil-tempered wire	A229	0.193	1,610
Hard-drawn wire	A227	0.201	1,510
Chrome vanadium	A232	0.155	1,790
Chrome silicone	A401	0.091	1,960

Table 6. Constants to Determine S_t for Spring Materials

The shear yield stress (used in linear helical springs) can be approximated as follows:

$S_{sy} = 0.45\ S_t$ for cold-drawn carbon steel

$S_{sy} = 0.50\ S_t$ for hardened and tempered carbon and low alloy steels

$S_{sy} = 0.35\ S_t$ for austenitic stainless steel and nonferrous alloys

The yield stress (used in helical torsion springs) can be approximated as follows:

$S_y = 0.78\ S_t$ for cold-drawn carbon steel

$S_y = 0.87\ S_t$ for hardened and tempered carbon and low alloy steels

$S_y = 0.61\ S_t$ for austenitic stainless steel and nonferrous alloys

Zimmerli discovered that the endurance limit in springs did not depend on size, surface finish, material, or tensile strength. For the materials above, the following endurance limits should be used:

For helical torsion springs (bending stress)

$S_e = k_a k_b k_c S'_e = 537$ MPa for unpeened springs

$S_e = k_a k_b k_c S'_e = 805$ MPa for peened springs

For linear helical springs (shear stress endurance limit)

$S_{se} = k_a k_b k_c S'_e = 310$ MPa for unpeened springs

$S_{se} = k_a k_b k_c S'_e = 465$ MPa for peened springs

As indicated, the product of the surface finish (k_a), size factor (k_b), load factor (k_c), and material endurance strength is a constant. The spring endurance limit must, however, be corrected for temperature (k_d), stress concentration (k_e), or for miscellaneous factors (k_f) when fatigue calculations are performed.

PROBLEM 6:

A helical torsion spring to be used to close a door approximately 90 cm wide. The design criterion is that when the spring is installed, the force at the door handle (80 cm from the hinge line) to open the door is at least 4 N and the force when the door is 90° open is a maximum of 8 N. The spring material is chrome-vanadium steel, hardened and tempered, and is round stock, 4 mm in diameter (E = 205 Gpa). The mean coil diameter is 24 mm. Determine: (a) the minimum number of coils necessary to meet the above criteria; (b) what the spring with the calculated coils will yield when the door is opened 90°; and (c) whether the spring will have an infinite fatigue life.

SOLUTION:

(a) First calculate the spring constant, as follows:

$$k = \frac{d^4 E}{64DN} = \frac{(0.004)^4(205 \times 10^9)}{64(0.024)N} = \frac{34.17}{N} \text{N-m/rad}$$

Using the given criterion:

$$\theta_{\text{cl}} = \frac{Fr}{k} = \frac{(4)(0.8)}{34.17} = 0.09365 \text{ N}$$

$$\theta_{op} = \frac{Fr}{k} = \frac{(8)(0.8)}{\frac{34.17}{N}} = 0.187299 \text{ N}$$

Combining equations:

$$\theta_{op} - \theta_{\text{cl}} = \frac{\pi}{2} = 0.09365 \ N$$

and

$$N = 16.77$$

Therefore, a minimum of 17 coils are needed.

(b) Determine the yield stress of the material and maximum stress on the spring at 90° open.

Using approximations in the text:

$$S_t = \frac{A}{d^m} = \frac{1,790}{(0.004)^{0.155}} = 4,212 \text{ MPa}$$

$$S_y = 0.50 S_t = (0.50)(4,212) = 2,106 \text{ MPa}$$

For our spring:

$$C = \frac{D}{d} = \frac{0.024}{0.004} = 6$$

$$K_i = \frac{4C^2 - C - 1}{4C(C-1)} = \frac{4(6)^2 - 6 - 1}{4(6)(6-1)} = \frac{137}{120} = 1.142$$

$$K_o = \frac{4C^2 + C - 1}{4C(C+1)} = \frac{4(6)^2 + 6 - 1}{4(6)(6+1)} = \frac{149}{168} = 0.887$$

The inside fiber is most limiting:

$$\sigma_i = K_i \frac{32Fr}{\pi d^3} = 1.142 \frac{32(8)(0.8)}{\pi(0,004)^3} = 1,164 \text{ MPa}$$

Yield will not occur.

(c) For fatigue use Soderburg criterion (because it is most limiting).

Stress on the inside fiber, when the door is closed, is:

$$\sigma_i = K_i \frac{32Fr}{\pi d^3} = 1.142 \frac{32(4)(0.8)}{\pi(0,004)^3} = 582 \text{ MPa}$$

$$\sigma_a = \frac{\sigma_{max} - \sigma_{min}}{2} = \frac{1,164 - 582}{2} = 291 \text{ MPa}$$

and

$$\sigma_a = \frac{\sigma_{max} + \sigma_{min}}{2} = \frac{1,164 + 582}{2} = 873 \text{ MPa}$$

For unpeened springs:

$$S_e = 537 \text{ MPa}$$

(No temperature (k_d), stress concentration (k_e), or miscellaneous (k_f) correction factors are indicated in the problem.)

$$\frac{1}{n} = \frac{\sigma_a}{\sigma_e} + \frac{\sigma_m}{S_y} = \frac{291}{537} + \frac{873}{2,106} = 0.96$$

and

$$n = 1.04$$

The spring should not fail due to fatigue if the loading is limited to 90°.

SQUARE THREAD POWER SCREWS

Square thread power screws are used to transfer angular motion (or torque) to linear motion (or force). Common examples include the screws on car jacks, vices, and presses. In the analysis, consider a square thread power screw with a single thread, a mean diameter (average of the root diameter and crest diameter) of d_m, and a pitch of p. The lead (l) of the screw is the pitch times the number of threads. If one were to imagine unrolling a single thread for one rotation, then draw a free-body diagram for raising and lowering the load F, is the result.

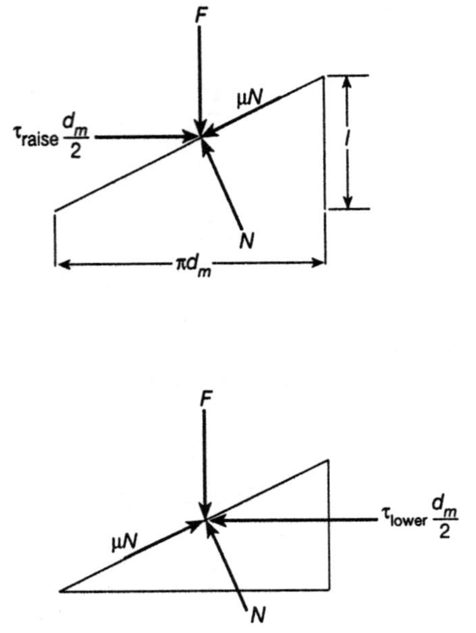

Figure 18. Unrolled Single Thread

Eliminating the normal force from the equations of statics and recognizing that the force P times the mean radius ($d_m/2$) equals the torque on the power screw, the following equations are derived:

$$T_{\text{lower}} = \frac{Fd_m}{2}\left(\frac{\pi\mu d_m - 1}{\pi d_n + \mu l}\right)$$

$$T_{\text{raise}} = \frac{Fd_m}{2}\left(\frac{\pi\mu d_m + 1}{\pi d_n - \mu l}\right)$$

Note that there are two components of this torque. The first is that required to overcome friction (the "$\pi\mu d$" term). The second is the torque caused by the force of the load on the inclined thread (the "l" term). If the lead (linear displacement per turn) "l" is large or the coefficient of friction

(μ) is small, the numerator of the lowering torque (T_{lower}) can become negative. The physical significance is that the power screw will lower under the weight of the load alone. It is generally desirable for an external torque to be required to lower the load (i.e., the numerator is positive). A power screw is said to be "self-locking" if an external torque in the lowering direction is required to lower the load.

Finally, a collar is required to support the power screw, and there is friction associated with the collar. The torque required to overcome the collar friction is:

$$T_c = \frac{F\mu_c d_c}{2}$$

and the final torque equations for a square thread power screw, including collar friction become:

$$T_{\text{lower}} = \frac{Fd_m}{2}\left(\frac{\pi\mu d_m - 1}{\pi d_n + \mu l}\right) + \frac{F\mu_c d_c}{2}$$

$$T_{\text{raise}} = \frac{Fd_m}{2}\left(\frac{\pi\mu d_m + 1}{\pi d_n - \mu l}\right) + \frac{F\mu_c d_c}{2}$$

where:

$T_{\text{lower}} = T_L =$ torque to lower the load

$T_{\text{raise}} = T_R =$ torque to raise the load

$d_c =$ mean collar diameter

$d_m =$ mean thread diameter

$l =$ lead

$F =$ load

$\mu =$ coefficient of friction for thread

$\mu_c =$ coefficient of friction for collar

The efficiency of the power screw is frequently used in the evaluation of a power screw. If we consider the ideal frictionless torque required to raise the load, we find:

$$T_0 = \frac{Fl}{2\pi}$$

Now define the efficiency of the power screw (η) as the frictionless raising torque divided by the actual torque required and the following equation is found:

$$\eta = \frac{Fl}{2\pi T}$$

PROBLEM 7:

A bench vice has a square thread screw, with a 4 mm pitch, a mean diameter of 16 mm, and a single thread. The collar has a mean diameter of 35 mm. The screw and collar are lubricated and have a coefficient of friction of 0.15. The vice handle is 250 mm long. A force of 200 N is used to tighten the vice. What is the force at the jaws, if all the force at the screw is transferred to the jaws, and what is the efficiency of the power screw.

SOLUTION:

In the vice, when tightening, both the force on the jaws and the force of friction must be overcome by the torque on the handle. Therefore, the torque for raising the load is used.

$$T_{\text{raise}} = T_R = \frac{Fd_m}{2}\left(\frac{\pi\mu d_m + l}{\pi d_m - \mu l}\right) + \frac{F\mu_c d_c}{2}$$

Rearrange to solve for force (F):

$$F = 2T_R\left[\frac{\pi d_m - \mu l}{d_m(\pi\mu d_m + l) + \mu_c d_c(\pi d_m - \mu l)}\right]$$

$$F = 2(.25)(200)\left[\frac{\pi(0.016) - 0.15(0.004)}{\begin{array}{l}0.016\{\pi(0.15)(0.016) + (0.004)\}\\ +0.15(0.035)\{\pi(0.016) - 0.15(0.004)\}\end{array}}\right]$$

$$F = 11.1 \ kN$$

$$\eta = \frac{Fl}{2\pi T} = \frac{11{,}100(0.004)}{2\pi(0.25)(200)} = 0.14$$

BOLTED JOINTS IN TENSION

Bolts are used to connect parts of an assembly when it is required that the assembly periodically be disassembled. In important bolted connections, the preload on the bolt is critical to provide the desired strength at the connection and in the selection of the proper bolts/fasteners. There are several methods to determine the preload in a specific fastener when it is installed; however, the easiest and most widely used method is to torque the fastener to a specified torque using a torque wrench. Torque values are specified on countless connections including wheel lug nuts and head bolts on cars. The engineer must be able to relate specified torque to bolt preload. The following is a frequently used relationship:

$$T = 0.2F_i d$$

where:

T = torque applied to the bolt

F_i = preload force on the bolt

d = nominal diameter of the bolt

Now let's consider a bolted connection, where the bolt has a preload F_i. The preload resists separation at the connection with axial forces and induces a normal force in the connection, which, because of friction, resists shear forces. The preload is critical to ensure that there is no motion of the bolted joint in the designed application.

Recall that in a bar the elongation (δ) is calculated as follows:

$$\delta = \frac{Fl}{AE}$$

and the stiffness constant for the bar is:

$$k = \frac{F}{\delta} = \frac{AE}{l}$$

where:

δ = deflection

A = cross-sectional area

F = force

E = Young's modulus (modulus of elasticity)

When an axial load is applied to the bolted joint, the bolt will stretch slightly and, in the connected members, the compressive forces induced by the preload of the bolt will decrease. As long as the external force is not too large (i.e., large enough to cause one of the connected members to separate from the other), the change in deflection of the bolt will be equal to the change in deflection of the bolted members, as follows:

$$\Delta\delta_b = \frac{p_b}{k_b} = \frac{P_m}{k_m} = \Delta\delta_m$$

Since:

$$P = P_b + P_m,$$

it follows that:

$$P_b = \frac{k_b P}{k_b + k_m}$$

The resultant forces on the bolt and the members can be calculated as follows:

$$F_b = P_b + F_i = \frac{k_b P}{k_b + k_m} + F_i$$

and

$$F_m = \frac{k_m P}{k_b + k_m} - F_i$$

Finally, if we define:

$$C = \frac{k_b}{k_b + k_m}$$

then the above equations become:

$$F_b = CP + F_i$$

and

$$F_m = (1 - C)P - F_i$$

when

$$F_m < 0$$

(connected members do not separate)

where:

F_b = resultant load in the bolt

F_m = resultant load in the connected members

P = external axial load

P_b = external axial load carried by bolt

P_m = external axial load carried by members

$\Delta\delta_b$ = change in bolt deflection due to load P

$\Delta\delta_m$ = change in member deflection due to load P

k_b = stiffness of bolt

k_m = stiffness of members

The stiffness of the clamped members, however, is a combined stiffness of each of the members included in the grip of the bolt and can be determined as follows:

$$\frac{1}{k_m} = \frac{1}{k_1} + \frac{1}{k_2} + \frac{1}{k_3} + \ldots + \frac{1}{k_n}$$

If one of the connected members is a gasket and the stiffness of the gasket is small compared to the other members, the gasket stiffness can be used directly as the member stiffness.

The stiffness of the bolt can be calculated by the following:

$$\frac{1}{k_b} = \frac{1}{k_t} + \frac{1}{k_s} = \frac{l_t}{A_t E} + \frac{l_d}{A_d E}$$

and

$$k_b = \frac{A_d A_t E}{A_d l_t + A_t l_d}$$

where:

k_b = stiffness of bolt

A_d = major diameter area (at shank)

A_t = tensile stress area (at threads)

l_d = length of unthreaded shank

l_t = length of threaded shank contained in grip

E = modulus of elasticity

We must now consider the bolt load factor and margin of safety to joint separation. For the bolt, the preload force must be greater than the external force or failure by joint separation will occur. Recall:

$$F_b = CP + F_i$$

Joint separation will occur when the bolt force caused by the external force (P term) is greater than the sum of the preload force (F_i term) and the force due to the elongation caused by the external force (CP term; therefore, the condition for joint separation is:

$$P \geq F_i + CP$$

The bolt load must also be less than the force necessary to yield the bolt or yield failure will occur. Therefore, F_i must be greater than CP and less than the force to yield (yield stress of bolt times the cross-sectional area of the bolts at the threads) as follows:

$$P < F_i + CP < S_{yb} A_t$$

where:

S_{yb} = minimum specified yield stress for the bolts

A_t = cross-sectional area of the bolts at the threads

If we define a bolt load factor, n_b, we can calculate the desired preload from the following:

$$F_1 = S_{yb} A_t - n_b CP$$

If the preload is known, then the bolt safety factor can be calculated as follows:

$$n_b = \frac{(S_{yb} A_t - F_i)}{CP}$$

Now recall that the force in the connected members is:

$$F_m = (1 - C)P - F_i$$

When F_m becomes zero, separation of the connected members occurs. Therefore, define the factor of safety against joint separation, n_s, and the equation becomes:

$$0 = n_s(1-C)P - F_i$$

or

$$n_s = \frac{F_i}{P(1-C)}$$

If the force P is applied and then removed in a cyclic manner, the bolt is subjected to fatigue loading. An example is the head bolts of an internal combustion engine, where the maximum load is the pressure in the cylinder caused by combustion. Recall that CP is the force on the bolt due to the external force P. If the external load cycles between 0 and P_{max}, the average force on the bolt due to the cyclic load is $CP_{max}/2$ and the amplitude of the cycle is also $CP_{max}/2$. Therefore, the mean stress and alternating stress on the bolt are:

$$\sigma_a = \frac{CP_{max}}{2A_t}$$

and

$$\sigma_m = F_i + \frac{CP_{max}}{2A_t}$$

The calculated values of mean and alternating stress can then be used in the Goodman, Solderberg or Gerber equations, which were discussed, previously.

PROBLEM 8:

A reciprocating air compressor has a single 0.10 m diameter piston. The pressure in the cylinder varies between 0 and 1 MPa (gage) on each stroke. The cylinder and head are made of cast iron and the area of contact between the cylinder and head is 0.005 m^2. The head is attached to the cylinder with 4-10 mm bolts with a 1.5 mm pitch (pitch diameter of 8.8 mm). The length of the bolts are 50 mm with 20 mm of the shaft unthreaded and 30 mm threaded. The bolts have a minimum yield stress S_{yb} of 400 MPa. The thickness of the head at the bold holes is 30 mm. Between the head and cylinder is a copper gasket 1.5 mm thick (E = 120 GPa). The manufacturer requires that the head bolts be torqued to 30 N-m. Determine: (a) the bolt safety factor and (b) the joint safety factor.

SOLUTION:

First, determine the stiffness of the components:

Stiffness of the head, gasket, and bolt:

$$k_{\text{head}} = \frac{AE}{l} = \frac{(0.005)(205 \times 10^9)}{0.03} = 34.2 \times 10^9 \text{ N/m}$$

$$k_{\text{gasket}} = \frac{AE}{l} = \frac{(0.005)(120 \times 10^9)}{0.0015} = 400 \times 10^9 \text{ N/m}$$

$$\begin{aligned} k_{\text{bolt}} &= \frac{A_d A_t E}{A_d l_t + A_t l_d} = \frac{\pi d_d^2 d_t^2 E}{4(d_d^2 l_t + d_t^2 l_d)} \\ &= \frac{\pi (0.01)^2 (0.0088)^2 (205 \times 10^9)}{4\left\{(0.01)^2(0.0115) + (0.0088)^2(0.02)\right\}} \\ &= 462 \times 10^6 \text{ N–m} \end{aligned}$$

Stiffness of the clamped material is:

$$\frac{1}{k_m} = \frac{1}{k_{\text{head}}} + \frac{1}{k_{\text{gasket}}} = \frac{1}{34.2 \times 10^9} + \frac{1}{400 \times 10^9} = 3.17 \times 10^{-11}$$

$$k_m = 31.5 \times 10^9 \text{ N–m}$$

Stiffness of all four bolts:

$$k_b = 4k_{\text{bolt}} = (4)(462 \times 10^6) = 1.85 \times 10^9 \text{ N–m}$$

Calculate C:

$$C = \frac{k_b}{k_b + k_m} = \frac{1.85 \times 10^9}{1.85 \times 10^9 + 31.5 \times 10^9} = 0.0555$$

Calculate the preload force on one bolt using torque relationship:

$$T = 0.2 F_i d$$

Rearrange:

$$F_i = \frac{T}{0.2d} = \frac{30}{0.2(0.01)} = 15{,}000 \text{ N}$$

for four bolts:

$$F_i = 4(15{,}000) = 60{,}000 \text{ N}$$

Calculate load P. This is due to the maximum pressure under the head times the area of the cylinder.

$$P = \frac{\pi}{4}(0.1)^2(1\times10^6) = 7{,}854 \text{ N}$$

(a) Calculate the bolt safety factor, n_b.

$$\begin{aligned} n_b &= \frac{(S_{yb}A_t - F_i)}{CP} \\ &= \frac{400\times10^6(4)\frac{\pi}{4}(0.0088)^2 - 60{,}000}{0.0555(7{,}854)} \\ &= 8.56 \end{aligned}$$

(b) Calculate the joint separation safety factor, n_s.

$$\begin{aligned} n_s &= \frac{F_i}{P(1-C)} \\ &= \frac{60{,}000}{7{,}854(1-0.0555)} \\ &= 8.1 \end{aligned}$$

BOLTED AND RIVETED JOINTS LOADED IN SHEAR

Frequently, tension members are joined by a lap joint, where one piece of plate is placed on the other, holes are drilled through both plates, and bolts or rivets are used to join the members. This type of joint is frequently seen on bridges, towers, steel shelving, etc. The joint is loaded in shear, where the axial load causes a shear stress between the two plates. There are several failure modes that must be analyzed to ensure that the joint will perform adequately. Three will be considered here.

First, the bolt or rivet will be placed in shear at the interface of the two members (much like a pair of scissors). Recall from mechanics of materials:

$$\tau = \frac{F}{A}$$

where:

τ = shear stress in the shank of the rivet/bolt

A = cross-sectional area of the rivet/bolt

F = shear force

The second failure mode is tensile failure of the plate, due to the decreased area where the holes are drilled for the bolts or rivets. Normally, the concern is the thinnest plate. The stress in the plate is given directly from mechanics of materials as follows:

$$\sigma = \frac{F}{A}$$

where:

σ = normal stress in plate ligaments

A = cross-sectional area of plate (thickness times width minus bolt hole diameters)

F = load

Finally, the joint can fail because the hole in the plate crushes (elongates) or the bolt/rivet crushes (becomes football-shaped). Again, directly from mechanics of materials:

$$\sigma = \frac{F}{A}$$

where:

σ = normal compressive stress in plate/fastener

A = projected cross-sectional area (thickness of plate times fastener diameter)

F = load

PROBLEM 9:

Consider the lap joint configuration shown in Figure 19. The rivets are 5 mm in diameter and made of 2024 – T4 aluminum (S_y = 325 MPa). The horizontal bar, which is riveted with the four rivets shown on the left end, is 8 mm thick aluminum (S_y = 90 MPa). Determine if the rivets or horizontal plate will fail at the riveted joint when the right end is loaded

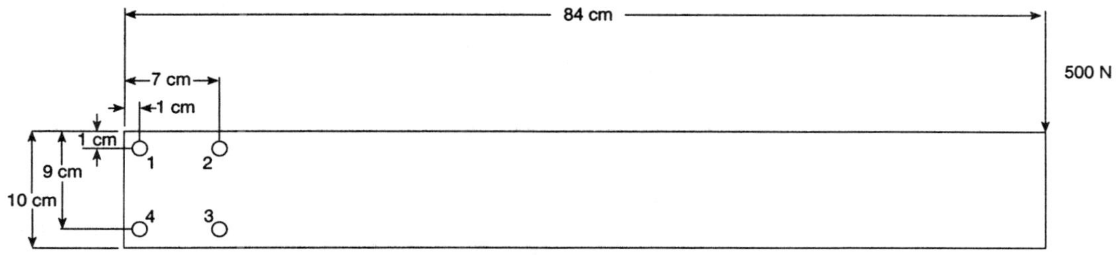

Figure 19

with a vertical load of 500 N. (There are other failure modes that may be of consideration in an actual design. For example, the horizontal plate is subject to buckling.)

SOLUTION:

The rivets must resist both the shear and the moment caused by this loading. The moment is resisted by rotation about the centroid of the rivet pattern; the force is proportional to the distance from the centroid and perpendicular to a line connecting the centroid and the center of the rivet. Because of symmetry, the centroid is located in the center of the pattern, with the distance to each rivet of 5 cm. It is further assumed that the shear force is distributed evenly between rivets.

The force on the rivets due to the moment is calculated as follows and shown on Figure 19a:

$$\sum M_{\text{centroid}} = 0 = 500(0.8) - (4)F_{\text{rivets}}(0.05)$$

$$F_{\text{rivets}} = \frac{500(0.8)}{4(0.05)} = 2{,}000 \text{ N}$$

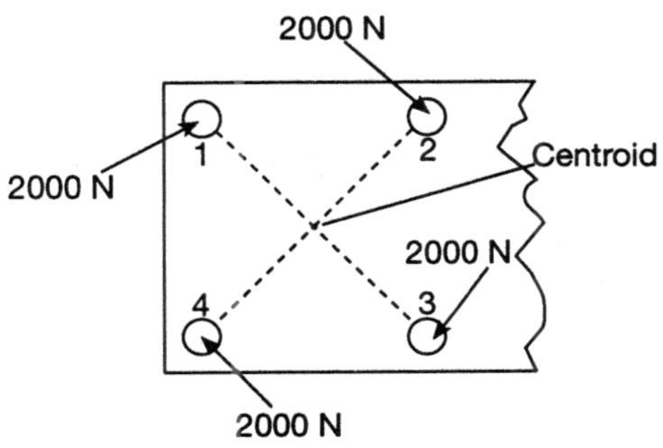

Figure 19a

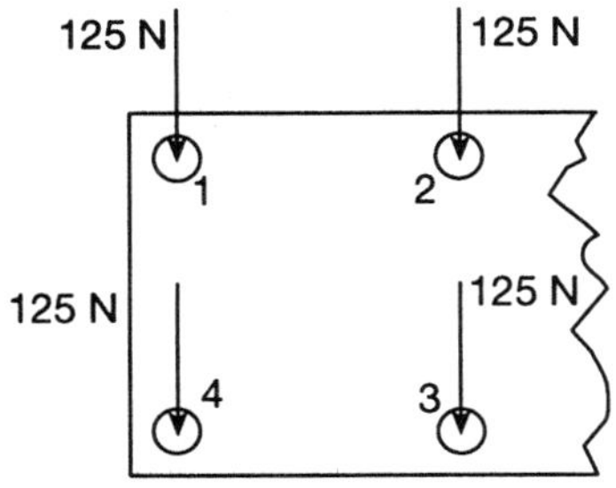

Figure 19b

Resolving the forces to vertical and horizontal forces produces:

Rivet 1: 1,600 N (right) and 1,200 N (up)

Rivet 2: 1,600 N (right) and 1,200 N (down)

Rivet 3: 1,600 N (left) and 1,200 N (down)

Rivet 4: 1,600 N (left) and 1,200 N (up)

The shear force acting down on each rivet is shown in Figure 19b:

$$F_{\text{shear}} = \frac{500}{4} = 125 \text{ N (down)}$$

The total force on the rivets are:

Rivet 1: 1,600 N (right) and 1,075 N (up)

Rivet 2: 1,600 N (right) and 1,325 N (down)

Rivet 3: 1,600 N (left) and 1,325 N (down)

Rivet 4: 1,600 N (left) and 1,075 N (up)

The most limiting rivets are #2 and #3, with a total force of:

$$F_2 = F_3 = \{(1{,}600)^2 + (1{,}325)^2\}^{\frac{1}{2}} = 2{,}080 \text{ N}$$

(Figure 19c shows forces on rivet 3.)

Check for rivet shear:

$$\tau = \frac{F}{A} = \frac{2{,}080}{\frac{\pi(0.005)^2}{4}} = 106 \text{ MPa}$$

$$\tau_y = 0.577 S_y = 0.577(325) + 188 \text{ MPa}$$

$$S.F. = \frac{\tau_y}{\tau} = \frac{188}{106} = 1.77$$

The rivets will not yield, but the safety factor is fairly small.

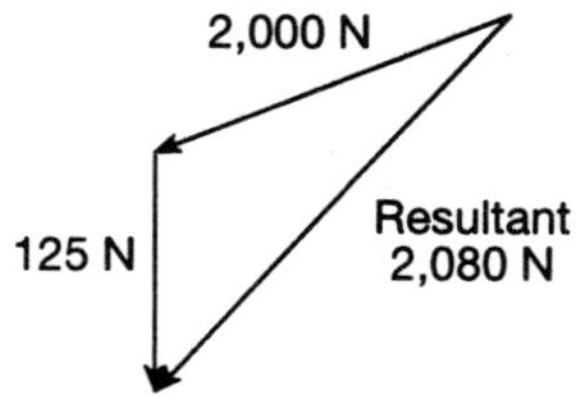

Figure 19c

Check for aluminum plate crush:

$$\sigma = \frac{F}{A} = \frac{2{,}080}{(0.008)(0.005)} = 52 \text{ MPa}$$

and

$$S.F. = \frac{S_y}{\sigma} = \frac{90}{52} = 1.73$$

The aluminum horizontal plate will not crush but the safety factor is fairly small.

Check for the aluminum plate failing between ligaments:

Because the ligaments are large (minimum of 55 mm) as compared to the rivet hole size (5 mm), the stress in the ligaments will be approximately a factor of 11 less than the crushing stress. It will not be limiting.

GEARING

Forces

Gear trains are always designed to transmit a specified power at a given input rotational speed (rpm) to an output that is normally at another speed. The tangential force on a gear times the radius gives a torque and that multiplied by the angular velocity is power.

$$P = T\omega$$

where:

P = power (watts)

T = torque in N-m

ω = angular velocity (rad/sec)

Gear Train Analysis

In analyzing gear trains, the fundamental concept is the velocity of the pinion and the gear are the same at the point of contact. Consider the pinion (1) driving a gear (2) in Figure 20.

By equating velocities at the teeth interface between gears, we can write:

$$n_2 N_2 = n_1 N_1$$

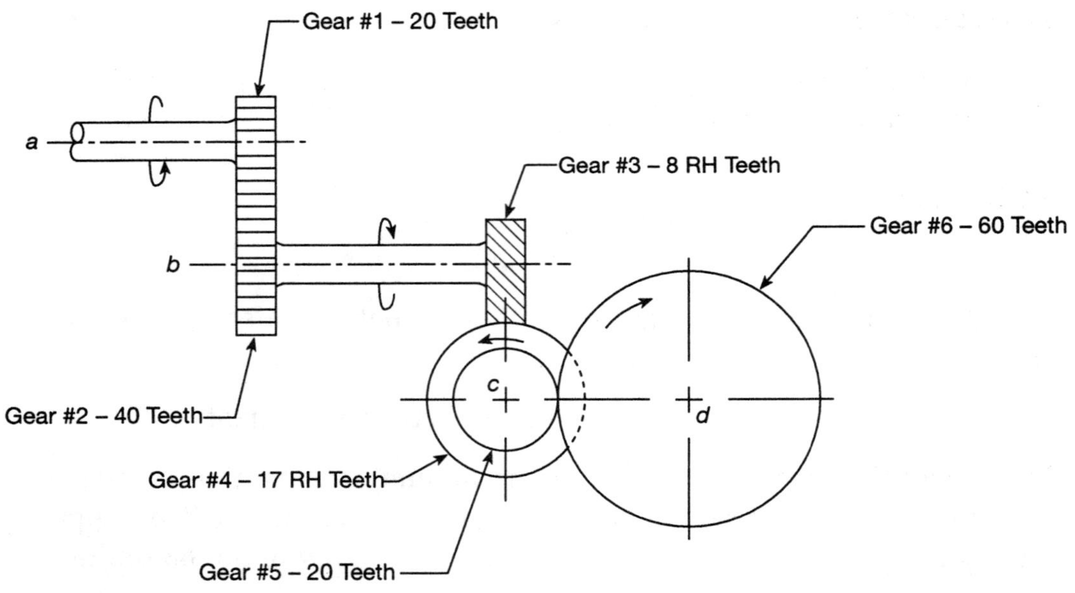

Figure 20. Typical Gear Train

Rearranged:

$$n_2 = \frac{N_1}{N_2} n_1 = \frac{d_1}{d_2} n_1$$

where:

n = rotational speed (rpm)

N = number of teeth

d = diametrical pitch

(subscripts are for gear 1, 2, etc.)

We did not consider direction of rotation, but in this case it is relatively simple. The rule of thumb is that if the gears are both external, the gears rotate in opposite directions. If one of the gears is an internal gear and the other external, the shafts rotate in the same direction. We can therefore define the gear train ratio (e) as:

$$e = \frac{\text{product of driving tooth numbers}}{\text{product of driven tooth numbers}}$$

Note: e is positive if the final gear turns in the same direction as the first and negative if the final gear turns in the opposite direction.

PROBLEM 10:

Shaft *a* in Figure 20 rotates at 1,200 rotations per minute (rpm) in the direction shown. Find the speed and direction of rotation of shaft *d*.

SOLUTION:

$$e = \frac{N_1 N_3 N_5}{N_2 N_4 N_6} = \frac{(20)(8)(20)}{(40)(17)(60)} = 0.0784314$$

$$\omega_6 = \omega_1(e) = 1{,}200(0.0784314) = 94.12 \text{ rpm CW}$$

Epicyclic (Planetary) Gear Trains

Epicyclic gear trains normally include a sun gear, a ring gear, and one or more planetary gears on an arm as shown in Figure 21. The difficulty with epicyclic gear trains is that any one of the three can be stationary while the other two provide the input and output.

The simplest way to solve a simple epicyclic gear train problem is the following:

1. Calculate the train value. (The train value will be negative when the sun and ring gears turn in opposite directions with a stationary arm.):

$$TV = \frac{N_{\text{ring}}}{N_{\text{sun}}}$$

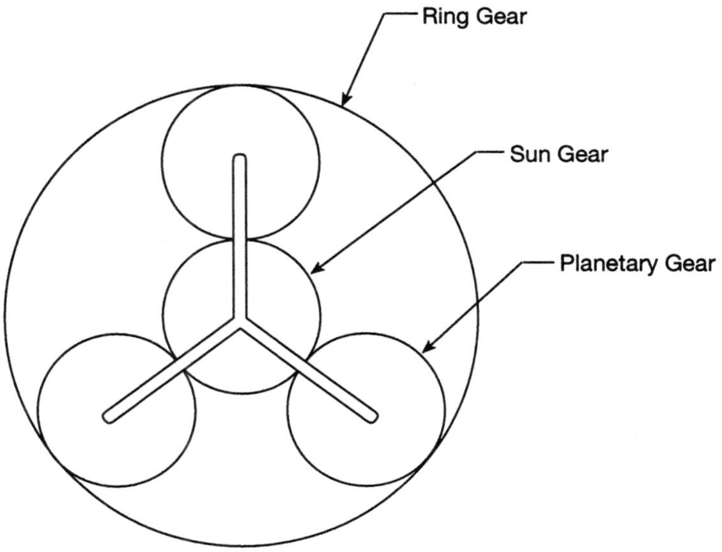

Figure 21. Typical Epicyclic (Planetary) Gear Train

Note: The train value is *not* the same as the gear train ratio (e).

2. Calculate the unknown angular velocities from the following: (Be sure to assign a positive sign for CW rotation and negative for CCW.)

$$\omega_{sun} = (TV)\omega_{ring} + \omega_{arm}(1 - TV)$$

3. The angular velocity of the planets can be found as follows: (The quantity is negative because the planet and sun gears turn in opposite directions.)

$$\frac{N_{planet}}{N_{sun}} = -\left(\frac{\omega_{sun} - \omega_{arm}}{\omega_{planet} - \omega_{arm}}\right)$$

PROBLEM 11:

An epicyclic gear train has a ring gear with 100 teeth, a sun gear with 50 teeth, and planetary gears with 25 teeth, as shown in Figure 22. If the ring gear is to be stationary and the sun gear is the input, calculate the input to output speed ratio of this gear train.

SOLUTION:

$$T.V. = \frac{N_{ring}}{N_{sun}} = -\frac{100}{50} = -2$$

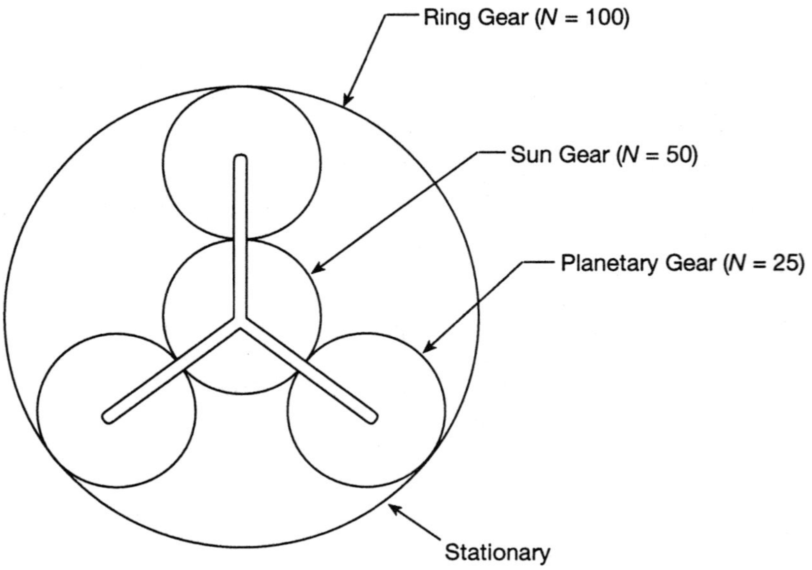

Figure 22

$$\omega_{sun} = (T.V.)\omega_{ring} + \omega_{arm}(1 - T.V.)$$

Because the ring gear is stationary, $\omega_{ring} = 0$; therefore,

$$\frac{\omega_{sun}}{\omega_{arm}} = (1 - T.V.) = 1 - (-2) = 3$$

SHAFT DESIGN

Shafting is normally designed after all the gearing and bearings are laid out. The questions are: What diameter must the shaft be? How must the gears be attached? What shoulders must be made for axial displacement? Several criteria must be met encompassing the following areas:

Deflection and rigidity (stiffness)

- Bending deflection
- Torsional deflection
- Slope at bearings and shaft-supported elements

Stress and strength

- Static strength
- Fatigue strength

Stiffness

Methods of analyzing beams and shafts must be employed to determine the shaft diameter required to meet stiffness requirements. In general, the equations can be solved for I or J and the diameter required can be calculated. Typical values for maximum deflections and slopes are:

Maximum slope at:

• cylindrical and roller bearings	0.001 rad
• deep groove ball bearings	0.0035 rad
• journal bearings	0.0002 rad
• gears	0.001 rad

Maximum deflections at:

• gears	0.002 inch

Fatigue Strength

A rotating shaft has a steady shear stress (from the torque transmitted by the shaft), a steady axial stress due to the thrust, and an alternating bending stress. Typically, the thrust component is negligible and is ignored. The shaft must be analyzed for static strength, to ensure the shaft does not yield at any point, and for fatigue strength, to ensure the shaft will not fail by fatigue. The Soderberg relationship combined with the maximum-shear-stress failure theory results in the formulas with two unknowns, shaft diameter and safety factor. The following formulas are used:

$$d = \left[\frac{32n}{\pi} \left(\left(\frac{M_m}{S_y} + \frac{K_f M_a}{S_e} \right)^2 + \left(\frac{T_m}{S_y} + \frac{K_{fs} T_a}{S_e} \right) \right)^{\frac{1}{2}} \right]^{\frac{1}{3}}$$

$$\frac{1}{n} = \frac{32}{\pi d^3} \left(\left(\frac{M_m}{S_y} + \frac{K_f M_a}{S_e} \right)^2 + \left(\frac{T_m}{S_y} + \frac{K_{fs} T_a}{S_e} \right)^2 \right)^{\frac{1}{2}}$$

where:

d = shaft diameter

n = factor of safety

K_f = stress concentration factor bending

K_{fs} = stress concentration factor in shear (torsion)

S_e = fatigue endurance strength

M_a = moment (subscript a stands for alternating)

M_m = moment (subscript m stands for mean)

T_m = torque (subscript m stands for mean or steady stress)

T_a = torque (subscript a stands for alternating)

S_y = yield strength

Similarly, we must still check for static strength because the fatigue strength may not be limiting. The following are used to calculate the maximum shear stress and the von Mises stress.

$$\tau_{max} = \frac{2}{\pi d^3}\left[(8M + Fd)^2 + 48T^2\right]^{\frac{1}{2}}$$

$$\sigma' = \frac{4}{\pi d^3}\left[(8M + Fd)^2 + 48T^2\right]^{\frac{1}{2}}$$

where:

τ_{max} = maximum shear stress in the shaft

σ' = von Mises stress

d = shaft diameter

M = bending moment on the shaft

F = axial thrust force

T = shaft torsion

Recall in maximum-shear-stress theory, $\tau_{fail} = 0.5S_y$ and we define the safety factor n as:

$$n = \frac{\tau_{fail}}{\tau_{max}}.$$

The following are derived and can be used to determine the minimum static safety factor or shaft diameter.

$$\frac{1}{n} = \frac{2}{\pi d^3 S_y}\left[(8M + Fd)^2 + (8T)^2\right]^{\frac{1}{2}}$$

and

$$d = \left\{\frac{4n}{\pi S_y}\left[(8M + Fd)^2 + (8T)^2\right]^{\frac{1}{2}}\right\}^{\frac{1}{3}}$$

Stress Concentration Factors

Stress concentration factors are particularly important in shafts, because shoulders, keyways, and shrunk-fit gears all cause a stress concentration. Keyways are in general most limiting. According to Shigley, conservative values of stress concentration on a standard end-milled keyway are $K_t = 2.14$ for bending and $K_t = 2.62$ for torsion.

PROBLEM 12:

A 5 cm solid steel shaft with two simple bearing supports carries two pulleys at a non-critical speed as shown in Figure 23. The pulleys are installed with square keys ($K_t = 2.14$). The shaft has the following characteristics: $S_t = 475$ MPa; $S_y = 330$ MPa; fully reversed stress fatigue limit, $S_e = 240$ MPa; and Poisson's ratio, $\mu = 0.3$. It is required that the deflection of the shaft not exceed 0.1 mm and the maximum angle of twist not exceed 0.3°. A safety factor of 1.75 is specified for fatigue and static loading. Does this shaft meet the requirements?

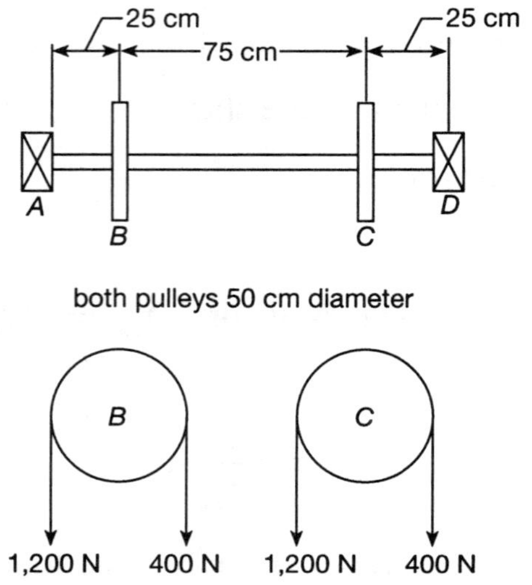

Figure 23

SOLUTION:

Deflection: Each pulley supports an unbalanced downward force of 1,600 N. Calculate the maximum deflection. For a beam of length L with two equal loads of F located at a distance a from each end, the maximum deflection is:

$$\gamma_{max} = \frac{Fa}{24EI}(3L^2 - 4a^2) = \frac{(1,600)(0.25)}{24(205 \times 10^9)\dfrac{\pi(0.05)^4}{4}}\left\{3(1,25)^2 - 4(0.25)^2\right\}$$

$$\gamma_{max} = 7.35 \times 10^{-5}\,\text{m} = 0.0735\text{ mm}$$

The deflection is acceptable.

Twist angle: A torque exists between the driver pulley and the driven pulley. Summing the moments about the center of the shaft on one pulley shows that the torque on the shaft is:

$$T = 1{,}200(0.25) - 400(0.25) = 200 \ \text{Nm}$$

$$\theta = \frac{TL}{JG} = \frac{(200)(0.75)}{\left(\dfrac{\pi(0.05)^4}{32}\right)(83\times10^9)} = 2.95\times10^{-3}\,\text{rad} = 0.169°$$

The twist angle is acceptable.

Static strength safety factor: Calculate the moment between the two pulleys:

$$M = (1{,}200 + 400)0.25 = 400 \ \text{Nm}$$

Now calculate the safety factor for static strength:

$$\frac{1}{n} = \frac{4}{\pi d^3 S_y}\left[(8M + Fd)^2 + (8T)^2\right]^{\frac{1}{2}}$$

$$= \frac{4}{\pi(0.05)^3(330\times10^6)}\left[(8(400) + 0(0.05)^2 + (8(200))^2\right]^{\frac{1}{2}}$$

$$= 0.110$$

$$n = 9.0$$

The static loading safety factor is satisfactory.

Fatigue safety factor:

$$\frac{1}{n} = \frac{32}{\pi d^3}\left(\left(\frac{M_m}{S_y} + \frac{K_f M_a}{S_e}\right)^2 + \left(\frac{T_m}{S_y} + \frac{K_{fs} T_a}{S_e}\right)^2\right)^{\frac{1}{2}}$$

$$\frac{1}{n} = \frac{32}{\pi(0.05)^3}\left(\left(\frac{0}{330\times10^6} + \frac{2.14(400)}{240\times10^6}\right)^2 + \left(\frac{200}{330\times10^6} + \frac{K_{fs}(0)}{240\times10^6}\right)^2\right)^{\frac{1}{2}}$$

$$= 0.295$$

$$n = 3.4$$

The fatigue safety factor is satisfactory.

The shaft meets the specifications.

BEARINGS

Antifriction Bearings

Antifriction bearings include ball, cylindrical, and tapered roller bearings. The standards for antifriction bearings are published by the Anti-Friction Bearing Manufactures Association (AFBMA) and in handbooks published by bearing manufactures. Basically, antifriction bearings can be selected for virtually any application but they have a finite life and must be periodically changed.

Journal Bearings

Journal bearings are widely used in heavy applications where long life is especially important. Internal combustion engines, gas and steam turbines, and marine propulsion line shaft bearings are a few of the applications. The journal is separated by a thick film of oil from the bearing surface by hydrodynamic lubrication. During startup the journal will tend to walk up the side of the bearing until the dynamic effects cause a wedge of oil to form under the journal. When the film is fully developed, the journal is off-center on the opposite side at an angle from the vertical.

FE/EIT

FE: PM Mechanical Engineering Exam

CHAPTER 12

Refrigeration and HVAC

CHAPTER 12

REFRIGERATION AND HVAC

The properties of "moist air" are essential to the analysis of HVAC systems. Thermodynamically, moist air is a combination of dry air and water vapor. The error in assuming ideal gas behavior is less than one percent and hence the use of ideal gas relations are adequate for most applications. Charts are also an important source of moist air data.

THE STANDARD ATMOSPHERE AND MOIST AIR

The *ASHRAE Handbook of Fundamentals* gives the following composition of dry air by volume fraction:

Table 1. Composition of Dry Air by Volume Fraction

Nitrogen	0.78084
Oxygen	0.20948
Argon	0.00924
Carbon dioxide	0.00031
Neon, helium, methane, sulfur dioxide, hydrogen, and other gases	0.00003

Based on this composition, the molecular weight, M_a, of dry air is 28.965 and the gas constant, R_a, for dry air is 53.35 ft-lb$_f$/lb$_m$–°R.

The basic medium in HVAC systems is a binary mixture of dry air and water vapor, which is assumed to behave as an ideal gas with gas constant, R_w, equal to 85.78 ft-lb$_f$/lb$_m$–°R.

MOIST AIR FUNDAMENTALS

Moist air up to about three atmospheres can be treated as an ideal gas for most engineering calculations. The Gibbs-Dalton law for a mixture of ideal gases states that the mixture pressure is the sum of the partial pressure of the constituents, which, for moist air, becomes:

$$P = P_a + P_w \tag{1}$$

where P is the total pressure of the mixture, P_a is the partial pressure of the air, and P_w is the partial pressure of the water vapor. Each constituent obeys the ideal gas equation of state:

$$\text{Dry air:} \qquad P_a V = m_a R_a T$$

$$\text{Water vapor:} \qquad P_w T = m_w R_w T \tag{2}$$

where V is the total volume of mixture, m_a is the mass of air, m_w is the mass of water vapor, and T is the absolute temperature of the mixture. This is referred to as the dry-bulb temperature in psychrometry since it represents the temperature registered by an ordinary thermometer.

The humidity ratio or specific humidity, W, is the ratio of the water vapor mass to the dry air mass:

$$W = \frac{m_w}{m_a} \tag{3}$$

By substituting Equation (2) into Equation (3) and using ideal gas relations:

$$W = 0.622\frac{P_w}{P_a} = 0.622\frac{P_w}{P - P_w} \tag{4}$$

The relative humidity, Φ, is the ratio of the mole fraction of the water vapor, x_w, in a mixture to the mole fraction, x_s, of water vapor in a saturated mixture at the same temperature and pressure. For an ideal gas, this ratio of mole fractions becomes the ratio of partial pressures, hence:

$$\phi = \left|\frac{x_w}{x_s}\right|_{T,P} = \frac{P_w}{P_g} \qquad (5)$$

where P_g represents the partial pressure of the water vapor in a saturated mixture. P_g is found in the steam tables at the mixture's temperature. For example, at a mixture temperature of 60°F, P_g would equal 0.2563 psia.

The degree of saturation, μ, is the ratio of the humidity ratio W to the humidity ratio of a saturated mixture at the same temperature and pressure W_s:

$$\mu = \left|\frac{W}{W_s}\right|_{T,P} \qquad (6)$$

The dew point temperature, T_{dp}, of moist air with humidity ratio W and P_w is the saturation temperature of water vapor corresponding to a saturation pressure of P_w. The dew point temperature is found by consulting the steam tables at the saturation temperature that corresponds to the partial pressure of the water vapor, P_w. The following equations can be used to approximate the dew point temperature:

$$T_{dp} = 79.047 + 30.5790\,\alpha + 1.8893\,\alpha^2 \qquad (7)$$

For temperatures below 32°F:

$$T_{dp} = 71.98 + 24.873\,\alpha + 0.8927\,\alpha^2 \qquad (8)$$

Where $\alpha = \ln(P_w)$. Note: P_w must be expressed in terms of inches of mercury (Hg), where 29.92 in. Hg being equivalent to 14.7 psia.

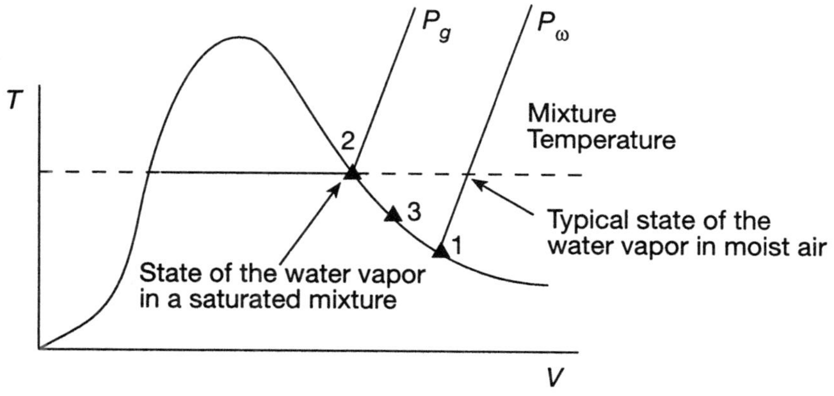

Figure 1. Moist Air Temperatures

The wet bulb temperature, T_{wb}, is the equilibrium temperature achieved by a thermometer fitted with a wetted wick over which the moist air sample flows. It is a widely used approximation of the "thermodynamic wet bulb temperature," which is required to describe the state of the moist air. The wet bulb temperature of a moist air sample will always be less than the dry bulb temperature and greater than the dew point temperature except in the case of saturated moist air at which point all three temperatures are identical. These temperatures are shown in Figure 1 corresponding to the moist air states. Point 1 corresponds to the dew point temperature of mixture, point 2 corresponds to the dry bulb temperature, and point 3 corresponds to the wet bulb temperature. The following relation can be derived from the above:

$$\phi = \frac{\mu}{1-(1-\mu)(P_g/P)} \tag{9}$$

The following expression is also used:

$$\phi = \frac{P_w}{P_{ws}} \tag{10}$$

where P_{ws} is saturation pressure of water vapor corresponding to dry bulb temperature.

The volume of the moist air mixture is usually expressed in terms of the dry air volume. Hence:

$$v_a = \frac{V}{m_a} = \frac{R_a T}{P_a} \tag{11}$$

By using Equation 4 this can be expressed as:

$$v_a = \frac{R_a T}{P}(1+1.6078\ \mathrm{W}) \tag{12}$$

The *actual* volume of the moist air would be $V/(m_a + m_w)$ and is given by the following expression:

$$v = \frac{V}{m_a + m_w} = \frac{v_a}{1+W} \tag{13}$$

Manipulation of the equations requires relationships between the wet bulb temperature and the other parameters. Two correlations are widely used, the first being Carrier's equation given by:

$$P_w = P_{sw,wb} - \frac{(P - P_{sw,wb}) - (T - T_{wb})}{2,831 - 1.43T_{wb}} \tag{14}$$

Where $P_{sw,wb}$ is the saturation pressure corresponding to the wet bulb temperature. The second is derived from an energy balance of the moist air during an adiabatic saturation process:

$$W = \frac{(1,093 - 0.556T_{wb})W_{s,wb} - 0.240(T - T_{wb})}{1,093 + 0.444T - T_{wb}} \tag{15}$$

The enthalpy of a mixture of ideal gases is the sum of the individual partial enthalpies of the components. The mixture enthalpy is expressed as unit energy per lb_m of air and is given by:

$$\begin{aligned} h &= h_a + Wh_g \\ h_a &= 0.240T \\ h_g &= 1,061 + 0.444T \end{aligned} \tag{16}$$

where h_a and h_g is the enthalpy of the dry air and water vapor, respectively, and T the dry bulb temperature in °F. The term h_g can be taken as the enthalpy of saturated water vapor at the dry bulb temperature. Combining the above, the moist air enthalpy becomes:

$$h = 0.240T + W(1061 + 0.444) \tag{17}$$

PROCEDURES FOR NUMERICAL CALCULATION OF MOIST AIR PROPERTIES

Table 2 that follows outlines the procedure for calculating moist air properties using ideal gas relations. While the chart methods are more convenient, they can only be used accurately for certain values of total pressure.

Table 2. Procedures to Calculate Moist Air Properties

Situation #1 Given:	***T*, T_{wb}, and *P***	**See Problem 1**
To Obtain	**Use above or other Eq.**	**Comments**
$P_{sw,wb}$	Steam tables	Sat. *P* for T_{wb}
$W_{s,wb}$	Eq. 4	Using $P_{ws,wb}$
W	Eq. 15	
$P_{ws,db}$	Steam tables	Sat. *P* for *T*
W_s	Eq. 4	Using $P_{ws,db}$
μ	Eq. 6	Using W_s
ϕ	Eq. 10	Find P_w from Eq. 4
v_a	Eq. 11 or 12	
h	Eq. 17	
T_{dp}	Steam tables or Eq. 7	
v	Eq. 13	

Situation #2 Given:	***T*, T_{dp}, and *P***	**See Problem 2**
To Obtain	**Use**	**Comments**
P_w	Steam tables	Sat. *P* for T_{dp}
W	Eq. 4	Use P_w
$P_{ws,db}$	Steam tables	Sat. *P* for *T*
W_s	Eq. 4	Use $P_{ws,db}$
μ	Eq. 6	Using W_s
ϕ	Eq. 10	
v_a	Eq. 11 or 12	
h	Eq. 17	
v	Eq. 13	
T_{wb}	Eq. 4 and Eq. 15 with steam tables	Requires trial-and-error solution

The following problem shows how the charts are implemented.

PROBLEM 1:

Determine the remaining moist air properties, given:

$$P = 14.0 \text{ psia}, T = 70°\text{F}, T_{wb} = 60°\text{F}$$

SOLUTION:

$P_{ws,wb} = 0.25611$ psia; steam tables at 60°F

$$W_{s,wb}\text{: } W_s = 0.622\frac{P_{ws}}{P - P_{ws}} = 0.622\left[\frac{0.25611}{14.0 - 0.25611}\right] = 0.01159\frac{\text{lbmv}}{\text{lbma}}$$

$$W = \frac{(1{,}093 - 0.556T_{wb})W_{s,wb} - 0.240(T - T_{wb})}{1{,}093 + 0.444T - T_{wb}}$$

$$= \frac{(1{,}093 - 0.556 \times 60) \times 0.01159 - 0.240 \times 10}{1{,}093 + 0.444 \times 70 - 60}$$

$$W = 0.009286\frac{\text{lbmv}}{\text{lbma}}$$

$P_{ws,db} = 0.36292$ psia; Steam tables at 70°F

$$W_s = 0.622\frac{P_{ws,db}}{P - P_{ws,db}} = \frac{(0.622) \times (0.36292)}{(14.0 - 0.36292)} = 0.01655\frac{\text{lbmv}}{\text{lbma}}$$

$$\mu = \frac{W}{W_s} = \frac{0.009286}{0.01655} = 0.561$$

$$\phi = \frac{P_w}{P_{ws}}; P_w = \frac{P \times W}{(0.622 + W)} = \frac{(14) \times (0.009286)}{(0.622 + 0.009286)} = 0.2059 \text{psia}$$

$$\phi = \frac{0.2059}{0.36292} = 0.567$$

$$v = \frac{R_a T}{P_a} = \frac{(53.35) \times (70 + 460)}{(14.0 - 0.2059) \times 144} = 14.23\frac{\text{ft}^3}{\text{lbma}}$$

$$h = 0.240T + W(1{,}061 + 0.444T)$$

$$= 0.240 \times 70 + (0.009286) \times (1{,}061 + 0.444 \times 70) = 26.94$$

$T_{dp} = T_{sat}$ at $P_{ws} = 0.2059$ psia; from the ASME steam tables

T_{sat}(°F)	P_{sat}(psia)	
54	0.20625	
T_{dp}	0.2059	Interpolating gives $T_{dp} = 53.99$°F
53	0.19883	

PROBLEM 2:

Determine the remaining moist air properties, given:

$P = 14.0$ psia, $T = 80$°F, $T_{dp} = 54$°F

SOLUTION:

$P_w = P_{sat}$ at 54°F $\Rightarrow P_w = 0.20625$ psia

$$W = 0.622\frac{P_w}{P - P_w} = \frac{(0.622)\times(0.20625)}{(14.0 - 0.20625)} = 0.009300\frac{\text{lbmv}}{\text{lbma}}$$

$P_{ws,db} = P_{sat}$ at 80°F $\Rightarrow P_w = 0.50683$ psia

$$W_s = 0.622\frac{P_{ws,db}}{P - P_{ws,db}} = \frac{(0.622)\times(0.50683)}{(14.0 - 0.50683)} = 0.02336\frac{\text{lbmv}}{\text{lbma}}$$

$$\mu = \frac{W}{W_s} = \frac{0.0093}{0.02336} = 0.3981$$

$$\phi = \frac{P_w}{P_{ws,db}} = \frac{0.20625}{0.50683} = 0.4069$$

$$v_a = \frac{R_a T}{P_a} = \frac{(53.35)\times(80+460)}{(14.0 - 0.20625)\times 144} = 14.5\frac{\text{ft}^3}{\text{lbma}}$$

$$h = .240T + W(1{,}061 + 0.444T)$$

$$= 0.240\times 80 + 0.0093\,(1061 + 0.444\times 80) = 29.40\ \frac{\text{Btu}}{\text{lbma}}$$

T_{wb} requires a trial-and-error solution, Note $80 > T_{wb} > 54$. Therefore, try 64°F.

$$P_{ws,wb} = 0.29497 \text{ psia; } W_{s,wb} = \frac{(0.622)\times(0.29497)}{(14.0-0.29497)} = 0.01339\frac{\text{lbmv}}{\text{lbma}}$$

$$W = \frac{(1{,}039 - 0.556\times 64)\times 0.01339 - .24\times(80-64)}{(1{,}093-(0.444\times 80)-64)}$$

$$= 0.00966\frac{\text{lbmv}}{\text{lbma}}$$

Since this is greater than W, try a lower guess for T_{wb}; say 62°F,

$$P_{ws,wb} = 0.27494 \text{ psia; } W_{s,wb} = 0.01246\frac{\text{lbmv}}{\text{lbma}}; W = 0.00823\frac{\text{lbmv}}{\text{lbma}}$$

Interpolating between these values:

$$T_{wb} \approx 63.5°\text{F}$$

PROBLEM 3:

Determine the remaining moist air properties, given:

$$P = 14.0 \text{ psia}, T = 70°\text{F}, \phi = 60\%$$

SOLUTION:

$$P_{ws,db} = P_{sat} \text{ at } 70°F \Rightarrow P_{ws,db} = 0.36292 \text{ psia}$$

$$\phi = \frac{P_w}{P_{ws}}; \Rightarrow P_w = \phi(P_{ws}) = (0.6)\times(0.36292) = .2176 \text{ psia}$$

$$W = 0.622\frac{P_w}{P-P_w} = \frac{(0.622)\times(0.2176)}{(14.0-0.2176)} = 0.009820\frac{\text{lbmv}}{\text{lbma}}$$

$$W_s = 0.622\frac{P_{ws,db}}{P-P_{ws,db}} = \frac{(0.622)\times(0.36292)}{(14.0-0.36292)} = 0.01655\frac{\text{lbmv}}{\text{lbma}}$$

$$\mu = \frac{W}{W_s} = \frac{0.009820}{0.01655} = 0.5933$$

$$v_a = \frac{R_a T}{P_a} = \frac{(53.35)\times(70+460)}{(14.0-0.2176)\times 144} = 14.25\frac{\text{ft}^3}{\text{lbma}}$$

$$h = 0.240T + W(1{,}061 + 0.444T)$$

$$= 0.240 \times 70 + 0.009820\,(1{,}061 + 0.444 \times 70)$$

$$= 27.52\ \frac{\text{Btu}}{\text{lbma}}$$

$$T_{dp} \Rightarrow T_{sat} \text{ at} \Rightarrow P_w = 0.2176 \text{ psia}$$

T_{sat} (°F)	P_w *(psia)*
56	0.22183
T_{dp}	0.2176
55	0.21392

By linear interpolation: $T_{dp} = 55.46$°F.

T_{wb} requires a trial-and-error solution. Note $70 > T_{wb} > 55.5$, therefore try 64°F.

$$P_{ws,wb} = 0.29497 \text{ psia;}\ \ W_{s,wb} = \frac{(0.622) \times (0.29497)}{(14.0 - 0.29497)} = 0.01339 \frac{\text{lbmv}}{\text{lbma}}$$

$$W = \frac{(1039 - 0.556 \times 64) \times 0.01339 - 0.24 \times (70 - 64)}{(1{,}093 - (0.444 \times 70) - 64)}$$

$$= 0.01202 \frac{\text{lbmv}}{\text{lbma}}$$

Since this is greater than W, try a lower guess for T_{wb}; say 60°F.

$$P_{ws,wb} = 0.25611 \text{ psia;}\ \ W_{s,wb} = 0.01159 \frac{\text{lbmv}}{\text{lbma}};\ \ W = 0.009237 \frac{\text{lbmv}}{\text{lbma}}$$

Interpolating between these values:

$$T_{wb} \approx 60.84\text{°F}$$

THE ASHRAE PSYCHROMETRIC CHART

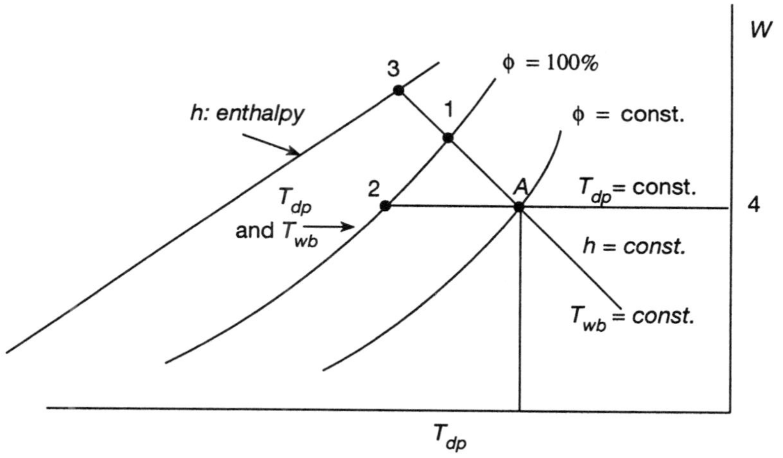

Figure 2. Schematic of Psychrometric Chart

To expedite engineering computations, graphical depictions of the moist air properties, known as psychrometric charts, have been developed. Figure 2 is a schematic of the ASHRAE Psychrometric Chart 1 and is typical of all available charts. Chart 1 has been designated to represent standard atmospheric pressure (29.92 inches Hg) and normal temperatures (32°F to 120°F). Other charts are available for standard pressures and both higher and lower temperature ranges as well as charts for normal temperatures and higher altitudes (lower pressures). The information on the psychrometric chart is as follows:

Dry-bulb temperature:	Straight, vertical lines.
Wet-bulb temperature:	Slanted lines within confines of chart
Humidity ratio:	Horizontal lines within confines of chart
Relative humidity:	Curved lines within confines of chart
Specific volume of air:	Slanted lines within the confines of chart
Enthalpy:	Area to the left of chart. Major lines of constant enthalpy also extend through the chart.
Saturation temperature:	Left boundary of chart; used for dew-point temperature.
Nomograph:	Upper-left corner; can be used to find enthalpy.
Protractor:	For determining the sensible heat ratio, SHR, and the ratio of ($\Delta h/\Delta W$).

PROBLEM 4:

Moist air at standard pressure has a dry-bulb temperature of 85°F and a wet-bulb temperature of 70°F. Determine all the moist air properties using the psychrometric chart.

SOLUTION:

The intersection of the 85°F dry-bulb line and the 70°F wet-bulb line defines the given state. All other properties will be determined with reference to this point.

Humidity ratio, W: From the reference point, move horizontally to the right and read $W = 0.0123$ lbmv/lbma.

Relative humidity, ϕ: Interpolate between the 40% and 50% relative humidity lines and read $\phi = 48\%$.

Specific volume, v_a: The reference point falls directly on 14.0 ft^3/lbma line.

Dew point temperature, T_{dp}: Move horizontally to the left from the reference point and read the temperature as 63.0°F.

Enthalpy, h: Follow the lines of constant enthalpy to find one that intersects the reference point. Since only the major enthalpy lines extend through the chart, a straightedge is required. Read $h = 34.0$ Btu/lbma.

An alternate method is available. From the reference point, follow the wet-bulb temperature line to the saturation line and read h_s = 34.2 Btu/lbma. Enter the nomograph at wet-bulb of 70°F and W = 0.0123 lbmv/lbma and read deviation, D, of -0.1 Btu/lbma. Since $h = h_s + D$, $h = 34.1$ Btu/lbma.

CLASSICAL MOIST AIR PROCESSES

The first law of thermodynamics or energy balance and the conservation of mass or mass balance are the basis for the analysis of moist air processes. For the problems that follow, steady-flow conditions are assumed as well as constant total pressure. All solutions will be done using the psychrometric charts.

Sensible Heating and Cooling of Moist Air

If heat is added to moist air with no change in the moisture content of the air, then this process is described as one of "sensible heating." An example would be moist air flowing over a heating coil or through a furnace. Figure 3 shows a schematic of the actual process and the resulting path on the psychrometric chart. The mass and energy balance for this process are as follows:

$$\dot{m}_{a,1}h_1 + {}_1\dot{q}_2 = \dot{m}_{a,2}h_2$$

$$\dot{m}_{a,1} = \dot{m}_{a,2}$$

$$\dot{m}_{a,1}W_1 = \dot{m}_{a,2}W_2$$

where the first equation is an energy balance, the second equation is a mass balance on the air, and the third equation is a mass balance on the water vapor. Since the mass flow rates are the same, the above equation can be written as:

$${}_1\dot{q}_2 = \dot{m}_a(h_2 - h_1)$$

where $\dot{m}_a$ is the mass flow rate of the air and $\dot{m}_v = \dot{m}_a W$ is the mass flow

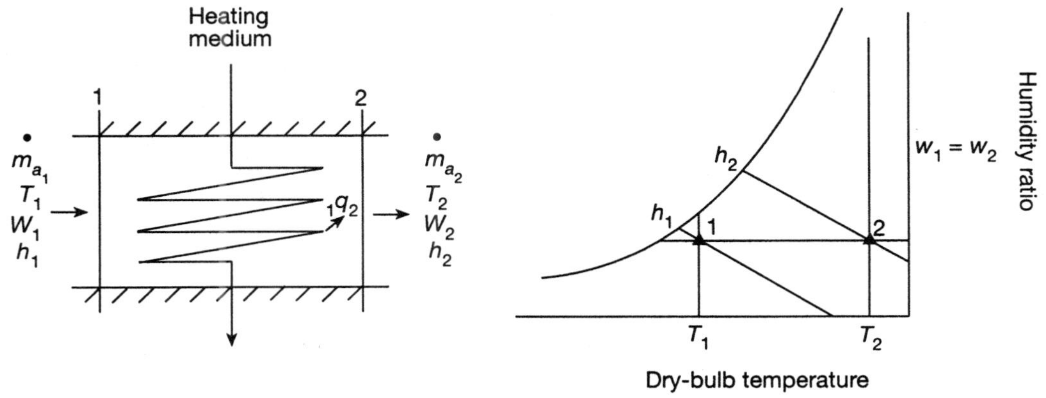

Figure 3. Sensible Heating and Cooling

rate of the water vapor. Since the humidity remains constant during this process, Δh can be approximated by $(0.24 + 0.45W) \times (T_2 - T_1)$.

Flows are often given in moist air processes in terms of volume flow rates rather than the mass flow rate required by the energy balance. The mass flow rate $\dot{m}$ is related to the volume flow rate, $\dot{Q}$ by:

$$\dot{m}_a = \frac{\dot{Q}}{v_a}$$

The most commonly used units for the volume flow rate is ft³/min or cfm, while the most commonly used units for the mass flow rate is lbm/hr; hence, a factor of 60 (to convert from minutes to hours or vice versa) is often required in the above equation.

PROBLEM 5:

Moist air enters a heating coil at 40°F dry-bulb and 36°F wet-bulb at a volume flow rate of 3,000 cfm at standard atmospheric pressure. The air leaves at 120°F dry-bulb. Determine the mass flow rate of steam required assuming the steam enters the coil as a saturated vapor at 220°F and leaves as a saturated liquid at 200°F.

SOLUTION:

This is a sensible heating problem. Properties will be determined by using a psychrometric chart.

$$W_1 = 0.0036\frac{\text{lbmv}}{\text{lbma}};\ h_1 = 13.5\frac{\text{Btu}}{\text{lbma}};\ v_1 = 12.7\frac{\text{ft}^3}{\text{lbma}};$$

$$W_2 = W_1;\ h_2 = 32.8\frac{\text{Btu}}{\text{lbma}}$$

$$\dot{m}_a = \frac{(3000) \times (60)}{12.7} = 14{,}170\frac{\text{lbma}}{\text{hr}}$$

$$_1q_2 = (14{,}170) \times (32.8 - 13.5) = 273{,}450\ \frac{\text{Btu}}{\text{hr}}$$

This can also be calculated using approximate equations:

$$_1q_2 = (14{,}170)[0.24 + (0.45) \times (0.0036)] \times (120 - 40)$$

$$=273{,}900\frac{\text{Btu}}{\text{hr}}$$

For the steam, h_g at 220°F = 1,153.4 Btu/lbmv and h_f at 200°F = 168.1 Btu/lbmv; therefore:

$$\dot{m}_s = \frac{{}_1q_2}{h_g - h_f} = \frac{273,450}{1,153.4 - 168.1} = 277.5\frac{\text{lbmv}}{\text{hr}}$$

COOLING AND DEHUMIDIFICATION OF MOIST AIR

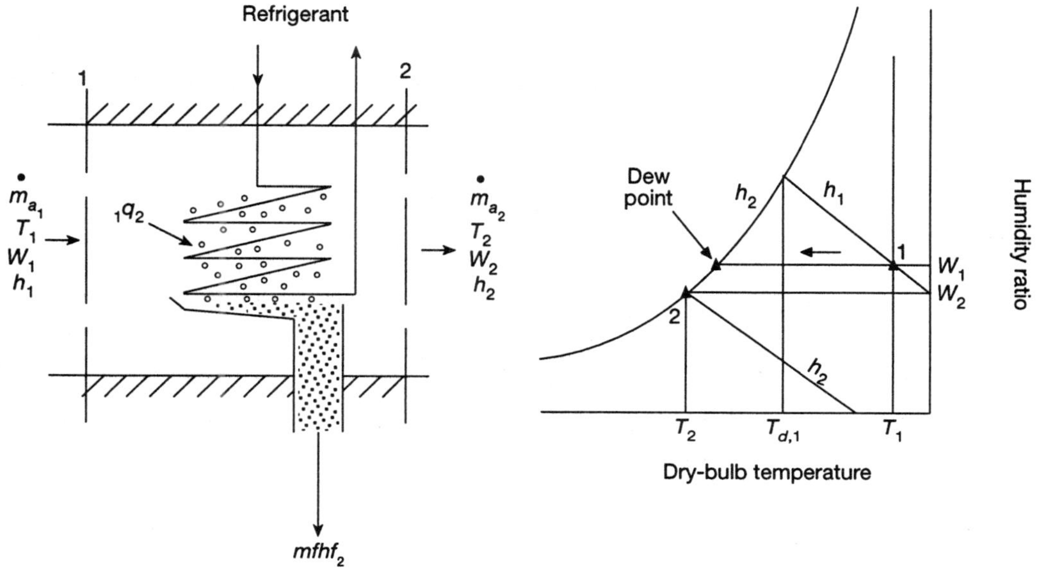

Figure 4. Cooling and Dehumidification of Moist Air

When moist air is cooled below its dew point, condensation of some of the water vapor will result. Figure 4 shows a schematic of the actual process and the path the process will follow on the psychrometric chart. Assume that all the moisture that condenses from the air is brought to a temperature T_2 before it is drained from the system. The conservation equations become:

$$\dot{m}_{a,1}h_1 = \dot{m}_{a,2}h_2 + {}_1q_2 + \dot{m}_f h_{f,2}$$

$$\dot{m}_{a,1} = \dot{m}_{a,2}$$

$$\dot{m}_{a,1}W_1 = \dot{m}_{a,2}W + \dot{m}_f$$

Therefore:

$${}_1q_2 = \dot{m}_a(h_1 - h_2) - \dot{m}_f h_{f,2}$$

$$\dot{m}_f = \dot{m}_a(W_1 - W_2)$$

PROBLEM 6:

Moist air at standard pressure enters a refrigeration coil at 80°F dry-bulb and 50% relative humidity at the rate of 200 lbma/min. The air leaves saturated at 50°F. Calculate the tons of refrigeration required.

SOLUTION:

The following properties are taken from the psychrometric chart and steam tables:

$$W_1 = 0.011\frac{\text{lbmv}}{\text{lbma}};\ h_1 = 31.3\frac{\text{Btu}}{\text{lbma}};\ W_2 = 0.0076\frac{\text{lbmv}}{\text{lbma}};$$

$$h_2 = 20.3\frac{\text{Btu}}{\text{lbma}};\ h_f = 18.04\frac{\text{Btu}}{\text{lbmv}}$$

$$_1q_2 = (200) \times [(31.3 - 20.3) - (0.011 - 0.0076)(18.04)]$$

$$= 2{,}212\ \frac{\text{Btu}}{\text{min}}$$

$$\text{tons of refrigeration} = \frac{2{,}212}{200} = 11.06 \text{ tons}$$

COMBINED HEATING AND HUMIDIFICATION OF MOIST AIR

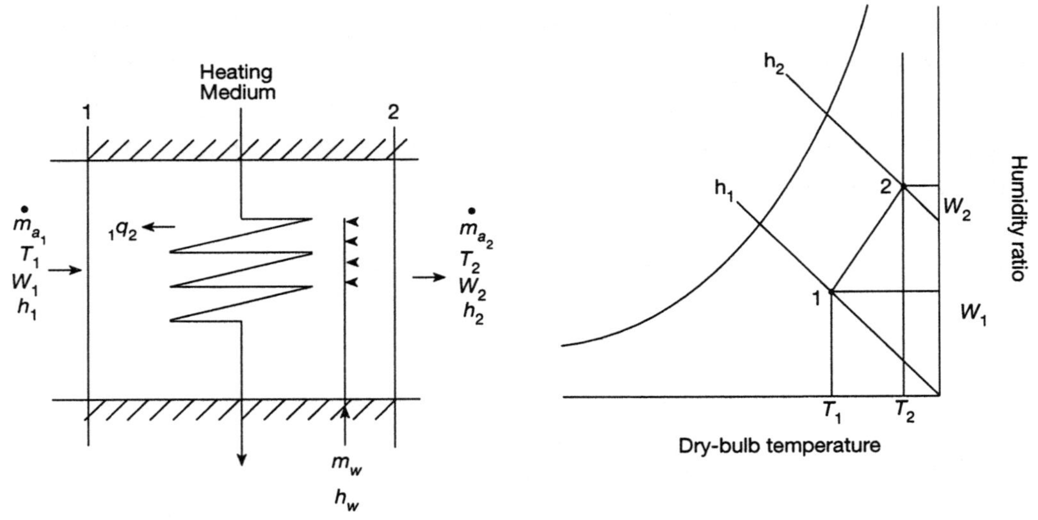

Figure 5. Combined Heating and Cooling

In winter, outside air must be heated and humidified before it is admitted into the conditioned space. In summer, air passing through the conditioned space absorbs the heat and moisture of the space and is heated and humidified. Figure 5 shows a schematic of a heating and humidifying apparatus and the corresponding path on the psychrometric chart. The conservation equations are as follows:

$$\dot{m}_{a,1}h_1 + \dot{m}_w h_w + {}_1q_2 + \dot{m}_{a,2}h_2$$

$$\dot{m}_{a,1} = \dot{m}_{a,2} \tag{1}$$

$$\dot{m}_{a,1}W_1 + \dot{m}_1 = \dot{m}_{a,2}W_2$$

Therefore:

$${}_1q_2 = \dot{m}_a(h_2 - h_1) - \dot{m}_w h_{w2}$$

$$\dot{m}_w = \dot{m}_a(W_2 - W_1)$$

By dividing the last two equations and rearranging the term:

$$\frac{\Delta h}{\Delta W} = \frac{h_2 - h_1}{W_2 - W_1} = \frac{{}_1q_2}{\dot{m}_w} + h_w \tag{2}$$

Equation (2) gives the value of the change of enthalpy divided by the change of specific humidity and as such must describe a straight line and fixed slope on the psychrometric chart. This characteristic is employed in the protractor that appears in the upper left-hand corner of the psychrometric chart. The following problem illustrates its use. Note that when both the dry-bulb and wet-bulb temperatures are given, it is common to express them as T/T_{wb}.

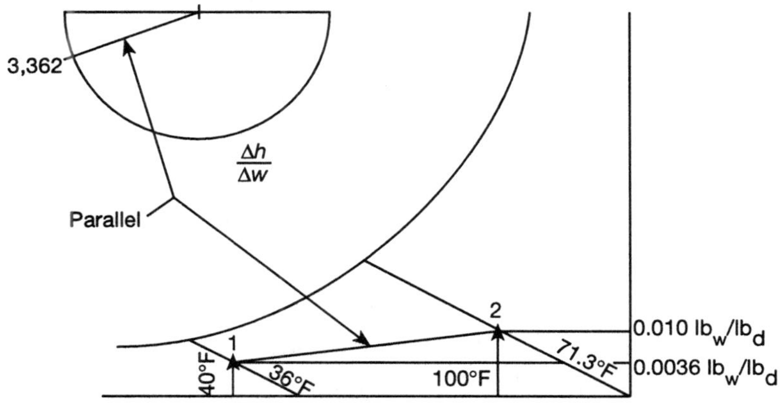

Figure 6. Using Protractor on Problem 7

PROBLEM 7:

Moist air at standard pressure enters a chamber at a dry-bulb temperature of 40°F and a wet-bulb temperature of 36°F at the rate of 5,000 ft^3/min. The air absorbs sensible heat at the rate of 333,000 Btu/hr and picks up 151 lbmw/hr of saturated steam at 230°F. Determine the dry-bulb and wet-bulb temperature of the exiting air.

SOLUTION:

From the psychrometric chart and steam tables, the values of the initial state and the water state are determined:

$$W_1 = 0.0036\frac{\text{lbmw}}{\text{lbma}};\ h_1 = 13.5\frac{\text{Btu}}{\text{lbma}};\ v_a = 12.7\frac{\text{ft}^3}{\text{lbma}};\ h_w = 1157\frac{\text{Btu}}{\text{lbmw}}$$

$$\dot{m}_a = \frac{5{,}000 \times 60}{12.7} = 23{,}620\frac{\text{lbma}}{\text{hr}}$$

From the equation:

$$W_2 = W_1 + \frac{\dot{m}_w}{\dot{m}_a} = 0.0036 + \frac{151}{23{,}620} = 0.010\frac{\text{lbmv}}{\text{lbma}}$$

$$h_2 = h_1 + \frac{{}_1q_2}{\dot{m}_a} + \frac{\dot{m}_w}{\dot{m}_a}h_w = 13.5 + \frac{333{,}000}{23{,}620} + \frac{(151)\times(1157)}{23{,}620}$$

$$= 4.99 \cong 35.0\frac{\text{Btu}}{\text{lbma}}$$

The intersection of W_2 and h_2 are used to find state 2, that is the conditions of the exiting air. Thus, it is found that T_2 = 100°F and $T_{2,wb}$ = 71.3°F.

The alternate procedure using the protractor on the psychrometric chart and Equation (2) determines:

$$\frac{\Delta h}{\Delta W} = \frac{333{,}000}{151} + 1{,}157 = 3{,}362\frac{\text{Btu}}{\text{lbmv}}$$

Two straightedges are used to draw a line parallel to the 3,362 line on the protractor through state (1), the initial conditions. The intersection of this line and W_2 = 0.010 lbmv/lbma locates state (2), the exiting conditions.

HUMIDIFICATION OF MOIST AIR

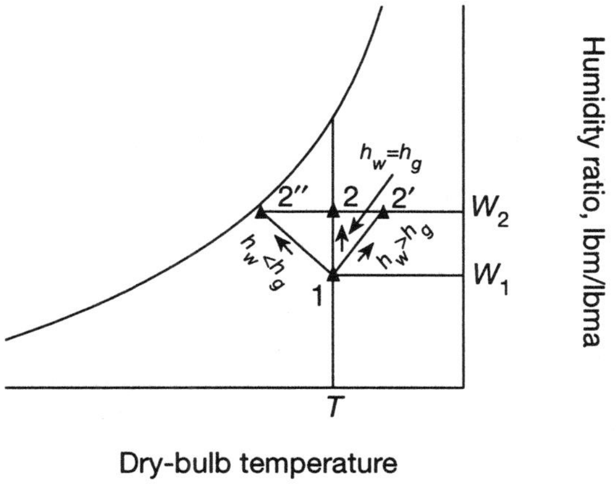

Figure 7. Humidification Processes

A frequent psychrometric process is the adiabatic humidification of moist air. The moisture may be added in either the liquid or vapor state. If the moisture is in liquid form, energy is required to vaporize the moisture prior to its absorption by the air. Since this energy must come from the moist air, a reduction in the dry-bulb temperature of the resulting moist air occurs. If the moisture is added as a vapor, say a superheated steam, the temperature of the moist air will increase in addition to humidification. The appropriate equations for this process can be found by setting ${}_1q_2 = 0$ in Equation (2), hence:

$$\frac{\Delta h}{\Delta W} = \frac{h_2 - h_1}{W_2 - W_1} = h_w$$

The direction of the condition line between (1) and (2) will depend on the enthalpy of the moisture. Two unique cases exist which are shown in Figure 7. If the moisture enthalpy equals h_g at the dry-bulb temperature, then the condition line will be vertical, i.e., the process is one of constant dry-bulb temperature. If the moisture enthalpy equals h_f at the wet-bulb temperature of the moist air, then the condition line will follow a line of constant wet-bulb temperature. From this we conclude that if $h_w > h_g$, then the air will be sensibly heated in addition to being humidified. If $h_w < h_g$, then the air will be sensibly cooled in addition to being humidified. The protractor method is quite useful for this type of problem. If the initial state and the moisture enthalpy are given, then the condition line can be determined and the end state determined from an end parameter, for example, specific humidity. Without the protractor, a trial-and-error problem would result. An important application in this area involves evaporative cooling, which is popular in the southwestern part of the country. Hot, dry air is drawn through a series of water saturated pads that results in adiabatic humidification at a moisture enthalpy h_f at the wet-bulb temperature of the air. The distance that the process follows the condition line will depend on the system design and initial conditions, but final relative humidities of 90% are often achievable with practical systems.

PROBLEM 8:

Five thousand cfm of moist air at 105°F and 10% relative humidity passes through an evaporative cooler and emerges at a relative humidity of 80%. Determine the outlet dry-bulb temperatures and the rate at which water is evaporated into the stream.

SOLUTION:

$$W_1 = 0.0048 \frac{\text{lbmv}}{\text{lbma}};\ T_{wb,2} = T_{wb,1} = 66^\circ F;\ v_a = 14.3 \frac{\text{ft}^2}{\text{lbma}}$$

Following the line of constant wet-bulb temperature to the 80% relative humidity line establishes the second state.

$$T_{db,2} = 70.5^\circ F;\ W_2 = 0.00128 \frac{\text{lbmv}}{\text{lbma}}$$

$$\dot{m}_a = \frac{(5{,}000) \times (60)}{14.3} = 20{,}980 \frac{\text{lbma}}{\text{hr}}$$

$$\dot{m}_w = \dot{m}_a (W_2 - W_1) = 20{,}980 \times (0.0128 - 0.0048) = 167.8\ \frac{\text{lbmw}}{\text{hr}}$$

ADIABATIC MIXING OF TWO STREAMS

In almost every air-conditioning system, the mixing of two or more air streams may occur. If the mixing occurs adiabatically, the following equations describe the process:

$$\dot{m}_{a,1} h_1 + \dot{m}_{a,2} h_2 = \dot{m}_{a,3} h_3$$

$$\dot{m}_{a,1} + \dot{m}_{a,2} = \dot{m}_{a,3}$$

$$\dot{m}_{a,1} W_1 + \dot{m}_{a,2} W_2 = \dot{m}_{a,3} W_3$$

By eliminating $\dot{m}_{a,3}$, we obtain:

$$\frac{\dot{m}_{a,1}}{\dot{m}_{a,2}} = \frac{h_2 - h_3}{h_3 - h_1} = \frac{W_2 - W_3}{W_3 - W_1}$$

Figure 8 shows the process on the psychrometric chart. Note that the mixture state must lie on a line that connects points 1 and 2. This allows for a graphical solution using line segment lengths.

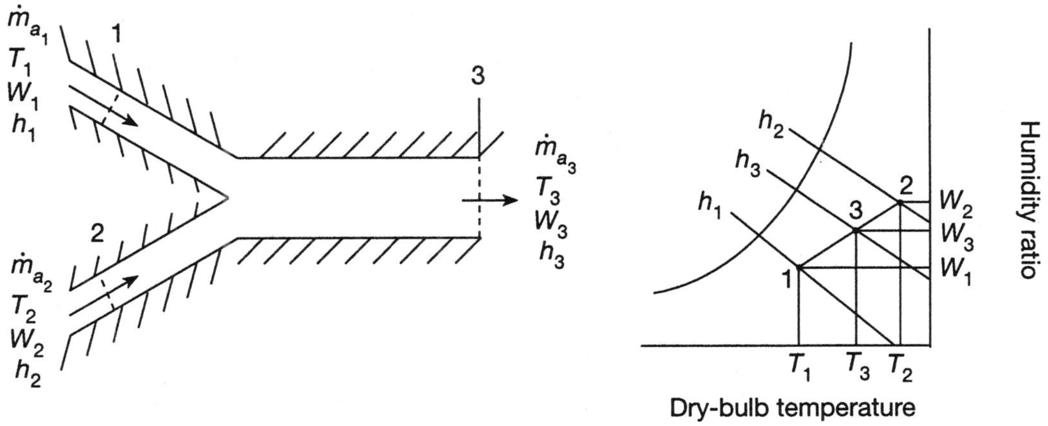

Figure 8. Adiabatic Mixing of Two Streams

PROBLEM 9:

Four thousand cfm of moist air at 100°F of dry-bulb, 75°F wet-bulb are mixed adiabatically with 2,000 cfm of moist air at 60°F of dry-bulb, 50°F wet-bulb in a steady flow device at sea level pressure. Determine the resulting outlet conditions of the air.

SOLUTION:

$$W_3 = W_1 + \frac{\dot{m}_{a,2}}{\dot{m}_{a,3}}(W_2 - W_1)$$

$$\dot{m}_{a1} = \frac{(2{,}000)\times(60)}{(13.22)} = 9077\frac{\text{lbma}}{\text{hr}};\ W_1 = 0.0053\frac{\text{lbmv}}{\text{lbma}}$$

$$\dot{m}_{a2} = \frac{(4{,}000)\times(60)}{(14.41)} = 16{,}650\frac{\text{lbma}}{\text{hr}};\ W_2 = 0.0128\frac{\text{lbmv}}{\text{lbma}}$$

$$W_3 = 0.0053 + \frac{16{,}650}{(9{,}077 + 16{,}650)}(0.0128 - 0.0053) = 0.0102\frac{\text{lbmv}}{\text{lbma}}$$

The intersection of W_3 with the line joining 1 and 2 locates point 3.

$$T_3 = 86.0°\text{F};\ T_{wb,3} = 67.3°\text{F}$$

Using complete graphical procedure:

$$\frac{\overline{13}}{\overline{12}} = \frac{\dot{m}_{a,2}}{\dot{m}_{a,3}} = \frac{16{,}650}{(16{,}650 + 9{,}077)} = 0.65$$

$$\overline{13} = 0.65\left(\overline{12}\right)$$

In the last example, once the ratio of line segments are known, the end state can be accurately determined by scaling the line segment.

CHAPTER 13

Stress Analysis

CHAPTER 13

STRESS ANALYSIS

All bodies are affected internally by forces that may act upon them. Contrary to the study of static forces, bodies cannot be considered perfectly rigid, instead, all bodies undergo deformations under different varieties of loads and conditions. Stress analysis is the study of the effects of forces on the elastic properties of structural materials.

STRESS AND STRAIN

Consider a straight metal bar of constant cross section loaded at its ends by co-linear forces directed in opposite directions at the longitudinal axis of the bar and acting through the centroid of each cross section. The magnitude of these forces must be equal in order to have static equilibrium. The bar is in *tension* if the forces are directed away from the bar. The bar is in *compression* when the forces are directed toward the bar (see Figure 1). These are the axial loads of tension and compression.

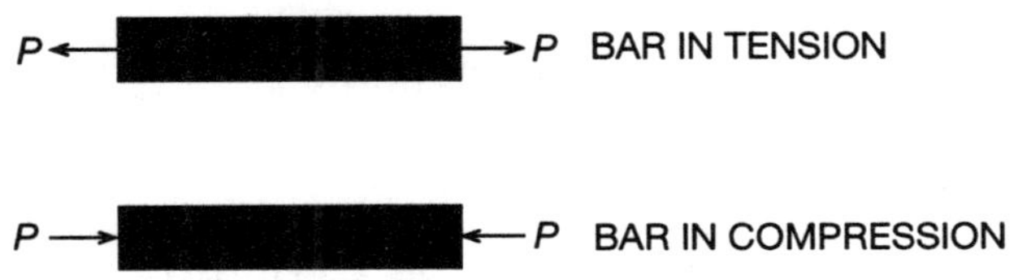

Figure 1. Bar in Tension and Compression

To analyze the effect of these forces along the length of the bar, we remove a section perpendicular to the horizontal axis of the bar. The section removed is replaced by the effects it exerts on the remaining section to the left. The original internal forces become external with regards to the remaining portion of the bar to the left. The force P for the left portion of the bar is equal to the resultant of distributed forces acting normal to and over the cross sectional area of the bar, for equilibrium (see Figure 2).

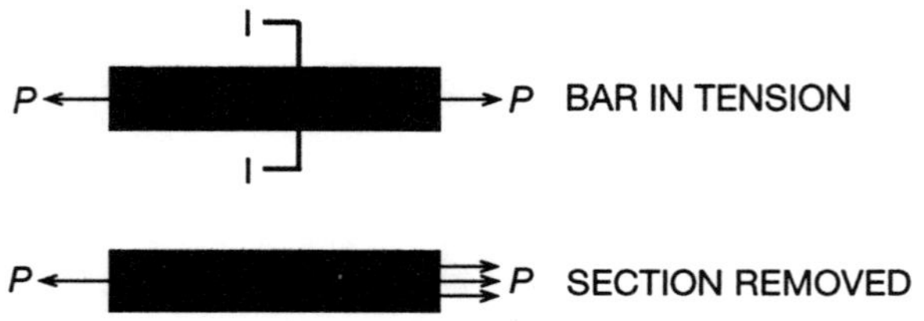

Figure 2. Bar with Section Removed

The intensity of the normal force acting over a unit area of the cross section of a body is the *normal stress*, which has the units of force over the area, N/m^2 or Pa. The tensile stress is achieved when forces are applied to the ends of a bar and the bar is in tension; if the forces are directed axially into the bar, the bar is in compression and has compressive stresses. The applied end forces pass through the centroid of each cross section of the bar.

Axial loading occurs often in structures and machine elements. In a laboratory test, this loading can be demonstrated by gripping a test specimen in an electrically driven gear or hydraulic type testing machine.

The American Society for Testing Materials (ASTM) issues specifications to standardize materials testing techniques that are used throughout the U.S. The test specimen is placed under axial load and the elongation is read anywhere along the gauge length at any increment of the load by a mechanical strain gauge. From these values, the normal strain, elongation per unit length, ϵ, can be found by dividing the total elongation, ΔL, by the gauge length, L_o. The units of strain are m per m and therefore dimensionless:

$$\Delta L = \text{elongation}$$

$$\epsilon = \text{strain}$$

$$\text{Strain} = \epsilon = \Delta L / L_o \ (\text{m/m})$$

A *stress-strain curve* is generated to reflect the loads placed on a particular structure and/or material. The axial load applied to the specimen is gradually increased. As each increment of the load increases, the elongation is measured and the normal stress and strains are noted until the specimen fractures (see Figure 3). This data is recorded and graphed. The shape that the graphs take for different materials under tension and compression are very different, as noted below, for mild steel, copper, and aluminum (see Figure 4).

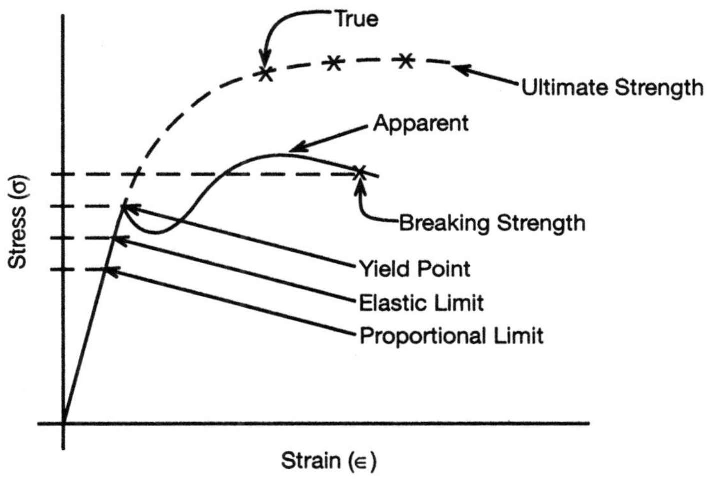

Figure 3. Stress-Strain Curve

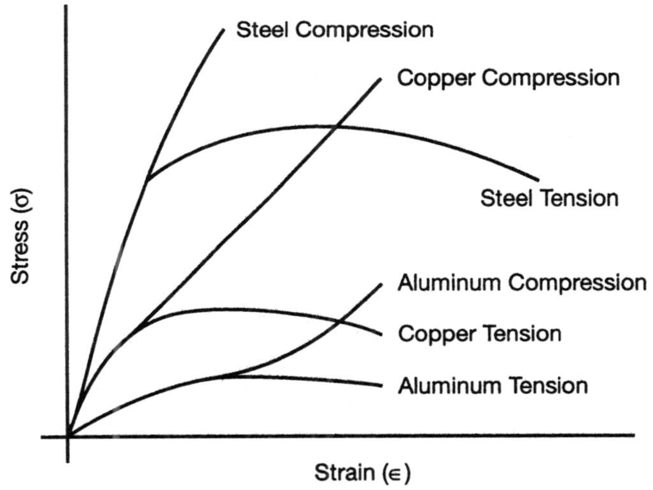

Figure 4. Stress-Strain Curve for Various Materials

Hooke's Law, named after Sir Robert Hooke, states that for some materials that have low values of strain, the relationship for stress and strain is linear and, therefore, reacts according to the following equation, where E is the slope of the straight-line portion of the stress-strain diagram:

$$\sigma = E\in$$

Also known as Young's Modulus, the *modulus of elasticity, E,* is the ratio of the unit stress to the unit strain of the material under tension and has the same units as stress, N/m^2 or Pa. E for materials under compression is close to the same value in tension. Values of E selected for design purposes fall along the linear portion of the stress-strain diagram. E is a constant for a given material up to the proportional limit.

PROBLEM 1:

A 7-meter-long steel rod has a cross-sectional area of 3.2 cm^2 and is carrying a load of 22,000 N (E = 207 GPa). What is the total elongation in the rod and load required for the rod to reach an elastic limit of 550 MPa?

SOLUTION:

Calculate the actual stress:

$$\sigma = \frac{P}{A} \frac{22,000\ \text{N}}{3.2\ \text{cm}^2} \times \frac{10^4\,\text{cm}^2}{\text{m}^2} = 68.8\ \text{MPa}$$

Solve for the actual strain. Substituting in the formula for stress:

$$\sigma = E\in$$

$$\in = \frac{\sigma}{E} = \frac{68.8 \times 10^6\ \text{N/m}^2}{207 \times 10^9\ \text{N/m}^2} = 3.32 \times 10^{-4}\ \text{m/m}$$

From the formula for strain, we get the total elongation:

$$\in = \frac{\Delta L}{L}$$

Therefore:

$$\Delta L = \in L = 3.32 \times 10^{-4}\,\text{m/m} \times 7\text{m} = 2.33 \times 10^{-3}\,\text{m} = .233\ \text{cm}$$

The load to reach the elastic limit is calculated as follows:

$$P = \sigma A = 550 \times 10^6 \frac{\text{N}}{\text{m}^2} \times 3.2\ \text{cm}^2 \times 10^{-4} \frac{\text{m}^2}{\text{cm}^2} = 176,000\ \text{N}$$

PROBLEM 2:

A 50 m long steel measuring tape is 1.25 cm wide and 0.05 cm thick ($E = 207 \times 10^9$ GPa). It is supported throughout its length and pulled with a 45 N force. What will be its length if it is pulled by a 180 N force?

SOLUTION:

Calculate the cross-sectional area of the tape:

$$A = \text{width} \times \text{thickness} = 1.25\ \text{cm}\ (0.05\ \text{cm}) = 0.0625\ \text{cm}^2$$

Determine the actual load—actual pull force:

$$P = 180\ \text{N} - 45\ \text{N} = 135\ \text{N}$$

Now, the added stress can be calculated:

$$\sigma = \frac{P}{A}$$

$$\frac{135\ \text{N}}{0.0625 \times 10^{-4}\ \text{m}^2} = 21.6\ \text{MPa}$$

The actual strain caused is determined:

$$\in = \frac{\sigma}{A} = \frac{21.6 \times 10^6\ \text{N/m}^2}{207 \times 10^9\ \text{N/m}^2} = 1.04 \times 10^{-4}\ \text{m/m}$$

The elongation caused by the 180 N pull is:

$$\Delta L = \in L = 1.04 \times 10^{-4}\ \text{m/m} \times 50\ \text{m} = 0.0052\ \text{m}$$

The total length at 180 N pull is:

$$L = 50\ \text{m} + \Delta L = 50\ \text{m} + 0.0052\ \text{m} = 50.0052\ \text{m}$$

The stress-strain diagram of Figure 3 reveals certain values and strength characteristics of the material known as *mechanical properties.*

The *proportional limit* is the maximum stress which can be developed during a tension test that remains a linear function of strain. The *elastic limit* is the maximum stress which can be developed under tension without causing permanent or residual deformation when the load is removed. The *elastic range* is the curve region from the origin to the proportional limit. The *plastic range* is the curve region from the proportional limit to the point of rupture.

The *yield point* is a point where there is an increase in the strain without a change in the stress. Some materials have two such points, the upper and lower yield points.

Resilience is the ability of a material to absorb energy in the elastic range. The *modulus of resilience* is the area under the curve from the origin to the proportional limit which represents the work done on a unit volume of material by a tensile force increased from the origin to the value of the proportional limit,

$$\frac{M-\text{N}}{m^2}.$$

Toughness is the ability of a material to absorb energy in the plastic range of the material. The *modulus of toughness* is the area under the curve from the origin to the point of rupture which represents the work done on a unit volume of material by a tensile force increased from the origin to rupture,

$$\frac{\text{N}-m}{m^3} \text{ or } Pa.$$

PROBLEM 3:

A specimen of medium-carbon steel having an initial diameter of 12.84 mm was tested using a gauge length of 50 mm. The following data was observed while the substance remained in the elastic state:

Load P kN	δ mm
0	0
5.5	0.005
7.8	0.015
13.6	0.025
18.9	0.035
26.9	0.05
33.2	0.065
35.6	0.08
36.8	0.11
37	0.145
37.1	0.2

Determine the modulus of elasticity and the yield strength.

SOLUTION:

The solution is most easily accomplished by using a spreadsheet, the results of which appear below.

Load P kN	δ mm	strain $\in$	stress Pa	E Pa
5.5	0.005	0.0001	4.26E+07	4.26E+11
7.8	0.015	0.0003	6.04E+07	2.01E+11
13.6	0.025	0.0005	1.05E+08	2.11E+11
18.9	0.035	0.0007	1.46E+08	2.09E+11
26.9	0.05	0.001	2.08E+08	2.08E+11
33.2	0.065	0.0013	2.57E+08	1.98E+11
35.6	0.08	0.0016	2.76E+08	1.72E+11
36.8	0.11	0.0022	2.85E+08	1.29E+11
37	0.145	0.0029	2.86E+08	9.88E+10
37.1	0.2	0.004	2.87E+08	7.18E+10

The strain is found by dividing the δ by the gage length of 50 mm. For line 2 of the data, the strain becomes

$$\frac{0.015\text{ mm}}{50\text{ mm}} = 0.003\text{ mm/mm}$$

The stress is found by dividing the load, P, by the cross-sectional area. Again for line 2 the stress becomes:

$$\sigma = \frac{P}{A} = \frac{7.8 \times 10^3\text{ N}}{\pi / 4(.01284\text{ m})^2} = 6.04 \times 10^7\text{ Pa}$$

Young's Modulus is found by dividing the stress by the strain. Again for line 2, Young's Modulus becomes:

$$E = \frac{\sigma}{\in} = \frac{6.04 \times 10^7\text{ Pa}}{.0003} = 2.01 \times 10^{11}\text{ Pa} = 201\text{ GPa}$$

From the data, E is found to be approximately 200 GPa.

The yield strength is found by locating the region where strain increases non-linearly. From the data, this occurs between data points 9 and 10. The yield strength is therefore 286 MPa.

The *working stress* will be selected within the elastic range of the material. The value is determined by selecting the yield or ultimate stress and dividing it by the safety factor. The safety factor used is selected by the designer based on experience and as specified in the governing building codes.

When metal deforms slowly over a long time of application of stress, well below its yield strength, the deformation of the metal is known as "creep" and as the material's *creep strength.* These materials are used for operation in high-temperature environments, such as turbine blades, where stresses are also high. See Figure 5 for idealized conditions.

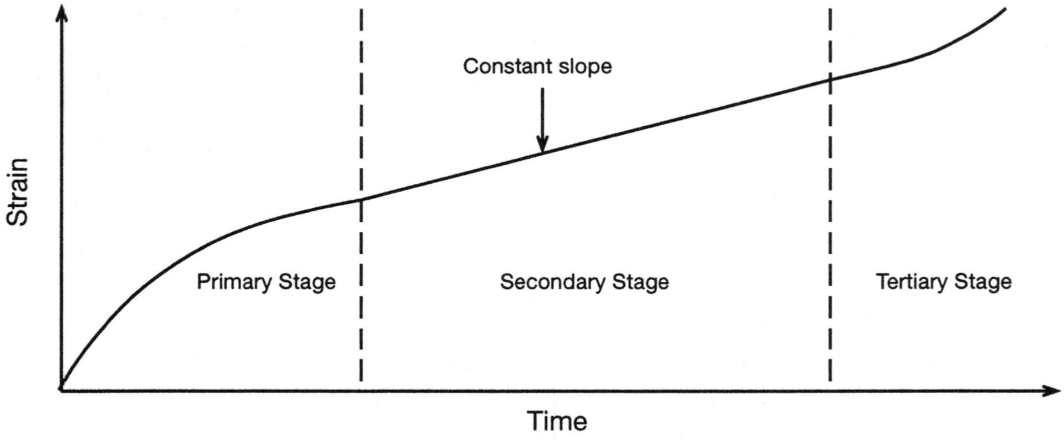

Figure 5. Creep

Metals are classed as *ductile* or *brittle materials.* A ductile material such as steel and aluminum has a high tensile strain up to the rupture point. A brittle material has a low strain to the rupture point. The strain separating these two classes of materials is accepted as 0.05 cm/cm. Cast iron and concrete are brittle materials.

Fatigue is the failure or rupture of a metal part under repeated application of a load which is well below permissible load.

The *endurance limit* is the amount of fatigue stress that will withstand 10 million cycles without breaking. For steels, the endurance limit is one-half the ultimate stress, S_u, and for non-ferrous materials, it is 20% to 40% of the tensile strength.

CYCLES OF FRACTURE

Stress concentration is critical when designing components. When designing members under cyclic loading, the concentration factor or notch sensitivity becomes very important. This occurs when there is a discontinuity or change of the section or cross-sectional area, such as a hole, groove, notch, or bend, which raises the stress. Localized stresses may reach three or more times the calculated average stress in the member. Rough machining produces localized stress concentration.

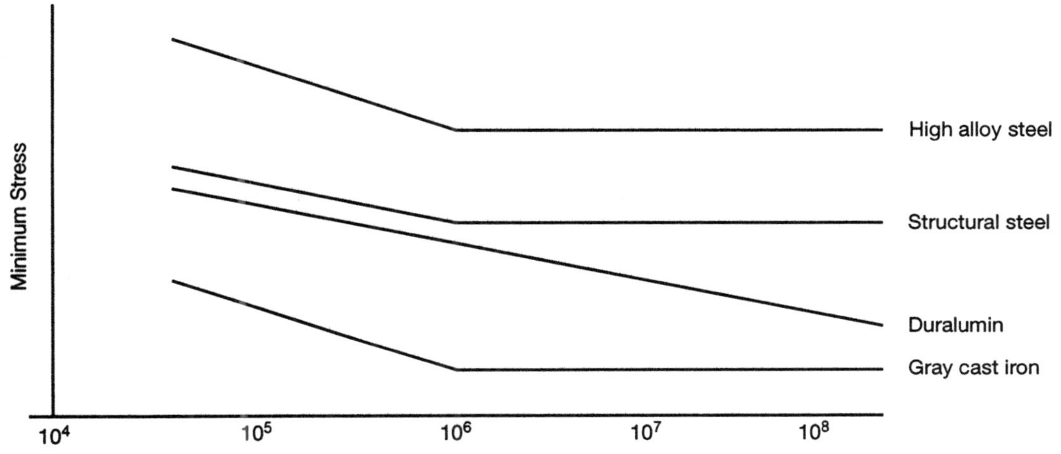

Figure 6. Stress Versus Cycles

When designing machine parts of varying cross-sectional areas, *Poisson's Ratio* becomes an important factor. A bar subjected to tension loads increases in length in the direction of the loads. But the bar undergoes a decrease in the width perpendicular to the load. The ratio of the strains in the axial and lateral directions is defined as Poisson's Ratio, μ, and is in the range of 0.25 to 0.35. Physical constants of materials are found in Table 1.

Another major concern of the design of any part and/or structural member is *Thermal Stress*. Temperature changes create statically indeterminate forces over the cross-sectional area of a body. Strain due to thermal expansion or contraction is linear in all directions and is added to strain due to stress algebraically. Every material has a coefficient of thermal expansion, cm/cm –°C. To calculate the stresses, it is assumed that the material remains rigid and straight and the yield point is never exceeded.

Table 1. Physical Constants of Materials

Material	E GPa	G GPa	μ	w kN/m^3
Aluminum	71	26.2	0.334	26.6
Brass	106	40.1	0.324	83.8
Carbon Steel	207	79.3	0.292	76.5
Cast Iron, gray	100	41.4	0.211	70.6
Copper	119	44.7	0.326	87.3
Douglas Fir	11	4.1	0.33	4.3
Glass	46.2	18.6	0.245	25.4
Lead	36.5	13.1	0.425	111.5
Nickel Steel	207	79.3	0.291	76
Stainless Steel (18–8)	190	73.1	0.305	76

PROBLEM 4:

A 1.2 m long copper bar that has a circular cross section of 5 cm in diameter is mounted between two walls with a gap at one end of 0.002 cm at room temperature. What is the stress in the bar, if the temperature is increased by 45°C and the coefficient of expansion and modulus of elasticity for copper are

$$\alpha = \frac{16.7 \times 10^{-6}\,\text{cm}}{\text{cm} \times {}^\circ\text{C}} \text{ and } E = 119 \text{ GPa, respectively?}$$

SOLUTION:

Determine the unrestrained increase of the length of the bar:

$$\Delta L = \alpha \Delta T L = \frac{16.7 \times 10^{-6}\,\text{cm}}{\text{cm} \times {}^\circ\text{C}} \times (45^\circ\text{C})\,(120\text{ cm})$$

$$= 9.02 \times 10^{-2}\,\text{cm} = 0.0902\text{ cm}$$

The bar expands 0.002 cm and then will be restrained. Stresswise, this is the same as if the bar were allowed to increase 0.0902 cm and then compressed 0.088 cm (0.0902 – 0.002 = 0.088 cm).

Calculate stress resulting in the bar.

Assume that the walls of the bar are perfectly rigid—no expansion or deformation, the bar remains straight and the yield point of copper is not exceeded.

Solve for stress:

$$\sigma = E\varepsilon = E\frac{\Delta L}{L} = \left(119 \times 10^9 \,\mathrm{N/m^2}\right)\frac{(0.088\ \mathrm{cm})}{120\ \mathrm{cm}}$$

$$= 8.73 \times 10^7 \,\mathrm{N/m^2} = 87.3\ \mathrm{MPa}$$

Materials are classified as homogeneous or isotropic. A material that has the same elastic properties (E, ν) throughout its body is homogeneous. A material that has the same elastic properties (E, ν) in all directions at one point is isentropic. Not all materials are isentropic. Some are anisotropic; they have over 21 elastic constants. A material is orthotropic if it has three mutually perpendicular planes of elastic symmetry.

STATICALLY INDETERMINATE FORCE SYSTEMS

A *statically determinate force system* exists when the external forces acting on the body can be calculated using the equations of static equilibrium.

Considering Figure 7, static equilibrium exists when all of the external forces on a body, acted upon by a force P, which causes the reactions R_1, R_2, and R_3, have a resultant force in all directions equal to zero. Since there are three equations of static equilibrium for a two-dimensional system, the values of each reaction can be determined.

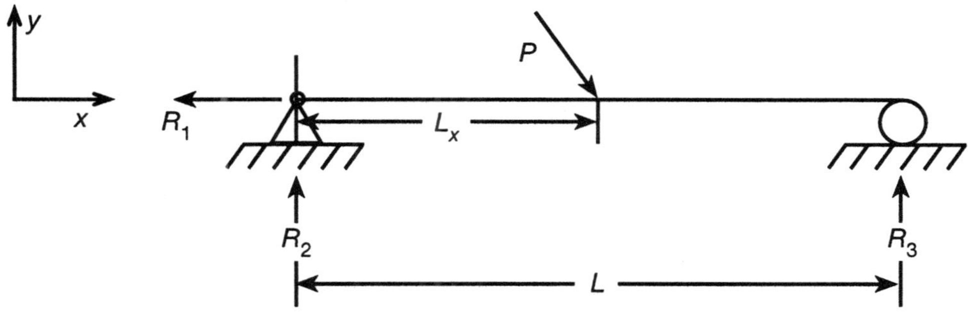

Figure 7. Statically Determine Force System

For static equilibrium, the sum of all forces acting on a body in the x and y directions equals zero. The three equations of force for equilibrium are:

$$+\rightarrow \Sigma F_x = -R_1 + P_x = 0$$

$$+\uparrow \Sigma F_y = -P_y + R_2 + R_3 = 0$$

and the sum of the moments about any point on a body equals zero.

$$+\curvearrowright \Sigma M(o) = +P_y L_x + -R_3 L = 0$$

When there are more unknown forces than equations of equilibrium, the unknown forces cannot be calculated using the equations of static equilibrium alone. These forces result in a *statically indeterminate system.*

A body is loaded by the force P and results in four reactions: R_1, R_2, R_3, and R_4 as shown in Figure 8c. Since there are only three equations of static equilibrium for the system, the values of each reaction cannot be determined. Such a system is indeterminate to the first degree (see Figure 8a below).

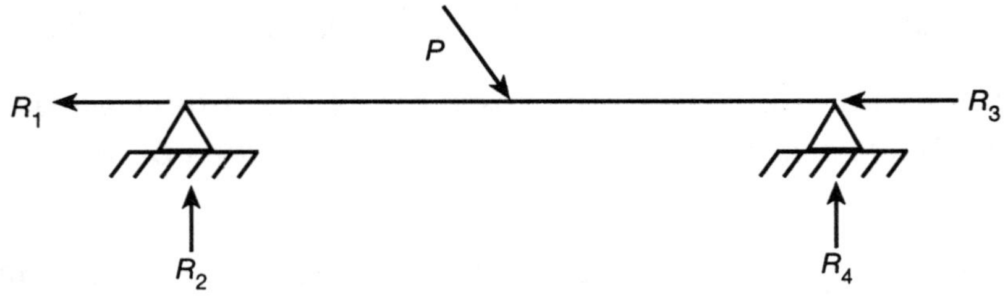

Figure 8a

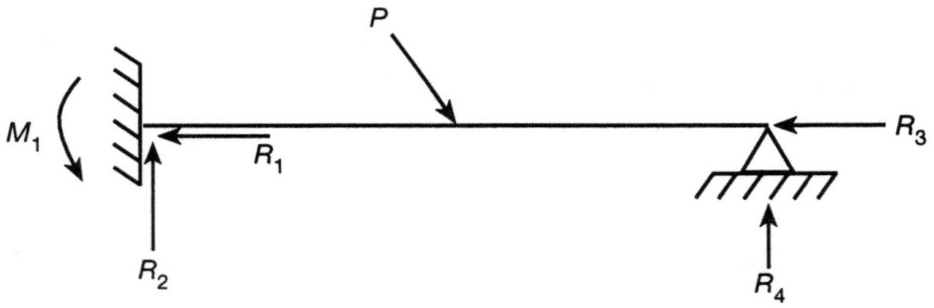

Figure 8b. Statically Indeterminate Force Systems

A body is loaded by the force P and results in five reactions: R_1, R_2, R_3, R_4, and M_1. Since there are only three equations of static equilibrium for the system, the values of each reaction cannot be determined. Such a system is indeterminate to the second degree.

The method of analysis used to find the reactions of indeterminate force systems is the deformation method. First, we write all of the equations of static equilibrium for the system and then add the equations based on the deformation of the structure. The total number of equations written for both statics and deformations equals the number of unknown forces.

A body with a constant cross-sectional area is held fixed between two walls. An axial force is applied to the body at a distance L_1 from one end as shown in Figure 8c.

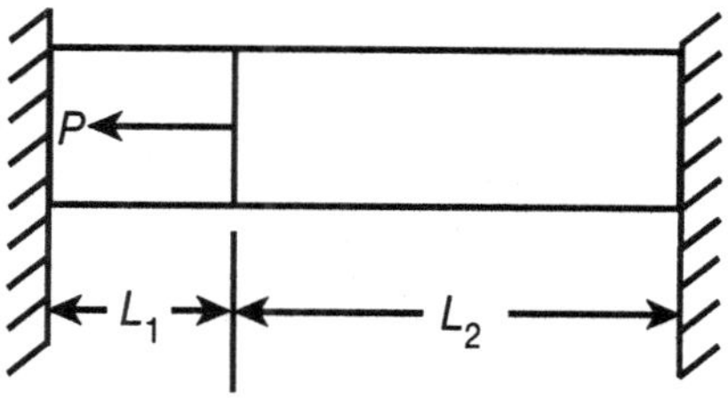

Figure 8c. Body Held Fixed Between Two Walls.

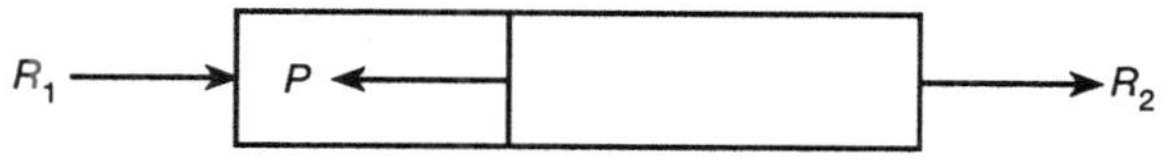

Figure 9. Free-Body Diagram of Body Held Fixed Between Two Walls.

To solve for the reactions, first draw a free-body diagram of the body as shown in Figure 9. The equations for static equilibrium are:

$$\Sigma F_x = R_1 - P + R_2 = 0$$

The equation has two unknowns, R_1 and R_2, and the problem is statically indeterminate to the first degree. To solve the problem, an additional equation based on the deformations of the body must be used.

Assume that the reduction of the body along length L_1 is equal to the elongation along the length L_2. The equation relating the deformation, δ, is $\sigma = P/A$ and $\in = \delta/L$.

$$E = \frac{\sigma}{\in} = \frac{P/A}{\delta/L}; \delta = \frac{PL}{AE}$$

$$\frac{R_1 L_1}{AE} = \frac{R_2 L_2}{AE}$$

Therefore, $R_1 L_1 = R_2 L_2$ and $R_1 = \frac{R_2 L_2}{L_1}$. Substituting this value into the static equation, the reactions can be determined:

$$R_1 = \frac{PL_2}{L_1 + L_2} \text{ and } R_2 = \frac{PL_1}{L_1 + L_2}$$

The elongation along L_2 is:

$$\delta_2 = \frac{R_2 L_2}{AE} = \frac{PL_1 L_2}{(L_1 + L_2)AE}$$

and the reduction is:

$$\delta_1 = \frac{-R_1 L_1}{AE} = \frac{-PL_1 L_2}{(L_1 + L_2)AE}$$

Therefore, $\delta_2 = \delta_1$.

PROBLEM 5:

A standard test specimen is 1.9 cm in diameter by 10 cm gauge length shows an increase in length of 0.002 cm when tension is increased from 1,800 N to 8,900 N. What is the modulus of elasticity?

SOLUTION:

1. Calculate the change in stress:

$$\sigma = \frac{P}{A} = \frac{(8{,}900 - 1{,}800)\text{ N}}{\pi(d^2/4)} = \frac{7{,}100\text{ N} \times 10^4\text{cm}^2/\text{m}^2}{(3.1416/4)(1.9\text{cm})^2}$$

$$= 2.5 \times 10^7 \text{N/m}^2 = 25\text{ MPa}$$

2. Calculate the change in strain:

$$\text{Strain} = \in = \frac{\Delta L}{L} = \frac{0.002\text{ cm}}{10\text{ cm}} = 0.0002\text{ cm/cm}$$

3. The modulus of elasticity is therefore:

$$\sigma = E \in$$

Therefore, if we solve for E:

$$E = \frac{\alpha}{\varepsilon}$$

$$E = \frac{25 \times 10^6 \text{N/m}^2}{(0.00020\text{ cm/cm})} = 1.25 \times 10^{11} \text{N/m}^2 = 125\text{ GPa}$$

PROBLEM 6:

Two parallel wires, 30 cm apart, one aluminum (E = 68.9 GPa) and one steel (E = 207 GPa), are used to support a load F as shown in Figure 10. The area of the aluminum wire is twice the area of the steel wire. What is the distance from the steel wire to the load that will permit the wires to be of equal length?

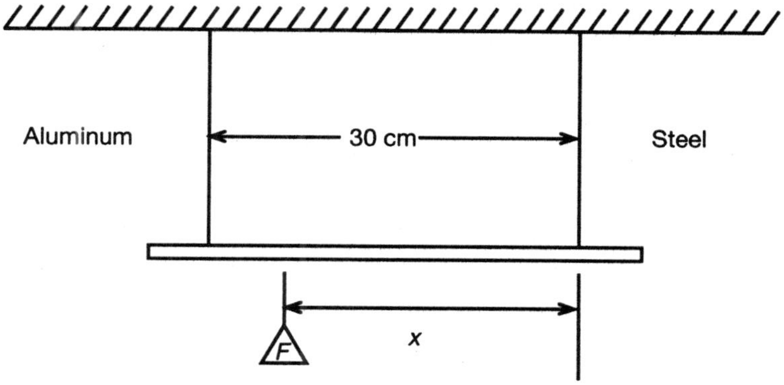

Figure 10

SOLUTION:

1. Calculate the relationship of the stress of the two wires.

Rearranging $\sigma = E \in$: $\in_{\text{alum}} = \in_{\text{steel}}$

$$\frac{\sigma_{\text{alum}}}{E_{\text{alum}}} = \frac{\sigma_{\text{steel}}}{E_{\text{steel}}}$$

$$\frac{\sigma_{\text{alum}}}{68.9 \times 10^9\,\text{N/m}^2} = \frac{\sigma_{\text{steel}}}{207 \times 10^9\,\text{N/m}^2}$$

Therefore,

$$\sigma_{\text{steel}} = 3.0\ \sigma_{\text{alum}}$$

2. Calculate the load, based on the length of the two wires being equal. Substituting in the equation $\sigma = F/A$, the load for the steel wire is:

(a) $\sigma = \dfrac{F}{A}$

$$\sigma_{\text{steel}} = \frac{F_{\text{steel}}}{A_{\text{steel}}};\quad \sigma_{\text{alum}} = \frac{F_{\text{alum}}}{A_{\text{alum}}}\quad \sigma_{\text{steel}} = 3.0\sigma_{\text{alum}}$$

$$\frac{F_{\text{steel}}}{A_{\text{steel}}} = 3.0 \frac{F_{\text{alum}}}{A_{\text{alum}}} \qquad F_{\text{steel}} = \frac{3.0\, F_{\text{alum}}}{(A_{\text{alum}}/A_{steel})}$$

$$F_{\text{steel}} = \frac{3}{2} F_{\text{alum}}$$

(b) Assume the system of forces is in equilibrium, therefore, the sum of the forces in the y-direction equal zero, as shown in Figure 11.

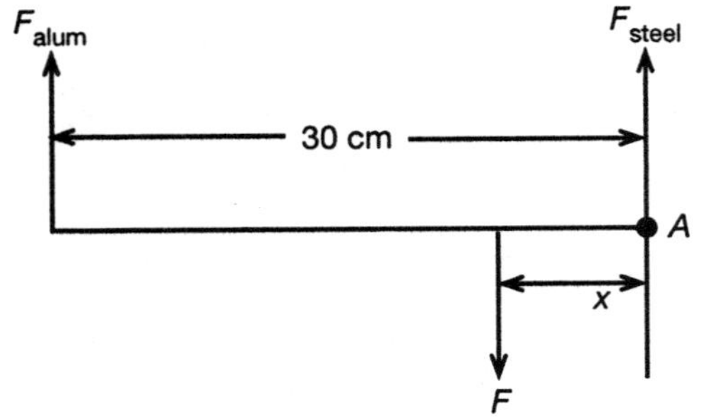

Figure 11

$$F_{\text{alum}} + F_{\text{steel}} = F$$

$$F_{\text{alum}} + \frac{3}{2} F_{\text{alum}} = F$$

$$\frac{5}{2} F_{\text{alum}} = F$$

Therefore,

$$F_{\text{alum}} = \frac{2}{5} F$$

$$F_{\text{steel}} = \frac{3}{5} F$$

3. Calculate the distance of the force from the steel wire.

Take the moments about point A (see Figure 11).

$$(L)\,(F_{\text{alum}}) = Fx$$

$$(30 \text{ cm})\left(\frac{2}{5} F\right) = Fx$$

$$\frac{60}{5} = x = 12 \text{ cm}$$

THIN-WALLED CYLINDERS AND SPHERES

Thin-walled cylinders and spheres under internal pressure, gas or liquid, are subjected to uniformly distributed normal stresses along the circumference and length—axial or longitudinal stresses and circumferential or tangential stresses—of the walls. Assume that the stresses acting on the walls are uniformly distributed along the thickness of the walls and symmetrical about the axis of the cylinder and the center of the sphere.

It is critical to maintain a ratio of the wall thickness to the curvature radius of less than 0.10 and that the structure have no discontinuities. Examples of thin-walled cylinders and spheres are liquid storage tanks and containers, water pipes, boilers, submarine hulls, and aircraft components.

Normal stresses are determined as follows:

$$\text{Hoop Stress} = \sigma_t = \frac{Pr}{t} = \frac{Pd}{\alpha t}$$

$$\text{Axial Stress of Tank} = \sigma_a = \frac{Pr}{2t} = \frac{Pd}{4t}$$

where:

P = uniform internal pressure

r = Internal radius of cylinder

t = wall thickness

d = internal diameter of cylinder

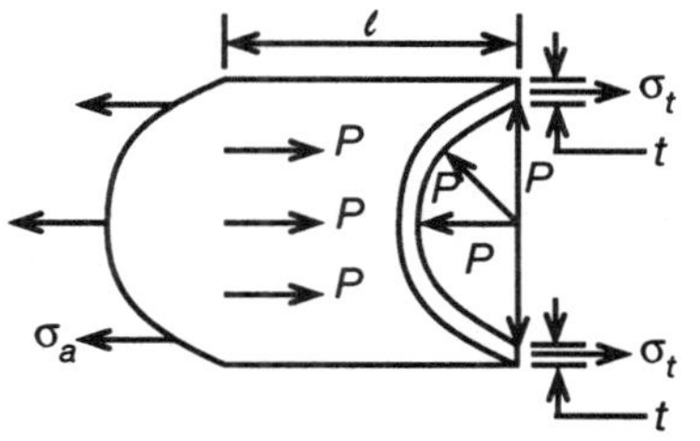

Figure 12. Stress in Thin-Walled Cylinders

PROBLEM 7:

What is the thickness of the wall of a tank cylinder containing oxygen at a pressure of 17.2 MPa? The inside diameter of the tank is 30 cm and the allowable stress is 138 MPa in the walls.

SOLUTION:

The variables are: t = thickness of shell; P = internal pressure; d = tank diameter; and σ_a = allowable stress.

Assume that this is a thin-walled cylinder. The formula for the wall thickness based on hoop stress that which is the greatest is as follows:

$$t = \frac{Pd}{2\sigma_t} = \frac{(17.2 \text{ MPa})(30 \text{ cm})}{(2)(138 \text{ MPa})} = 1.87 \text{ cm}$$

TORSION

A bar rigidly held at one end and twisted at the other end with a torque, T, resulting from forces, F, applied at a distance, d, as shown in Figure 13, is said to be in torsion.

$$T = Fd$$

Pure torsional loading—couple only, no bending—produces a shearing stress in a bar with the stress in any cross section being proportional to the distance from the bar center as shown in Figure 14 and an angle displacement with respect to the supported end.

A twisting moment or *moment of torsion* is created by couples, forces in opposite directions along the length of the bar and perpendicular to the diameter of the bar. The *polar moment of inertia* for a hollow circular bar is:

$$J = \frac{\pi}{32}\left(\left(D_{\text{outer}}\right)^4 - \left(D_{\text{inner}}\right)^4\right)$$

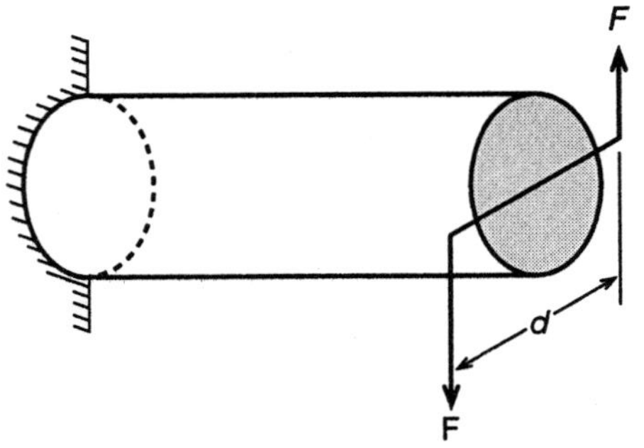

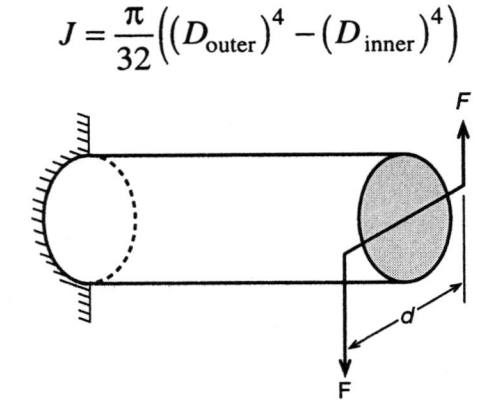

Figure 13. Bar in Torsion

This equation also applies to a solid bar with the inner diameter equal to zero.

$$J = \frac{\pi}{32}(D_{\text{outer}})^4$$

For any cylindrical bar being acted upon by a twisting moment, T, there is *torsional shearing stress*, τ, at any distance from the center, ρ, equal to:

$$\tau = \frac{T\rho}{J}$$

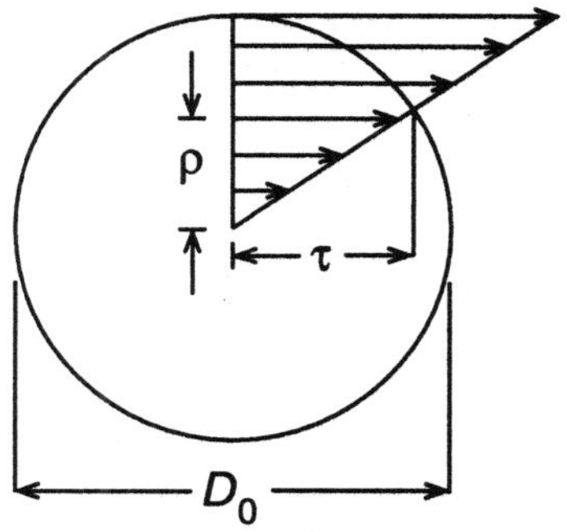

Figure 14. Stress Developed by Bar in Torsion

Referring to Figure 15a, *a-b* is a line along the surface prior to twisting while *a-b′* is the same line after twisting. The shearing strain at the surface of the bar is equal to the angle, γ, in radians that results from the twisting force on the unsupported end.

The ratio of shear stress, τ, to shear strain, γ, is called the modulus of rigidity and is given by:

$$G = \frac{\tau}{\gamma}$$

Referring to Figure 15b, the angle of twist for a round bar is the angle ϕ through which one end of the bar will twist relative to the other and is given by:

$$\phi = \frac{Tl}{GJ}$$

where:

T = torque

l = length

G = modulus of rigidity

J = polar area moment of inertia

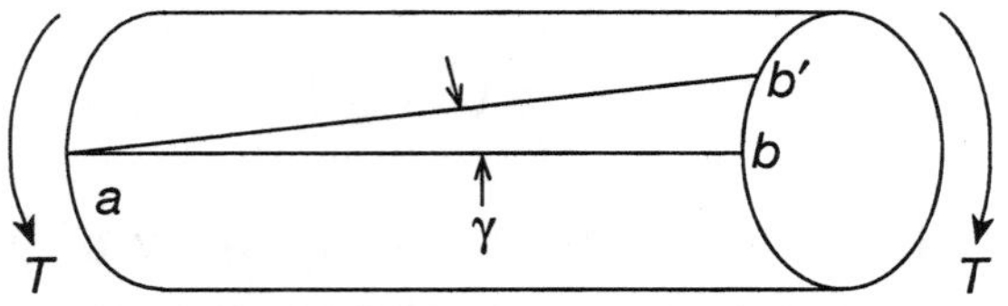

Figure 15a. Shear Strain for a Bar in Torsion

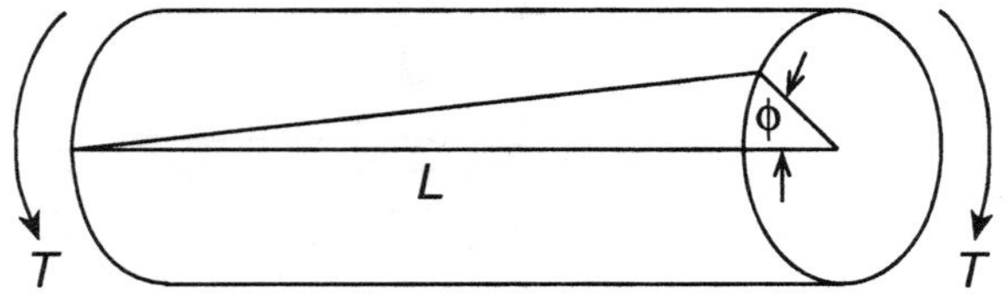

Figure 15b. Angle of Twist for Bar in Torsion

It is often necessary to obtain the torque T from the power and speed of a rotating shaft. In SI units the applicable equation is:

$$H = T\omega$$

where:

H = power, Watts

ω = angular velocity, radians/s

T = torque, $N \times m$

T/ϕ gives the twisting moment per radian of twist and is designated as the torsional stiffness, k or c.

PROBLEM 8:

A round shaft 10 cm in diameter transmits 150 kW at 100 rpm. What is the torsional stress in the outer fibers? If the shaft is made hollow with a 5 cm inside diameter, what is the increase in stress and the stress at the inner fibers?

SOLUTION:

From the power transmitted by the shaft, the torque, T, can be found as follows:

$H = T \times \omega$ where H is the power and ω is the angular velocity in rad/s. Since there are 2π radians in a revolution and there are 60 seconds in a minute, 100 rpm = $(2\pi/60) \times 100$ rad/s.

$$T = \frac{H}{\omega} = \frac{150{,}000 \text{ W}}{(20\pi / 6) \text{ rad/s}} = 1.43 \times 10^4 J$$

$$(J = \text{Joules or N} - \text{m})$$

The polar moment, J, for the solid shaft is given by:

$$J = \frac{\pi D^4}{32} = \frac{\pi (10 \text{ cm})^4}{32} = 983 \text{ cm}^4$$

The shear stress is found as follows:

$$\tau = \frac{T\rho}{J} = \frac{(1.43 \times 10^4 J) \times (5 \text{ cm})}{982 \text{ cm}^4} \times \frac{10^6 \text{ cm}^3}{\text{m}^3}$$

$$= 72.8 \times 10^6 \text{Pa} = 72.8 \text{ MPa}$$

For the hollow shaft, J is determined as follows:

$$J = \frac{\pi}{32}\left(D_0^4 - D_i^4\right) = \frac{\pi}{32}\left(10^4 - 5^4\right) \text{cm}^4 = 920 \text{ cm}^4$$

$$\tau = \frac{T\rho}{J} = \frac{\left(1.43 \times 10^4 J\right) \times (5 \text{ cm})}{920 \text{ cm}^4} \times \frac{10^6 \text{ cm}^3}{\text{m}^3}$$

$$= 77.7 \times 10^6 \text{ Pa} = 77.7 \text{ MPa}$$

The increase in stress is 77.7 – 72.8 = 4.9 MPa.

The stress at the inner fiber is found using the above equation with $\rho = 2.5$ cm:

$$\tau = \frac{T\rho}{J} = \frac{\left(1.43 \times 10^4 J\right) \times (2.5 \text{ cm})}{920 \text{ cm}^4} \times \frac{10^6 \text{ cm}^3}{\text{m}^3}$$

$$= 38.9 \times 10^6 \text{ Pa} = 38.9 \text{ MPa}$$

SHEARING FORCE AND BENDING MOMENT

A beam is a bar that is subjected to forces (concentrated force or uniformly distributed loads) or couples, acting perpendicular, along its longitudinal axis. A *cantilever beam*, shown in Figure 16, is supported or restrained on one end and that end is prevented from rotating. The other end is free to deflect or bend. The restrained end transmits a vertical force and a moment along the plane of the beam.

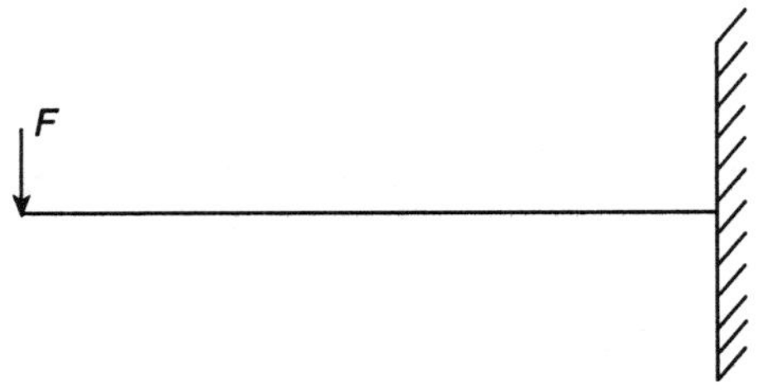

Figure 16. Cantilever Beam

A *simple beam* is shown in Figure 17 freely supported at both ends and allowed to exert only forces on the beam, not moments. One of the end supports has to be allowed to move.

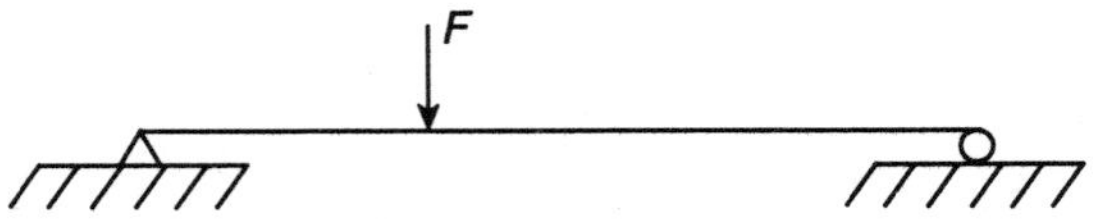

Figure 17. Simple Supported Beam

An *overhanging beam* is supported at two points and has end(s) that extend past the supports.

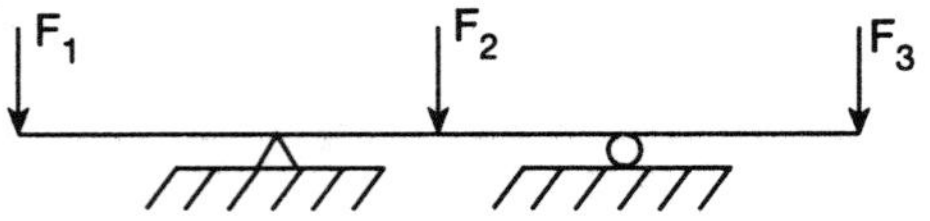

Figure 18. Overhanging Beam

The cantilever, simple, and overhanging beams are *statically determinate beams*; the reactions of the supports can be determined using the equations of static equilibrium. The reactions are independent of the beam deformations as shown in Figure 19, a cantilever beam supported at the free end or rigidly fixed at both ends and a beam having three or more supports are examples of *statically indeterminate beams*. To solve for the reactions, the equations of static equilibrium have to be supplemented by the equations based on the beam deformations.

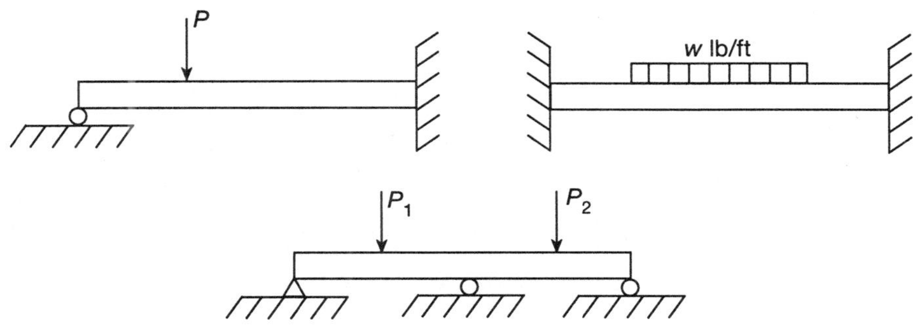

Figure 19. Statically Indeterminate Beams

Common *load types* applied to beams are concentrated forces at a point, uniformly distributed loads, force per unit of length, uniformly varying loads, and by a couple. *Internal stresses*, both normal and shearing, develop in beams due to loads by forces and couples. Using the resultant force and moment acting at any point on a beam allows the determination of the magnitude of the stresses at any section on the beam. The stresses are found by applying the equations of static equilibrium as shown in Figure 20.

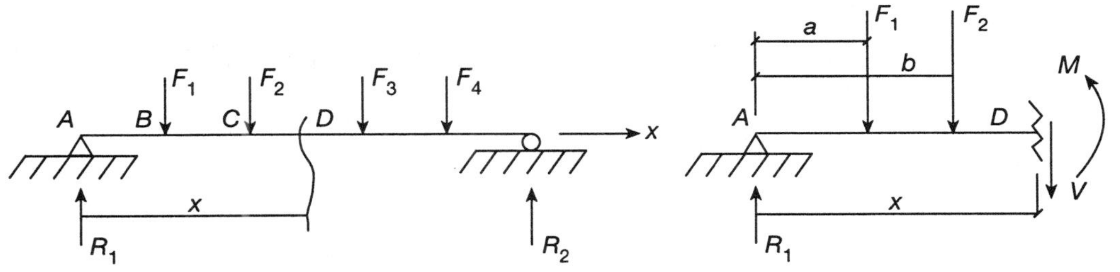

Figure 20. Free-Body Diagram of a Beam

To find the value of the *resisting moment*, *M*, the sum of all of the moments about any point must equal zero for static equilibrium (Sum of Moments = 0). Taking moments about point *D*.

$$\circlearrowleft + \Sigma Mo = M - R_1x + F_1(x - a) + F_2(x - b) = 0$$

$$M = R_1 x - F_1(x - a) - F_2(x - b)$$

The resisting moment equals the resultant couple caused by the stresses created over the vertical section and acting horizontally at *D*. The stresses are compressive and tensile along different cross sections of the beam.

The *resisting shear, V*, is the vertical force at section *D*. To solve for *V*, use the equation for static equilibrium in the *y*-direction.

$$\Sigma Fy = R_1 - F_1 - F_2 - V = 0$$

$$V = R_1 - F_1 - F_2$$

The *bending moment* at D is equal to the algebraic sum of the moments of the external forces on one side of D about an axis through D and opposite in direction to the resisting moment and of the same magnitude.

$$M_b = R_1\,x - F_1(x - a) - F_2(x - b)$$

The shearing force at D is equal to the algebraic sum of the vertical forces on one side of D and opposite in direction to the resisting shear and of the same magnitude.

$$V_s = R_1 - F_1 - F_2$$

In order to determine the proper value of bending moment and shear, a *sign convention* is established as follows:

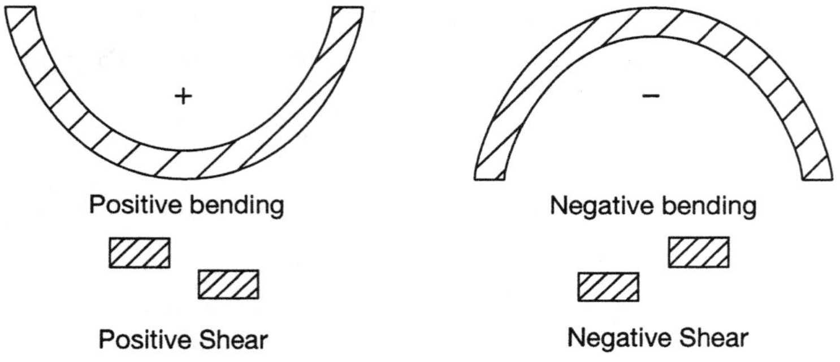

Figure 21. Sign Convention for Shear and Bending of a Beam

A force that causes the beam to bend concave upward and tends to shear the left portion of the beam upward produces a positive bending moment and a positive shearing force. Upward external forces produce positive bending moments and downward forces result in negative bending moments.

The pictorial representation of V and M are called *shearing force and bending moment diagrams*. The x-axis represents the location of the beam section and the y-axis is the magnitude of the shearing force and bending moment, respectively. The diagrams show the shearing force is equal to the rate of change of the bending moment with respect to x—the differential of the bending moment.

$$dM = Vdx$$

$$V = \frac{dM}{dx}$$

The following example demonstrates how to calculate the shear and bending moment for a simply supported beam subject to a distributed load of 15 kN/m, as shown in Figure 22.

By considering the entire beam as a free-body (Figure 23), the equations of equilibrium are applied to obtain:

$$W = \frac{\text{force}}{\text{unit length}}$$

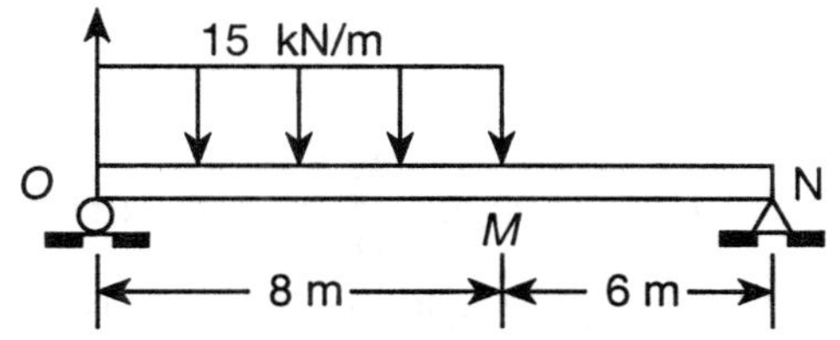

Figure 22. Beam Subjected to Uniform Load

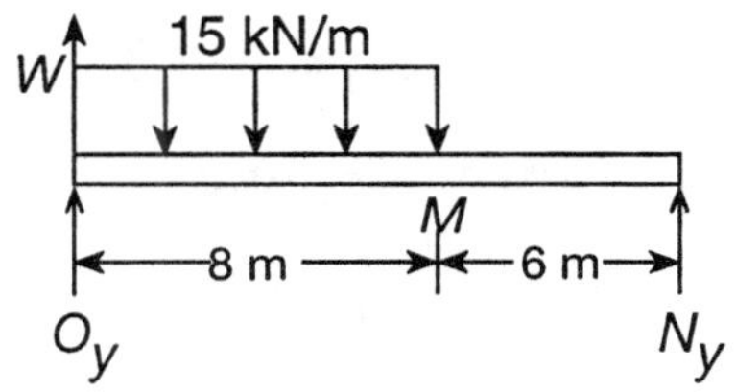

Figure 23. Free-Body Diagram of Beam

$$+\uparrow \Sigma F_y = 0;\ O_y - (15\ \text{kN/m})\ (8\ \text{m}) + N_y = 0$$

$$O_y + N_y = 120\ \text{kN}$$

$$+\circlearrowleft\ \Sigma M = 0_N;\ -O_y(14\ \text{m}) + 15\,\frac{\text{kN}}{\text{m}}\ (8\ \text{m})\ (10\ \text{m}) = 0$$

$$O_y = 85.7\ \text{kN}$$

Therefore,

$$N_y = 34.3\ \text{kN}$$

$$\frac{dS}{dx} = -w \qquad (1)$$

or

$$S_x - S_1 = -\int_{x_1}^{x} w\,dx = -w(x - x_1) \qquad (2)$$

The shear at M is given as:

$$S_2 - 85.7 \text{ kN} = -\int_0^{8m} 15\frac{\text{kN}}{\text{m}},$$

or

$$S_2 - 85.7 \text{ kN} = -15\frac{\text{kN}}{\text{m}}\ (8 \text{ m} - 0)$$

$$S_m = -34.3 \text{ kN}$$

Since w is a constant, the slope $\frac{dS}{dx}$ is just linear (from Equation (1)) between points 0 and M. To the right of M there is no load; therefore, the shear does not change in that region.

Hence:

$$S_N = S_M = -34.3$$

The shear diagram is shown in Figure 24.

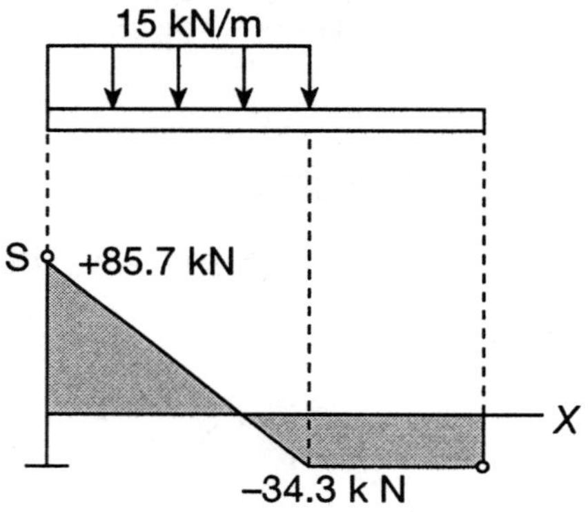

Figure 24. Shear Diagram

Bending Moment:

The bending moment is related to the shear by:

$$\frac{dM}{dx} = \text{S} \tag{3}$$

Between O and M, the equation for the shear may be obtained by analyzing Figure 25.

$$S = \frac{85.7 - (-34.3)}{0 - 8}\frac{\text{kN}}{\text{m}}x + 85.7 \text{ kN}$$

or

$$S = -15x + 85.7 \quad 0 \leq x \leq 8 \tag{4}$$

It can be seen from Equations (3) and (4) that the moment will be parabolic with its maximum value occurring at:

$$S = \frac{dM_x}{dx} = -\ 15x + 85.7 = 0$$

or

$$x = \frac{85.7}{15} = 5.71 \text{ m}$$

It is now possible to construct the bending moment diagram by looking at the shear diagram and summing areas.

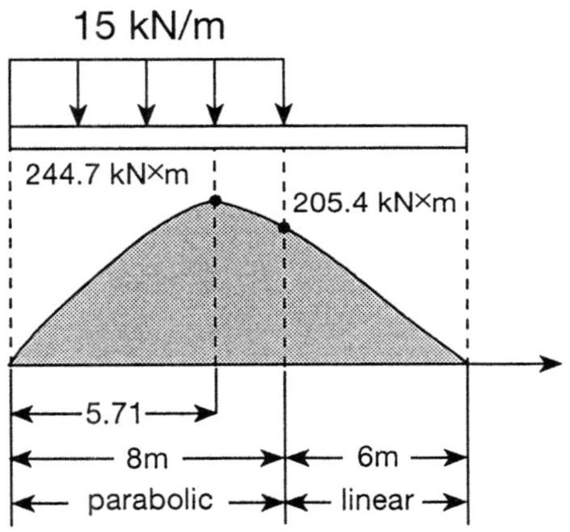

Figure 25. Bending Moment Diagram

At $x = 5.71$ m, $M = \frac{1}{2}(85.7 \times 5.71) = 244.7$ kN—m

At $x = 8$ m, $M = 244.7$ kN—m $+ \frac{1}{2}[-34.3 \times (8 - 5.71)]$kN—m

$= 205.7$ kN—m

As a check,

at $x = 14$ m, $M = 205.7$ kN-m $+ [-34.3 \times 6]$ kN-m $= 0$.

With the shear and the moment determined at any section of the beam, the stresses in the beam can now be determined. Let ρ = the radius of curvature of the deflected axis of the beam and y the distance from the neutral axis to the longitudinal fiber in question. The axial stress, σ_x, is given by the following:

$$\sigma_x = -\frac{Ey}{\rho} = -\frac{My}{I} = \pm\frac{Mc}{I} = \pm\frac{M}{S}$$

where:

M = moment at the section

I = moment of inertia of the cross-section

c = distance from the neutral axis to the outermost fiber of beam

$S = I/c$ = section modulus

The shear stress on the surface, τ_{xy}, is given by the following:

$$\tau_{xy} = \frac{VQ}{Ib}$$

where:

V = shear force at the section

b = width of the cross section

Q = first moment of the area of the vertical face about the neutral axis

DEFLECTION OF BEAMS

Using the relationship for the curvature of a beam,

$$\frac{1}{\rho} = \frac{M}{EI}$$

the differential equation of the deflection curve is:

$$EId^2y = \frac{M}{dx^2}$$

By double integration of the above equation and application of the boundary conditions, the deflection of a beam is determined. Table 2 is a list of various beam deflections.

An important technique in the solution of the beam deflection problems is the use of superposition. Since the governing differential equation of beam defection is linear, problems can be divided into a series of simpler problems for which solutions exist. The final solution is the sum of these sub-problems. This procedure will be demonstrated in Problem 9.

PROBLEM 9:

Consider the beam pictured in Figure 26 which is made of steel (E = 207 GPa). The beam has a moment of inertia, I, equal to 87.2 cm^4 and is simply supported. Determine the deflection of the beam at point B directly under the 1.8 kN force.

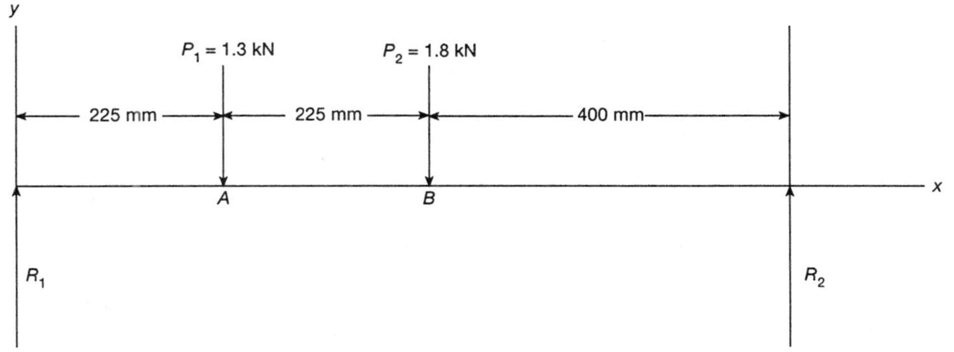

Figure 26

SOLUTION:

The problem will be solved by using superposition. First a beam will be analyzed with just the 1.3 kN force and then the beam will be analyzed with just the 1.8 kN force. In each case the deflection at point B will be found. The summation of these two deflections will give the deflection for the original problem.

For case one, only consider the 1.3 kN force. The formulas from Table 2 are given by:

$$\delta_1 = \frac{Pb}{6LEI}\left[\frac{L}{b}(x-a)^3 - x^3 + \left(L^2 - b^2\right)x\right], \text{ for } x > a$$

with a = 225 mm, b = 625 mm, and x = 450 mm. For computation, all the lengths are converted to meters.

$$\delta_1 = \frac{(1300 \text{ N} \times .625 \text{ m}) \times \left(10^8 \text{ cm}^4/\text{m}^4\right)}{6 \times .85 \text{ m} \times 207 \times 10^9 \text{ N/m}^2 \times 87.2 \text{ cm}^4}$$

$$\left[\frac{0.85 \text{ m}}{0.625 \text{ m}}(0.45 \text{ m} - 0.225 \text{ m})^3 - (0.45 \text{ m})^3\right.$$

$$\left.+\left((0.85 \text{ m})^2 - (0.625 \text{ m})^2\right) 0.45 \text{ m}\right]$$

$$\delta_1 = 65.0 \times 10^{-6} \text{ m}$$

Table 2
Beam Deflection Formulas – Special Cases
(d is positive downward.)

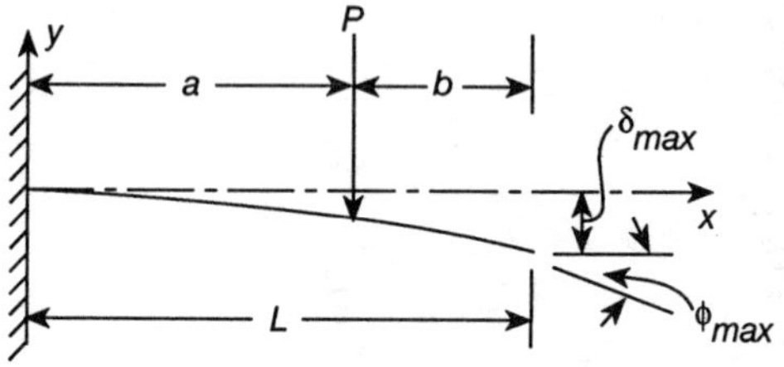

$$\delta = \frac{Pa^2}{6EI}(3x - a), \text{ for } x > a$$

$$\delta = \frac{Px^2}{6EI}(-x + 3a), \text{ for } x \le a$$

$$\delta_{\max} = \frac{Pa^2}{6EI}(3L - a)$$

$$\phi_{\max} = \frac{Pa^2}{2EI}$$

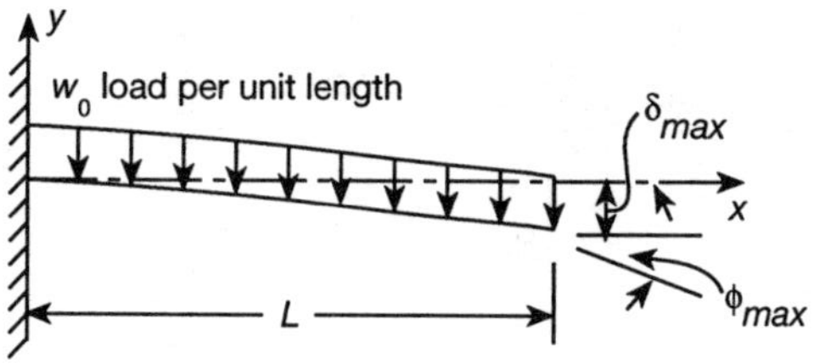

$$\delta = \frac{w_0 x^2}{24EI}\left(x^2 + 6L^2 - 4Lx\right)$$

$$\delta_{\max} = \frac{w_0 L^4}{8EI}$$

$$\phi_{\max} = \frac{w_0 L^3}{6EI}$$

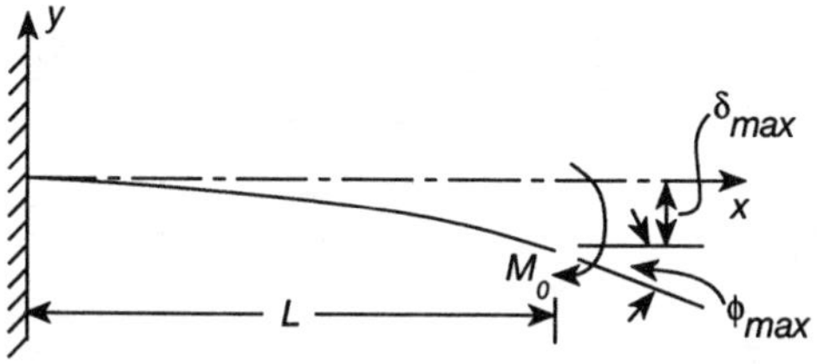

$$\delta = \frac{M_0 x^2}{2EI}$$

$$\delta_{\max} = \frac{M_0 L^2}{2EI}$$

$$\phi_{\max} = \frac{M_0 L}{EI}$$

Table 2 (continued)
Beam Deflection Formulas – Special Cases
(d is positive downward.)

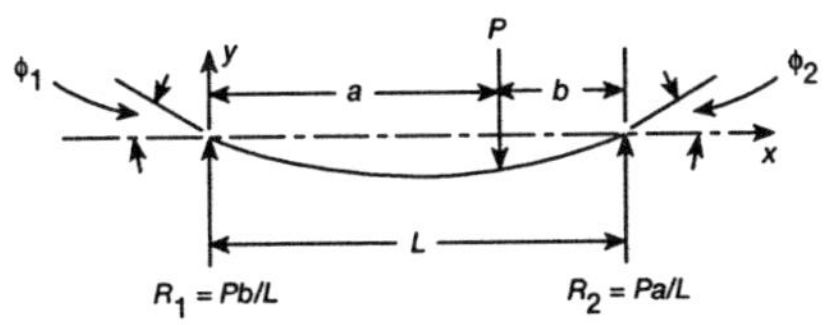	$\delta = \frac{Pb}{6LEI}\left[\frac{L}{b}(x-a)^3 - x^3 + \left(L^2 - b^2\right)x\right]$, for $x > a$ $\delta = \frac{Pb}{6LEI}\left[-x^3 + \left(L^2 - b^2\right)x\right]$, for $x \le a$	$\delta_{max} = \frac{Pb\left(L^2 - b^2\right)^{3/2}}{9\sqrt{3}\,LEI}$ at $x = \sqrt{\frac{L^2 - b^2}{3}}$	$\phi_1 = \frac{Pab(2L-a)}{6LEI}$ $\phi_2 = \frac{Pab(2L-b)}{6LEI}$
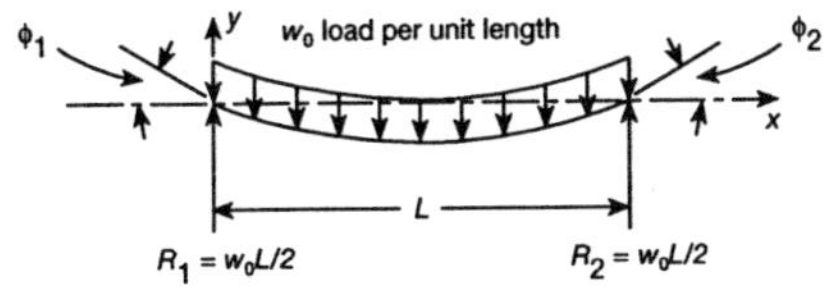	$\delta = \frac{w_0 x}{24EI}\left(L^3 - 2Lx^2 + x^3\right)$	$\delta_{max} = \frac{5w_0L^4}{384EI}$	$\phi_1 = \phi_2 = \frac{w_0L^3}{24EI}$
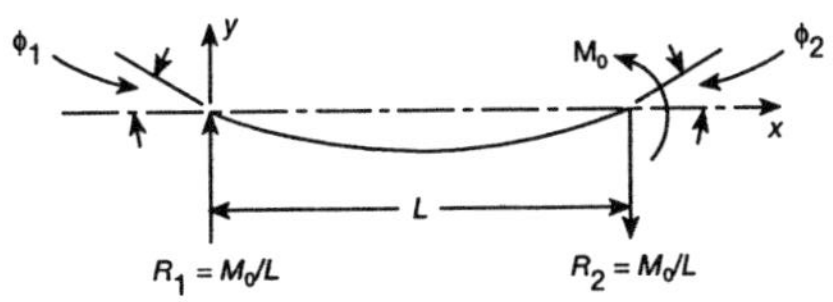	$\delta = \frac{M_0Lx}{6EI}\left(1 - \frac{x^2}{L^2}\right)$	$\delta_{max} = \frac{M_0L^2}{9\sqrt{3}\,EI}$ at $x = \frac{L}{\sqrt{3}}$	$\phi_1 = \frac{M_0L}{6EI}$ $\phi_2 = \frac{M_0L}{3EI}$

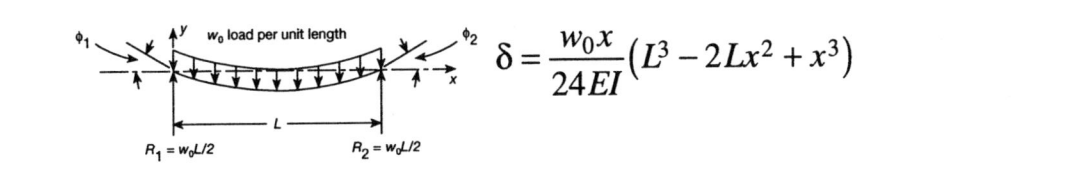

For case 2, only the 1.8 kN force will be considered. The appropriate formula from Table 2 is:

$$\delta_2 = \frac{Pb}{6LEI}\left[-x^3 + \left(L^2 - b^2\right)x\right], \text{ for } x < a$$

with $a = x = 450$ mm and $b = 400$ mm.

$$\delta_2 = \frac{\left(1800\ \text{N} \times 0.4\ \text{m}\right) \times \left(10^8\ \text{cm}^4/\text{m}^4\right)}{6 \times 085\ \text{m} \times 207 \times 10^9\ \text{N/m}^2 \times 87.2\ \text{cm}^4}$$

$$\left[-(0.45\ \text{m})^3 + \left((0.85\ \text{m})^2 - (0.4\ \text{m})^2\right)(0.45\ \text{m})\right]$$

$$\delta_2 = 197 \times 10^{-6}\ \text{m}$$

$$\delta = \delta_1 + \delta_2 = 262 \times 10^{-6}\ \text{m}$$

Mohr's Circle

The stresses on a specified plane surface can be determined from the stresses on two other surfaces that are perpendicular to each other. The Mohr's circle is a graphical method of determining these combined stresses.

FE/EIT

FE: PM Mechanical Engineering Exam

CHAPTER 14

Thermodynamics

CHAPTER 14

THERMODYNAMICS

BASIC LAWS OF THERMODYNAMICS

Zeroth Law—Two bodies, each in thermal equilibrium with a third body, are in equilibrium with each other.

Consequences—permits the use of temperature measuring devices such as thermometers, thermocouples, etc.

First Law—Energy can neither be created nor destroyed, but can only be changed from one form to another.

$$\text{Energy}_{\text{in}} = \text{Energy}_{\text{out}} + \Delta\text{Energy}_{\text{stored}}$$

Consequences—forbids perpetual motion machines of the first kind since energy cannot be created.

Second Law—It is impossible to construct a device that operates in a cycle and whose sole effect is to transfer heat from a cooler body to a hotter body (Clausius statement).

This law can also be stated that it is impossible to construct a device that operates in a cycle and produces no other effect than the production of work and the exchange of heat with a single reservoir (Kelvin-Planck).

Consequences—forbids the operation of devices that while not in violation of the First Law are still impossible to build. An example would be building an adiabatic steam turbine with water exiting the system.

Third Law—The entropy of any pure substance in thermodynamic equilibrium approaches zero as the absolute temperature approaches zero.

Consequences—establishes an absolute scale for entropy as opposed to the arbitrary scale in the gas and steam tables.

BASIC DEFINITIONS

Closed System—specified region of space containing a fixed amount of mass. Energy but not mass can cross a system's boundary.

Open System—specified region of space that allows energy and mass to cross its boundaries.

Surroundings—everything that is not part of the system.

Units and Dimensions—Newton's Second Law of Motion is used to relate the units of force, mass, length, and time.

F, force, is proportional to the time rate of change of momentum:

$$F \propto \frac{d(mV)}{dt}$$

To change the proportionality to an equality requires the insertion of a constant into the equation.

$$F = \frac{ma}{g_c}$$

where g_c is the proportionality constant. While g_c may have a numerical value of 1 in some unit systems, it is still "required" to make the equations dimensionally homogeneous. Table 1 on the following page summarizes the various unit systems. g_c should be considered as a conversion factor that can be used to adjust the units of an equation.

OTHER UNITS

Since both work and heat are forms of energy, they have units of ft-lbf (force × distance) in the U.S. customary system and N–m in the SI system.

1 N–m = 1 Joule = 1 J	1 cal = 4.1855 J
1 kcal = 1 Cal = 1,000 cal	1 Btu = 778 ft-lbf

Temperature—The measure of the average kinetic energy of molecular motion. Table 2 lists the most widely used temperature scales.

Table 1. Summary of Unit Systems

System	Length	Mass	Time	Force	g_c
U.S. customary	foot, ft	pound mass, lbm	second, sec	pound force, lbf	$32.2 \dfrac{\text{ft–lbm}}{\text{lbf–sec}^2}$
English	foot, ft	slug	second, sec	pound force, lbf	$1 \dfrac{\text{ft–slug}}{\text{lbf–sec}^2}$
SI	meter, m	kilogram, kg	second, sec	Newton, N	$1 \dfrac{\text{kg–m}}{\text{N–sec}^2}$

Table 2. Temperature Scales and Conversions

Scale	Type	System	Lowest possible temp.
Fahrenheit	Arbitrary	U.S. customary	–459.69°F
Rankine	Absolute	U.S. customary	0°R
Celsius	Arbitrary	Metric	–273.15°C
Kelvin	Absolute	Metric	0 K

Absolute temperatures are required for division and multiplication.

For conversion from one temperature scale to another:

$$°F = 32.0 + 9/5\ °C$$

$$°R = °F + 459.69$$

$$K = °C + 273.15$$

For conversion of temperature for ΔT, dT, or units such as specific heat or entropy:

$$\Delta T \text{ of } 1°F = \Delta T \text{ of } 1°R$$

$$\Delta T \text{ of } 1°C = \Delta T \text{ of } 1\ K$$

$$\Delta T \text{ of } 1°F = \Delta T \text{ of } 5/9°C$$

$$\Delta T \text{ of } 1°C = \Delta T \text{ of } 9/5°F$$

Pressure—The normal force per unit area caused by molecular impact on a surface. *Absolute pressure is required in most thermodynamic calculations.*

Pressure Conventions—Expressed as force per unit area (actual definition) or as height of a fluid column (pseudo-definition). Pressure is given as absolute or gage (which is the pressure relative to the atmospheric pressure). Most pressure measuring devices (gage, manometer) measure gage pressure. The atmospheric pressure must be added to the gage pressure to find the absolute pressure.

P_a or P = absolute pressure

P_g = gage pressure

P_{atm} = atmospheric pressure

$P_a = P_g + P_{atm}$

Standard Atmospheric Pressure:

$$114.696 \text{ lbf/in}^2 = 4.696 \text{ psia} = 0 \text{ psig} = 1 \text{ atm} =$$

$$2116 \text{ lbf/f}t^2 = 29.92 \text{ in Hg}$$

$$101{,}325 \text{ N/m}^2 = 101{,}325 \text{ Pa} =$$

$$101.325 \text{ kPa} = 760 \text{ mm hg}$$

Properties of Single-Component Systems—A single component system is one that has only one relevant work mode. As a result, only two independent properties are required to determine all the remaining. The only pair of properties that are sometimes not independent are pressure and temperature (P, T). (P, T) are independent in a single-phase state but not independent in a multiple-phase region. All other pairs of thermodynamic properties in all regions are independent.

T = absolute temperature	(°R or K)
P = absolute pressure	(lbf/in^2 or Pa)
V = total volume	(ft^3 or m^3)
U = total internal energy	(Btu or kJ)
H = total enthalpy	(Btu or kJ)
S = total entropy	(Btu/°R or kJ/K)
C_P = total heat capacity at constant pressure	(Btu/°R or kJ/K)
C_v = total heat capacity at constant volume	(Btu/°R or kJ/K)

Properties are intensive (independent of mass) or extensive (proportional to mass of the system). P and T are intensive. All the other properties defined above are extensive. An extensive property can be made intensive by dividing by the mass.

$v = V/m$ = specific volume (ft^3/lbm or m^3/kg)

$u = U/m$ = specific internal energy (Btu/lbm or kJ/kg)

$h = H/m$ = specific enthalpy (Btu/lbm or kJ/kg)

$s = S/m$ = specific entropy (Btu/lbm-°R or kJ/kg-K)

$c_P = C_P/m$ = specific heat capacity at constant P (Btu/lbm-°R or kJ/kg-K)

$c_v = C_V/m$ = specific heat capacity at constant V (Btu/lbm-°R or kJ/kg-K)

$h = u + Pv$

$H = U + PV$

Specific heat capacity at constant P, $c_p = \left(\frac{\partial h}{T}\right)_P$

Specific heat capacity at constant V, $c_v = \left(\frac{\partial u}{\partial T}\right)_v$

Properties in a Two-Phase, Liquid-Vapor Region—Quality x (applies to liquid-vapor systems at saturation) is defined as the mass fraction of the vapor phase.

$$x = \frac{m_g}{\left(m_g + m_f\right)}$$

where:

m_g = mass of vapor

m_f = mass of liquid

The specific volume of a two-phase region can be written as follows:

$$v = xv_g + (1 - x)\,v_f$$

or

$$v = xv_{fg} + v_f$$

where:

v_f = specific volume of saturated liquid

v_g = specific volume of saturated vapor

v_{fg} = specific volume change under vaporization = $v_g - v_f$

Similar expressions exist for u, h, and s:

$$u = xu_g + (1 - x)\, u_f$$

$$h = xh_g + (1 - x)\, h_f$$

$$s = xs_g + (1 - x)\, s_f$$

The term saturation in thermodynamics refers to where a substance exists in a multiple phase region where T and P are linked and no longer independent. From the steam table at a temperature of 150°C the corresponding or saturation pressure is 0.4758 kPa. In the two-phase, liquid-vapor region, the quality x plus either P or T will allow determination of thermodynamic properties.

PROBLEM 1:

Determine the specific volume and enthalpy of water at a temperature of 50°C and a quality of 50%. Determine the total volume, V, if $m = 0.25$ kg.

SOLUTION:

$$v = xv_g + (1 - x)\, v_f = (0.5) \times (12.03) + (1 - 0.5) \times (0.001012)$$

$$= 6.02 \text{ m}^3/\text{kg}$$

$$h = h_f + xh_{fg} = 209.33 + (0.5) \times (2{,}382.8) = 1{,}401 \text{ kJ/kg}$$

$$V = vm = (6.02) \times (0.25) = 1.51 \text{ m}^3$$

A superheated vapor is defined as a vapor whose $T > T_{sat}$. Since a superheated vapor is a single-phase substance, T and P are independent and can be used to fix the state of the system.

PROBLEM 2:

Determine the internal energy and entropy of water at $P = 0.40$ MPa and $T = 500$°C. What would be the total entropy if $m = 0.5$ kg?

SOLUTION:

From the superheated water tables, $u = 3{,}129.2$ kJ/kg and $s = 8.1913$ kJ/kg-K. Thus, $S = sm = (8.1913) \times (0.5) = 4.10$ kJ/K.

A compressed liquid or sub-cooled liquid is a pure substance at $T < T_{sat}$. Normally, the properties of compressed liquids are assumed to be that of a saturated liquid at the substance temperature.

PROBLEM 3:

Estimate the internal energy of water at $P = 1$ atm and $T = 60°C$.

SOLUTION:

At $P = 1$ atm, $T_{sat} = 100°C$, since $T < T_{sat}$, the water is a compressed liquid. From the saturated water tables, $u \cong u_f = 251.11$ kJ/kg.

The equation of state of a pure substance is the relationship between the properties *P*, *T*, and *V*. An ideal gas is a pure substance whose equation of state is given by the following:

$$Pv = RT \text{ or } PV = mRT \text{ or } \frac{P_1 v_1}{T_1} = \frac{P_2 v_2}{T_2}$$

where:

m = mass of gas (kg or lbm)

R = particular gas constant

$$\frac{\overline{R}}{(\text{mol.wt.})}$$

$\overline{R}$ = the universal gas constant

$\overline{R}$ = 1,545 ft-lbf/(lbmole-°R) = 8,314 J/(kmol-K)

= 8.314 kJ/(kmol-K)

For ideal gases, $c_P - c_v = R$.

For a thermodynamic substance in general, $u(T, X)$, where X is any thermodynamic coordinate. For an ideal gas, $u\ (T)$ (i.e., the internal energy of an ideal gas is solely a function of temperature). Since $h = u + Pv$ and $Pv = RT$ for an ideal gas, $h = u + RT$; hence, $h(T)$ is determined as an ideal gas also. Values of u and h are tabulated as a function of temperature. The change in u or h can be calculated using the values of c_p and c_v. Both c_p and c_v are functions of temperature only for an ideal gas. For cold air standard, c_p and c_v are assumed to be constant at their room temperature values. These room temperature values are presented in the Table 3 at the end of this chapter. In this case the following are true:

$$\Delta u = c_v\, \Delta T \quad \Delta h = c_P\, \Delta T$$

$$\Delta s = c_p \ln\left(\frac{T_2}{T_1}\right) - R\ln\left(\frac{P_2}{P_1}\right)$$

$$\Delta s = c_v \ln\left(\frac{T_2}{T_1}\right) + R\ln\left(\frac{v_2}{v_1}\right)$$

For reversible, adiabatic processes:

$$\frac{P_2}{P_1} = \left(\frac{v_1}{v_2}\right)^k;\ \frac{T_2}{T_1} = \left(\frac{P_2}{P_1}\right)^{\frac{k-1}{k}};\ \frac{T_2}{T_1} = \left(\frac{v_1}{v_2}\right)^{k-1}$$

PROBLEM 4:

A rigid container with a volume of 0.5 m^3 contains oxygen at an initial pressure of 100 kPa and an initial temperature of 27°C. The oxygen is heated to a final temperature of 127°C. Determine the oxygen's:

(a) specific volume,

(b) final pressure,

(c) change of enthalpy, and

(d) change of entropy,

SOLUTION:

(a)

$$Pv = RT;\quad R = \frac{\overline{R}}{(\text{Mol.wt.})} = \frac{(8{,}314 \text{ J/kmol–K})}{32 \text{ kg/kgmol}} = 260\text{J/kg–K}$$

$$v = \frac{RT}{P} = \frac{(260 \text{ J/kg–K}) \times (27 + 273 \text{ K})}{100{,}000 \text{ N/m}^2} = 0.779 \text{ m}^3\text{/kg}$$

(b)

$$\frac{P_1 v_1}{T_1} = \frac{P_2 v_2}{T_2} \text{ and } v_1 = v_2$$

$$P_2 = \frac{T_2}{T_1} P_1 \text{ or } P_2 = \frac{T_2}{T_1} P_1 \text{ or } P_2 = \frac{(127 + 273 \text{ K})}{(27 + 273 \text{ K})} \times 100 \text{ kPa} = 133 \text{ kP}$$

(c)

$$\Delta h = c_P\,\Delta T = c_P\,(T_2 - T_1)$$

$$= 0.918 \text{ kJ/kg-K} \times (127 - 27 \text{ K}) = 91.8 \text{ kJ/kg}$$

(d)

$$\Delta s = c_p \ln\left(\frac{T_2}{T_1}\right) - R\ln\left(\frac{P_2}{P_1}\right)$$

$$= (0.918 \text{ kJ/kg-K})\ln\left(\frac{400K}{300K}\right) - (0.260 \text{ kJ/kg-K})\ln\left(\frac{133 \text{ kPa}}{100 \text{ kPa}}\right)$$

$$\Delta s = 0.189 \text{ kJ/kg-K}$$

FIRST LAW OF THERMODYNAMICS

First Law Analysis of a Closed System

$$Q - W = \Delta U \text{ or } q - w = \Delta u$$

Heat, Q, is energy transferred as a result of a temperature difference. Q is considered positive if it is inward to the system and negative if it is outward from the system.

Work, W, is considered positive if it is produced by the system and negative if it is added to the system. For a closed system, the *reversible* or ideal work is given by:

$$w_{rev} = \int P\,dv$$

In general, if a simple compressible substance expands in a closed system, work is *produced by* the system. If the substance is compressed, work is being *performed on* the system.

Special Cases

$$P = \text{constant},\; w_{rev} = P(v_2 - v_1)$$

$$v = \text{constant},\; w_{rev} = 0$$

Polytropic Process

A process is polytropic if the equation of state can be written in the form Pv^n = constant. Any simple compressible substance can behave in a polytropic manner.

$$w_{rev} = \frac{P_2 v_2 - P_1 v_1}{1-n}$$

$$n \neq 1$$

Ideal Gas Cases

For an ideal gas and a polytropic process:

$$w_{rev} = \frac{R(T_2 - T_1)}{1-n}$$

$$n \neq 1$$

For an ideal gas and an isothermal (constant temperature) process:

$$w_{rev} = P_1 v_1 \ln\frac{v_2}{v_1} = RT\ln\frac{P_1}{P_2} = RT\ln\frac{V_2}{V_1}$$

Open Thermodynamic Systems: First Law Energy Balance

Flow work is the energy input to a system when mass enters an open system and is the energy output from an open system when mass leaves a system. The flow work is handled by using enthalpy in the energy balance instead of internal energy.

$$\sum \dot{m}_i\left[h_i + \frac{V_i^2}{2\alpha} + gZ_i\right] - \sum \dot{m}_e\left[h_e + \frac{V_e^2}{2\alpha} + gZ_e\right] + \dot{Q}_{in} - \dot{W}_{net} = \frac{d(m_s u_s)}{dt}$$

where:

i = inlet

e = exit

$\dot{W}_{net}$ = rate of net work produced by the system

$\dot{m}$ = mass flow rate

m_s = fluid mass in the system

u_s = specific internal energy of fluid in system

$\dot{Q}$ = rate of heat transfer into the system

g = gravitational acceleration

Z = elevation

V = velocity

α = kinetic energy conversion factor
(1 for turbulent flow, 0.5 for laminar flow)

Steady-State Systems

Such a system is defined as a system that does not change its state with time. This assumption is valid for the steady operation of turbines, pumps, compressors, throttling valve nozzles, and heat exchangers including boilers and condensers.

$$\sum \dot{m}_i \left[h_i + \frac{V_i^2}{2\alpha} + gZ_i \right] - \sum \dot{m}_e \left[h_e + \frac{V_e^2}{2\alpha} + gZ_e \right] + \dot{Q}_{in} - \dot{W}_{net} = 0$$

$$\sum \dot{m}_i = \sum \dot{m}_e$$

For many open system problems, the kinetic and potential energy terms are ignored. Exceptions are listed below for special cases of the steady-flow energy equation. Also listed below are the isentropic efficiencies of some of the processes. This is the actual system performance compared to the maximum system performance predicted by the second law of thermodynamics.

Nozzles, Diffusers

Velocity terms are significant. Elevation changes are ignored and work and heat assumed to be zero. The steady-flow energy equation becomes:

$$h_i + \frac{V_i^2}{2} = {}_e + \frac{V_e^2}{2};\ V_e = \sqrt{2(h_i - h_e) + V_i^2}$$

$$\eta_s \text{ (nozzle)} = \frac{(h_i - h_e)}{(h_i - h_{es})}$$

h_{es} = enthalpy at an isentropic exit state.

Turbines, Pumps, and Compressors

These devices are often considered to operate adiabatically (i.e., without adding or losing heat). Kinetic and potential energy is also ignored. The work that is input or produced is most important.

$$h_i = h_e + w \text{ or } \dot{m}(h_i - h_e) = \dot{W}$$

$$\eta_s \text{ (turbine)} = \frac{h_i - h_e}{h_i - h_{es}}$$

$$\eta_s \text{ (pump, compressor)} = \frac{h_{es} - h_i}{h_e - h_i}$$

Throttling Valves

With a throttling valve it is normal to assume that there is no work, no heat transfer, and no kinetic or potential energy involved.

$$h_i = h_e$$

Boilers, Condensers, Evaporators, and One Side in a Heat Exchanger

Only heat transfer is significant for these devices. Work, kinetic and potential energies are ignored.

$$h_i + q = h_e \text{ or } \dot{Q} = \dot{m}(h_e - h_i)$$

Heat Exchangers

No heat or work as heat transfer occurs internally. No kinetic or potential energy is involved. There are two separate flow rates, $\dot{m}_e$ and $\dot{m}_i$:

$$\dot{m}_1(h_{1i} - h_{1e}) = \dot{m}_2(h_{2e} - h_{2i})$$

Mixing Devices, Separators, Open and Closed Feedwater Heaters

For these type devices:

$$\sum \dot{m}_i h_i = \sum \dot{m}_e h_e \text{ and } \sum \dot{m}_i = \sum \dot{m}_e$$

Compressors

Work is input to a compressor operating on an ideal gas. Kinetic and potential energy is ignored. Three cases are considered: isentropic (Pv^k = constant), polytropic (Pv^n = constant), and isothermal (Pv = constant).

Isentropic: $$w_{comp} = \frac{nR(T_1 - T_2)}{n-1} = \frac{nRT_1}{n-1}\left[1 - \left(\frac{P_2}{P_1}\right)^{\frac{(n-1)}{n}}\right]$$

Polytropic: $$w_{comp} = \frac{nR(T_1 - T_2)}{n-1} = \frac{nRT_1}{n-1}\left[1 - \left(\frac{P_2}{P_1}\right)^{\frac{(n-1)}{n}}\right]$$

Isothermal: $$w_{comp} = RT \ln \frac{P_1}{P_2}$$

Ideal Pump Work

Neglecting kinetic and potential energy, the reversible (ideal) work is given by the following:

$w_{rev} \equiv \int v dP = v(P_1 - P_2)$ for an incompressible substance.

$\eta_s = w_{rev} / w$

CYCLES

For a thermodynamic device to operate in a practical manner, it must operate continuously by following a cycle. A heat engine is a thermodynamic system that takes in energy from a high temperature source, rejects energy to a low temperature source, and produces work. Denoting Q_H as the energy input at T_H and Q_L as the energy rejected to T_L, the efficiency of a heat engine is given as follows:

$$\eta = \frac{W}{Q_H} = \frac{Q_H - Q_L}{Q_H}$$

The most efficient heat engine is the Carnot heat engine whose efficiency is given as follows:

$$\eta c = \frac{T_H - T_L}{T_H} = 1 - \frac{T_L}{T_H}$$

Note: T_H and T_L must be expressed in terms of absolute temperatures (K or °R).

The Rankine Cycle

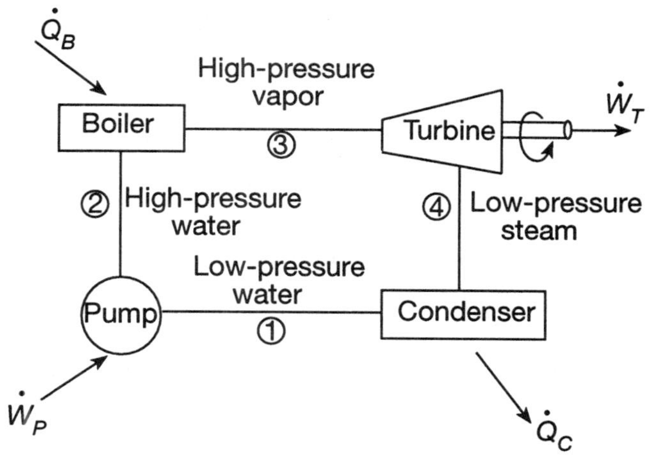

Figure 1. The Ideal Rankine Cycle

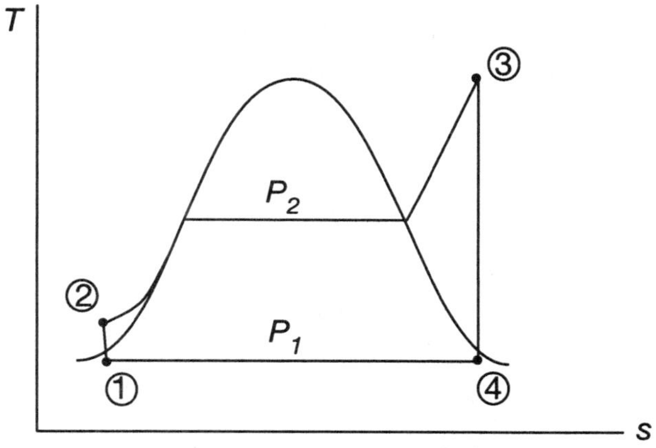

Figure 2. *T-s* Diagram for Ideal Rankine Cycle

Figure 1 shows the ideal Rankine Cycle and Figure 2 is the corresponding temperature-entropy (T-s) diagram. The process 3–4 through the turbine is assumed to be reversible and adiabatic and appears as a vertical line. State 1 is assumed to be a saturated liquid and the pump is assumed to be adiabatic. If we neglect kinetic and potential energy effects, the cycle efficiency is given by:

$$\eta_s = \frac{(h_3 - h_4) - (h_2 - h_1)}{h_3 - h_2}$$

PROBLEM 5:

Steam enters the turbine of an ideal Rankine Cycle at a pressure of 0.6 MPa and a temperature of 300°C and exits at a pressure of 70.14 kPa. Determine the efficiency of the cycle.

SOLUTION:

The cycle efficiency is given by the following where the h's must be determined:

$$\eta_s = \frac{(h_3 - h_4) - (h_2 - h_1)}{h_3 - h_2}$$

$h_3 = 3{,}061.6$ kJ/kg   Superheated; $P = 0.6$ MPa, $T = 300$ °C

$s_3 = 7.3724$ kJ/kg-K

$s_2 = s_3$

$P_2 = 70.14$ kPa

$s_f = 1.1925$ kJ/kg-K

$s_g = 7.4791$ kJ/kg-K, therefore state 2 is 2-phase and the quality x is required.

$$x = \frac{s_2 - s_f}{s_g - s_f}$$

$$= \frac{7.3724 - 1.1925}{7.4791 - 1.1925} = 0.983$$

$h_2 = h_{f2} + x_2\, h_{fg2} = 376.92 + (.983)\,(2{,}283.2) = 2{,}620$ kJ/kg

$h_1 = h_f$ at 70.41 kPa; $h_1 = 376.92$ kJ/kg

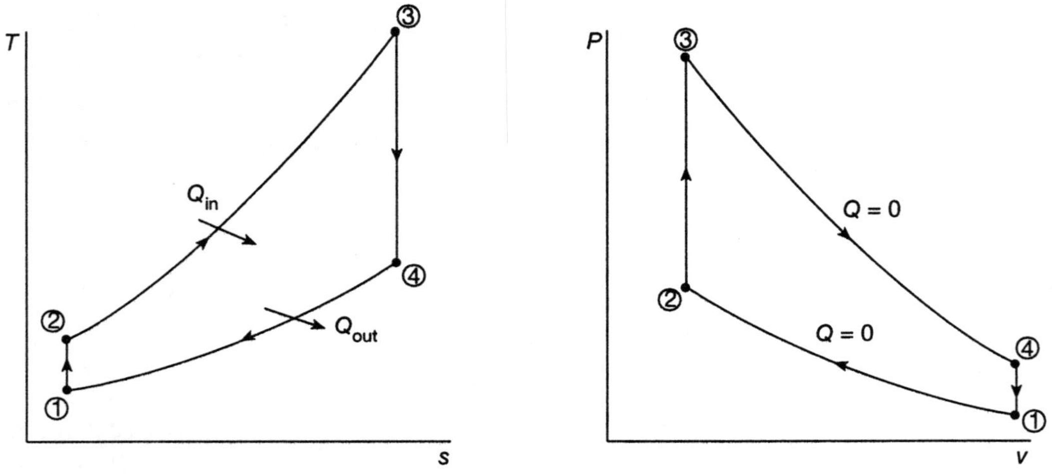

Figure 3. The Otto Cycle

The pump work is given by $v_1(P_2 - P_1)$ since the pump is assumed to be isentropic.

$$w_p = 0.001036 \text{ m}^3/\text{kg } (600 - 70.14)\text{kPa} = 0.549 \text{ kJ/kg} = h_2 - h_1$$

$$h_2 = h_1 + w_p$$

$$h_2 = 376.92 + 0.549 = 377.5 \text{ kJ/kg}$$

$$\eta_s = \frac{(3{,}061.6 - 2{,}620) - (0.549)}{3{,}061.6 - 377.5} = 0.164$$

The Otto Cycle

The Otto Cycle is the ideal cycle on which the internal combustion engine operates. In the air-standard cycle shown in Figure 3, the working substance is assumed to be air. The efficiency of the Otto cycle is given as follows:

$$\eta_s = 1 - r^{1-k}; \; r = \frac{v_1}{v_2}$$

Reverse or Refrigeration Cycle

The purpose of the reverse or refrigeration cycle is to transfer heat from a lower temperature to a higher temperature. A refrigerator and a heat pump are examples of reversed cycles.

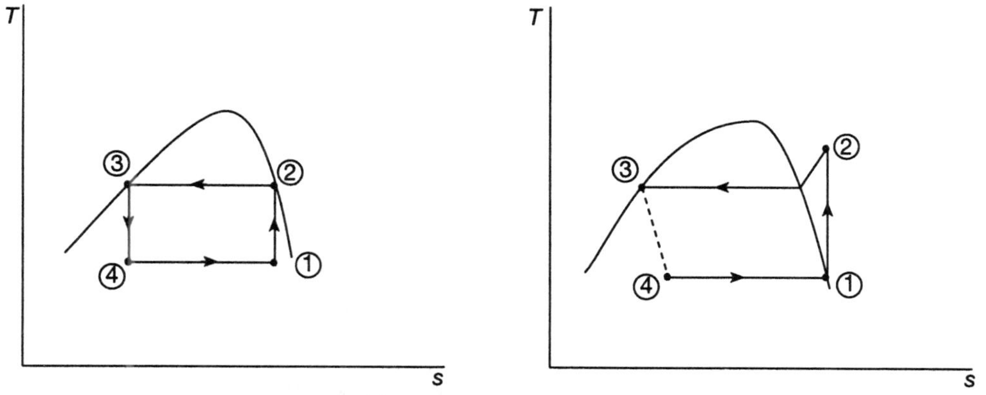

Figure 4. Refrigeration Cycle

The ideal reversed cycle is shown in Figure 4. The performance of reversed cycles is measured by the coefficient of performance, *COP*, which is the ratio of energy transfer to energy input.

$$COP_{ref} = \frac{h_1 - h_4}{h_2 - h_1}$$

$$COP_{HP} = \frac{h_2 - h_3}{h_2 - h_1}$$

PROBLEM 6:

An ideal heat pump cycle using refrigerant R-134a is operating at an upper pressure and temperature of 1 MPa and 100°C and a lower pressure of 0.2 MPa. Determine the *COP* and the mass flow rate required to transfer energy at the rate of 40 kW. Estimate the quality of point 4.

SOLUTION:

The enthalpies of the various state points must be determined. The Mollier (P–v) diagram for R-134a will be used to find the properties.

$h_2 = 482$ kJ/kg from the superheated region of the Mollier diagram

$s_2 = s_1 = 1.90$ kJ/kg-K

By following the line of constant entropy back to 0.2 MPa:

$h_1 = 440$ kJ/kg

$h_3 =$ saturation liquid at 1.0 MPa = 260 kJ/kg

$h_4 = h_3 = 260$ kJ/kg

$$COP_{HP} = \frac{h_2 - h_3}{h_2 - h_1} = \frac{483 - 260}{482 - 440} = 5.3$$

$\dot{Q}_{2\text{-}3} = \dot{m}(h_2 - h_3)$; 40 kJ/s $= \dot{m}(482 - 260)$ kJ/kg

$\dot{m} = 0.18$ kg/s = 649 kg/h

Since $h_3 = h_4$, point 4 is located by dropping a vertical line from point 3 to 0.2 MPa. By inspection the quality of point 4 is 0.35.

SECOND LAW OF THERMODYNAMICS

A thermal energy reservoir is an idealized mass that receives or gives up energy with no change in temperature. As a result:

$$\Delta S_{reservior} = \frac{Q}{T_{reservoir}}$$

where Q is measured with respect to the reservoir (i.e., it is positive if the flow is into the reservoir).

Kelvin-Planck Statement of the Second Law—No heat engine can operate in a cycle while transferring heat with a single heat reservoir. Application of the Kelvin-Planck statement leads to the following Corollary: No heat engine can have a higher efficiency than a Carnot cycle operating between the same reservoirs.

Clausius Statement of the Second Law—No refrigerator or heat pump can operate without a net work input. Application of the Clausius statement leads to the following Corollary: No refrigerator or heat pump can have a higher coefficient of performance (*COP*) than a Carnot cycle refrigerator or heat pump.

ENTROPY

The concept of entropy is centered about the principle of reversibility. A thermodynamic process is either reversible or irreversible. A reversible process is an ideal process; one that does not exist in nature, but represents the theoretical limit that is possible. An analogy from mechanics is the frictionless pulley system that gives optimum performance. Actual systems though are modeled using these ideal systems and by determining an efficiency that is related to the theoretical limit.

Entropy is a thermodynamic property that is defined as follows:

$$ds = \frac{1}{T}\,\delta Q_{rev}$$

which upon integration yields:

$$s_2 - s_1 = \int \frac{1}{T}\,\delta Q_{rev}$$

For tabulated substances such as water, the entropy values can be found in tables or charts. For ideal gases, the change in entropy can be found from the following equations:

$$s_2 - s_1 = c_v \ln \frac{T_2}{T_1} + R \ln \frac{v_2}{v_1}$$

$$s_2 - s_1 = c_p \ln \frac{T_2}{T_1} + R \ln \frac{P_2}{P_1}$$

For an isothermal reversible process: $\Delta s = s_2 - s_1 = Q/T$

For an isentropic process: $\Delta s = 0$; $ds = 0$

By definition, a process that is reversible and adiabatic is isentropic.

For an irreversible, adiabatic process: $\delta Q = 0$; $\Delta s > 0$

This is an important consequence of the second law since it determines if a process is possible. An *adiabatic* process that results in a *decrease* in entropy is impossible!

Increase of Entropy Principle—An isolated system is defined as a "system plus its surroundings." The Second Law requires that the entropy of an isolated system must increase or in the limit of a reversible process remain the same. This is why hot water cools and ice melts when exposed to room temperatures. The first law would allow the ice to get colder or the hot water to get hotter, but these occurrences would be contrary to the second law.

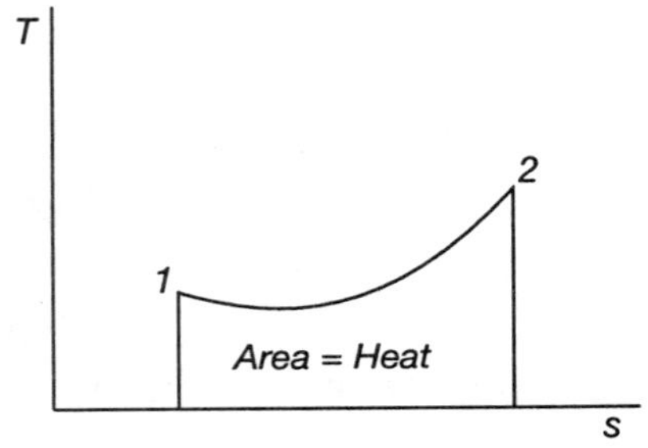

Figure 5. Temperature-Entropy Diagram

$$\Delta s_{total} = \Delta s_{system} + \Delta s_{surroundings} \geq 0$$

$$\Delta s_{total} = \sum \dot{m}_{out} s_{out} = \sum \dot{m}_e s_e - \sum \left(\frac{\dot{Q}_{external}}{T_{external}} \right) \geq 0$$

The temperature-entropy diagram has two fundamental uses in thermodynamics. For a reversible process, the area under the curve represents the heat transfer. For an irreversible process, the departure of actual processes from reversible processes can be shown.

$$Q_{rev} = \int T ds$$

Entropy Change for Solids and Liquids

$$ds = c\frac{dT}{T}$$

$$s_2 - s_1 = \int c\frac{dT}{T} = c_{mean}\ln\left(\frac{T_2}{T_1}\right)$$

Irreversibility:

$$I = W_{rev} - W_{actual}$$

Closed-system Availability:

$$\Phi = (u - u_0) - T_0(s - s_0) + P_0(v - v_0)$$

$$w_{reversible} = \Phi_1 - \Phi_2$$

Open-system Availability:

$$\Psi = (h - h_0) - T_0(s - s_0) + \frac{V^2}{2} + gz$$

$$w_{reversible} = \Psi_1 - \Psi_2$$

PROBLEM 7:

A manufacturer makes the following claims concerning the operation of a steam turbine: under adiabatic conditions, steam will enter the turbine at an absolute pressure of 3 MPa and 350°C and exit as a saturated vapor at 75 kPa. If the mass flow rate of the steam is 1 kg/s:

(a) prove that the above turbine can operate as advertised,

(b) determine the power output of the turbine in kW, and

(c) determine the isentropic efficiency of the turbine.

SOLUTION:

Since state 1 and state 2 are known, their properties can be determined from the steam tables.

State 1: $P = 3$ MPa, $T = 350$°C, superheated; $h_1 = 3{,}115.3$ kJ/kg, $s_1 = 6.7428$ kJ/kg-K

State 2: saturated vapor at 75 kPa; $h_2 = 2{,}663.0$ kJ/kg, $s_2 = 7.4564$

(a) $\Delta s = s_2 - s_1 = 7.4564 - 6.7428 = 0.7136$. The process is *possible* since the change of entropy is positive and the process is adiabatic.

(b) $\dot{W} = \dot{m}\,(h_1 - h_2) = (1 \text{ kg/s})(3{,}115.3 - 2{,}663) \text{ kJ/kg} = 452.3 \text{ kW}$

(c) The isentropic efficiency of a turbine is the ratio of the actual to reversible work. To find the reversible work, the new outlet condition must be determined. The reversible work for an adiabatic process results in an isentropic (constant entropy process). $s_2 = s_1 = 6.7428$ and since $s_2 < s_g$ at 75 kPa, the new state 2, 2*s*, is in the liquid-vapor region. The quality *x* must be found.

$$x_{2s} = \frac{s_{2s} - s_f}{s_{fg}};$$

the values of s_f and s_{fg} are found in the saturated tables at 75 kPa

$$x_{2s} = \frac{6.7428 - 1.2130}{6.2434} = 0.886$$

$h_{2s} = h_f + x_{2s}\,h_{fg,}$ where h_f and h_{fg} are again found in the saturated tables at 75 kPa

$$h_{2s} = 384.39 + (0.886)(2{,}278.6) = 2{,}403 \text{ kJ/kg}$$

$$\eta_s = \frac{w_{act}}{w_{rev}} = \frac{h_1 - h_2}{h_1 - h_{2s}} = \frac{(3{,}115.3 - 2{,}663)}{(3{,}115.3 - 2{,}403)} = 0.63 \text{ or } 63\%$$

PROBLEM 8:

A cylinder-piston apparatus contains 0.5 kg of air initially at 27°C and 100 kPa. The air is compressed polytropically to a final pressure of 500 kPa and final temperature of 127°C.

(a) Find the initial and final total volume of the air.

(b) Find the polytropic exponent *n* for this process.

(c) Find the change of internal energy of the air.

(d) Find the work done on or by the air as a result of this process.

(e) Find the heat transfer into or out of the air during the process.

(f) Find the change of entropy of the air.

(g) If the surroundings are at a temperature of 25°C, what will be the total change of entropy (system + surroundings).

(h) Based on part (g), was this process reversible, irreversible, or impossible?

SOLUTION:

Assume air to be an ideal gas with constant specific heats.

$R = 0.287$ kJ/kg-K or

0.287 kPa-m³/kg-K

$c_P = 1.00$ kJ/kg-K

$c_v = 0.718$ kJ/kg-K

$T_1 = 27°\text{C} = 300$ K

$T_2 = 127°\text{C} = 400$ K

(a) $PV = mRT; \quad V_1 = \dfrac{mRT_1}{P_1}$

$= (0.5 \text{ kg})(0.287 \text{ kPa–m}^3/\text{kg–K})(300 \text{ K})/(100 \text{ kPa}) = 0.431 \text{ m}^3$

$$V_2 = \frac{mRT_2}{P_2}$$

$= (0.5 \text{ kg})(0.287 \text{ kPa–m}^3/\text{kg–K})(400 \text{ K})/(500 \text{ kPa}) = 0.115 \text{ m}^3$

(b) For a polytropic process, $P_1V_1{}^n = P_2V_2{}^n$

Solving for n:

$$n = \frac{\ln(P_1 / P_2)}{\ln(V_2 / V_1)} = \frac{\ln(100 / 500)}{\ln(.115 / .431)} = 1.22$$

(c) $\Delta U = mc_v\Delta T = (0.5 \text{ kg})(0.718 \text{ kJ/kg-K})[(400 - 300)\text{K}] = 35.9 \text{ kJ}$

(d) For a polytropic process:

$$W = \frac{P_2V_2 - P_1V_1}{1-n} = \frac{(500)(0.115)-(100)(0.431)}{1-1.22} = -65.5 \text{ kJ}$$

(e) Using a first law energy balance:

$Q - W = \Delta U; \; Q = W + \Delta U = -65.5 + 35.9 = -29.6 \text{ kJ}$

(f) For an ideal gas with constant specific heats:

$$s_2 - s_1 = c_P \ln\frac{T_2}{T_1} + R\ln\frac{v_2}{v_1}$$

$= (1.00 \text{ kJ/kg-K}) \ln[(400/300)] + (0.287 \text{ kJ/kg-K})$

$\ln[(0.115)/(0.431)]$

$= -0.0915 \text{ kJ/kg-K}$

Note: $\frac{v_2}{v_1} = \frac{V_2}{V_1}$; $\Delta S = m\Delta s = -0.0457$ kJ/K

(g) $\Delta S_{Total} = \Delta S_{Sys} + \Delta S_{Surr}$; $\Delta S_{Surr} = Q_{Surr} / T_{Surr}$; Q_{Surr}

$= -Q_{Sys} = +29.6$ kJ; $T_{Surr} = 298$ K

$\Delta S_{Total} = (-0.0457 \text{ kJ/K}) + (29.6 \text{ kJ})/(298 \text{ K}) = +0.054 \text{ kJ/K}$

(h) Since $\Delta S_{Total} > 0$, the process was irreversible.

PROBLEM 9:

Steam enters a turbine with a pressure of 3 MPa, a temperature of 400°C , and a velocity of 160 m/s. The steam exits as a saturated vapor at 100°C with a velocity of 100 m/s. At steady-state conditions, the turbine develops work at the rate of 540 kJ/kg. Assuming that T_0 = 25°C and P_0 = 1 atm, and neglecting changes in potential energy, determine:

(a) the stream availability entering the turbine, Ψ_1,

(b) the stream availability leaving the turbine, Ψ_2,

(c) the rate of heat loss from the turbine,

(d) the reversible work, w_{rev}, and

(e) the irreversibility.

SOLUTION:

The initial, final, and dead state properties must be found using the steam tables.

State 1: $P_1 = 3$ MPa, $T_1 = 400$°C (superheated); $h_1 = 3{,}230.9$ kJ/kg,

$s_1 = 6.9212$ kJ/kg-K

State 2: Saturated vapor at 100°C; $h_2 = 2{,}676.1$ kJ/kg,

$s_2 = 7.3549$ kJ/kg-K

Dead State: $h_0 = h_{f\,@\,25°C} = 104.89$ kJ/kg, $s_{f\,@\,25°C} = 0.3674$ kJ/kg-K

The stream availability function (neglecting potential energy) is:

$$\Psi = (h - h_0) - T_0\,(s - s_0) + \frac{V^2}{2}$$

(a) $$\Psi_1 = (h_1 - h_0) - T_0\,(s_1 - s_0) + \frac{V_1^2}{2}$$

$$= (3{,}230.9 - 104.89) - 298\text{ K}\,(6.9212 - 0.3674) + 160^2/(2 \times 10^3)$$

$$= 1{,}186\text{ kJ/kg}$$

(b) $$\Psi_2 = (h_2 - h_0) - T_0\,(s_2 - s_0) + \frac{V_2^2}{2}$$

$$= (2{,}676.1 - 104.89) - 298\text{ K}\,(7.3549 - 0.3674) + 100^2/(2 \times 10^3)$$

$$= 493.9\text{ kJ/kg}$$

(c) First law energy balance on system:

$$q - w = h_2 - h_1 + \frac{V_2^2 - V_1^2}{2}$$

$$q - 540 = (2{,}676.1 - 3{,}230.9) + (100^2 - 160^2)/(2 \times 10^3)$$

$$q = -22.6\text{ kJ/kg}$$

(d) $w_{rev} = \Psi_1 - \Psi_2 = 1{,}186 - 493.9 = 692$ kJ/kg

(e) $I = w_{rev} - w_{act} = 692 - 540 = 152$ kJ/kg

Note: The 10^3 factor used in the above solution is required to convert J to kJ.

Table 3. Specific Heat (at Room Temperature)

Substance	Mol wt	c_P kJ/kg-K	c_P Btu/lbm-°R	c_V kJ/kg-K	c_V Btu/lbm-°R	k
Gases						
Air	29	1.00	0.240	0.718	0.171	1.40
Argon	40	0.520	0.125	0.312	0.0756	1.67
Butane	58	1.72	0.414	1.57	0.381	1.09
Carbon dioxide	44	0.846	0.203	0.657	0.158	1.29
Carbon monoxide	28	1.04	0.249	0.744	0.178	1.40
Ethane	30	1.77	0.427	1.49	0.361	1.18
Helium	44	5.19	1.25	3.12	0.753	1.67
Hydrogen	2	14.3	3.43	10.2	2.44	1.4
Methane	16	2.25	0.532	1.74	0.403	1.30
Neon	20	1.03	0.246	0.618	0.148	1.67
Nitrogen	28	1.04	0.248	0.743	0.177	1.40
Octane vapor	114	1.71	0.409	1.64	0.392	1.04
Oxygen	32	0.918	0.219	0.658	0.157	1.40
Propane	44	1.68	0.407	1.49	0.362	1.12
Steam	18	1.87	0.445	1.41	0.335	1.33

Table 3. Specific Heat (cont.)
(at Room Temperature)

Substance	c_P		Density	
	kj/kg-K	Btu/lbm-°R	kg/m^3	lbm/ft^3
	Solids			
Ammonia	4.80	1.146	602	38
Mercury	0.139	0.033	13,560	847
Water	4.18	1.000	997	62.4
	Solids			
Aluminum	0.900	0.215	2,700	170
Copper	0.386	0.092	8,900	555
Ice (0°C, 32°F)	2.11	0.502	917	57.2
Iron	0.450	0.107	7,840	490
Lead	0.128	0.030	11,310	705

FE/EIT

FE: PM Mechanical Engineering Exam

Practice Test 1

FUNDAMENTALS OF ENGINEERING EXAMINATION

TEST 1

(Answer sheets appear in the back of this book.)

TIME: 4 Hours
60 Questions

DIRECTIONS: For each of the following questions and incomplete statements, choose the best answer from the four answer choices. You must answer all questions.

1. What is the function of a macro in a spreadsheet?

   (A) To do repetitive operations

   (B) It is required for large spreadsheets.

   (C) It must precede any table larger than 10×10.

   (D) It is not used in a spreadsheet.

2. For the following programming segment, the value of S will be equal to:

   INPUT N
   $S = 0; P = 1$
   FOR $I = 1$ TO N
   $P = P \times I$
   $S = S + 1/P$
   NEXT I

   (A) $1 + \frac{1}{1!} + \frac{1}{2!} + \frac{1}{3!} + \ldots + \frac{1}{n!}$

(B) $1+\frac{1}{2}+\frac{1}{3}+\frac{1}{4}+\ldots+\frac{1}{n}$

(C) $1+\frac{1}{2!}+\frac{1}{3!}+\ldots+\frac{1}{n!}$

(D) $S=\frac{1}{1-n}$

3. $^{1}/_{3} \times 3$ yields a value of 0.99999999 on a computer that is capable of storing data to eight significant figures. Could this present problems in engineering computations?

   (A) No, since three significant figures is the acceptable accuracy of most engineering calculations.

   (B) No, since most modern computers are programmed to spot this kind of problem and would correctly return the value of 1.0000000.

   (C) Yes, since most engineering calculations require at least eight significant figures.

   (D) Yes, if the computer is being used to perform iterations, the errors will become cumulative and could become quite significant.

Questions 4–6 refer to the following situation.

An air compressor operates at steady-state with air entering at $T_1 = 25°\text{C}$, $P_1 = 100$ kPa and exiting at $P_2 = 500$ kPa. The air undergoes a reversible polytropic process with $n = 1.3$. Make the appropriate assumptions using the cold air standard for ideal gases.

4. The work output of the compressor is closest to

   (A) −167 kJ/kg.

   (B) +167 kJ/kg.

   (C) −79.9 kJ/kg.

   (D) −14.0 kJ/kg.

5. The heat transfer per unit mass passing through the compressor is closest to

   (A) +33.0 kJ/kg.

   (B) −33.0 kJ/kg.

   (C) −120 kJ/kg.

   (D) +120 kJ/kg.

6. What would be the required work if an ideal two-stage compressor with intercooling is used?

   (A) –80 kJ/kg (C) –152 kJ/kg

   (B) +80 kJ/kg (D) +152 kJ/kg

Questions 7–10 relate to the beam pictured below.

The loads to which the beam is subjected and the beam's cross section are shown. The moment of inertia about the neutral axis (*NA*) equals:

$$I_{NA} = 1.46 \times 10^{-4}\ \text{m}^4$$

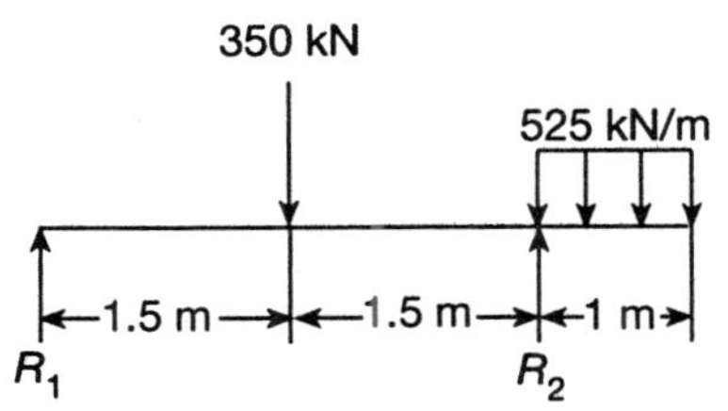

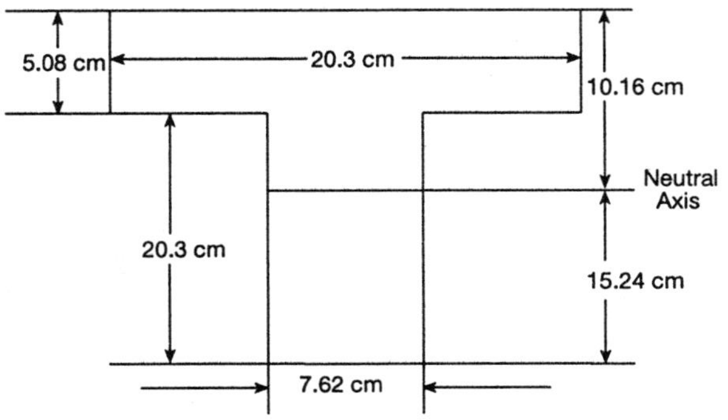

7. The maximum transverse shear force in the beam is nearly

   (A) 100 kN. (C) 525 kN.

   (B) 225 kN. (D) 825 kN.

8. The maximum stress (tensile or compressive) in the beam is

   (A) 172 MPa. (C) 472 MPa.

   (B) 271 MPa. (D) 872 MPa.

9. At a point in the beam where the fiber stress is 55 MPa tension and the transverse shearing stress is 21 MPa, the maximum normal stress (principle stress) is

   (A) 34 MPa. (C) 62.1 MPa.

   (B) 55 MPa. (D) 76 MPa.

10. If the beam is made of nodular cast iron having a modulus of elasticity E = 172 GPa, the deflection at the 350 kN load is

(A) 0.192 cm up.
(B) 0.192 cm down.
(C) 1.92 cm down.
(D) zero.

Questions 11–13 refer to the following situation.

The figure below represents an example of a low-head hydroelectric generation plant. Two constant height reservoirs are connected by 30.5 m of riveted steel pipe (e = 1.82 mm). A 0.61 m gate hinged at the top is used to control the water flowing through the turbine. The water temperature is 17°C and the barometric pressure is 724 mm Hg. Assume the kinematic viscosity, ν, to be 1.08×10^{-6} m²/s.

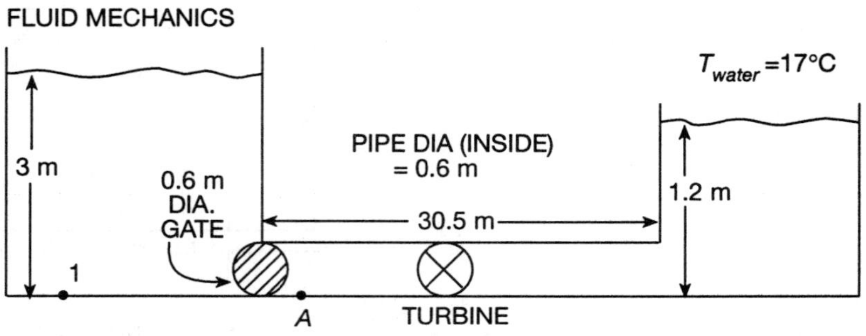

11. The force F required at the bottom of the gate to open it (assuming the pipe is not present and there is only atmospheric air) is closest to

(A) 3.86 kN.
(B) 3.97 kN.
(C) 7.72 kN.
(D) 10 kN.

12. If the pipe was severed just downstream from the gate and the gate was completely open, the outlet velocity of the water would be

(A) 2.67 m/s.
(B) 5.67 m/s.
(C) 7.67 m/s.
(D) 13.9 m/s.

13. If the volume flow rate, $\dot{Q}$, through the pipe is 17 m³/min, the horsepower output of the turbine is

(A) 2.76 kW.
(B) 4.77 kW.
(C) 6.76 kW.
(D) 8.76 kW.

Questions 14–16 refer to the sketch below.

A bowling ball with a mass of 7.26 kg is free to slide down a frictionless, curved surface as shown. The ball at position 1 has a velocity of 3 m/s and upon leaving the track impacts a block having a mass of 14.5 kg. The block is connected to a spring having a spring constant k = 1,750 N/cm. It is observed that after impact with the block, the bowling ball comes completely to rest.

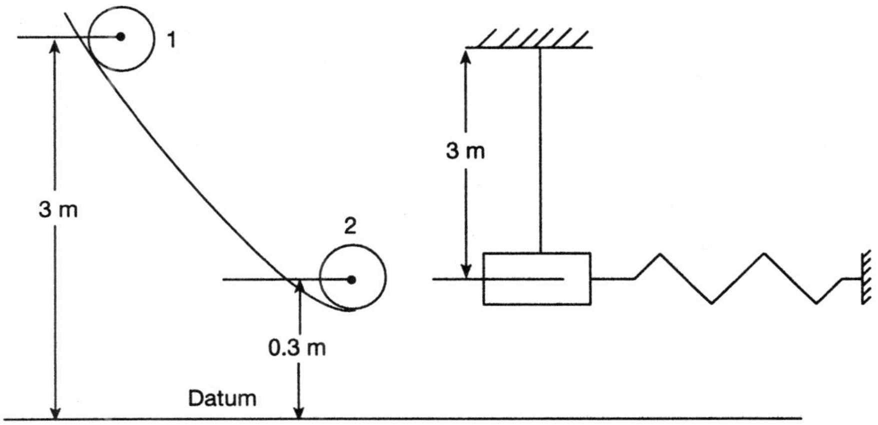

14. The velocity of the bowling ball immediately after impact is

    (A) 0.5 m/s.
    (B) 3.94 m/s.
    (C) 7.86 m/s.
    (D) 15.7 m/s.

15. If the spring is initially compressed 5 cm from its unstretched length, the total deflection of the spring is

    (A) 0.5 cm.
    (B) 2.5 cm.
    (C) 6.2 cm.
    (D) 7.5 cm.

16. If the spring were missing, the tension in the cable immediately after impact is

    (A) 14.5 N.
    (B) 142 N.
    (C) 185 N.
    (D) 217 N.

Questions 17–19 refer to the following situation.

It is desired to heat cold water at a pressure 600 kPa from 4°C to 49°C at a rate of 0.9 kg/s in a thin-walled copper tube (*ID* = 26.04 mm, *OD* = 28.58 mm) by condensing waste process steam at 276 kPa gage on the outside of the tube.

17. The convection coefficient inside the tube is

    (A) 2,300 W/m^2-K.
    (B) 5,300 W/m^2-K.
    (C) 7,300 W/m^2-K.
    (D) 9,300 W/m^2-K.

18. The convection coefficient outside the tube is closest to

    (A) 2,700 W/m^2-K.
    (B) 4,700 W/m^2-K.
    (C) 6,300 W/m^2-K.
    (D) 8,700 W/m^2-K.

19. The length of a single tube that would accomplish this temperature rise is

    (A) 5.2 m.
    (B) 16.3 m.
    (C) 163 m.
    (D) 263 m.

Questions 20–22 refer to a vapor-compression refrigeration cycle using HFC-134a as the refrigerant.

The evaporator temperature is –10°C and the condenser temperature is 50°C.

20. If the vapor entering the compressor is saturated and the compression process is adiabatic, the minimum work required is

    (A) 15 kJ/kg.
    (B) 28 kJ/kg.
    (C) 39 kJ/kg.
    (D) 50 kJ/kg.

21. If the refrigerant leaves the condenser as a saturated liquid, the COP (coefficient of performance) of the cycle is

    (A) 0.5.
    (B) 1.5.
    (C) 4.3.
    (D) 6.2.

22. If this system is to provide two tons of refrigeration, the required mass flow rate is

    (A) 1.5 kg/min. (C) 3.5 kg/min.

    (B) 2.5 kg/min. (D) 6.0 kg/min.

Questions 23–25 refer to the following situation.

An air-conditioning system is to take in outdoor air at 40°F and 80% relative humidity at a steady rate of 500 ft^3/min and to condition it to 100°F and 60% relative humidity. The outdoor air is first heated to 80°F in the heating section and then heated and humidified by the injection of steam.

23. The heat added in the heating section is closest to

    (A) 4,780 Btu/hr. (C) 8,780 Btu/hr.

    (B) 5,780 Btu/hr. (D) 21,900 Btu/hr.

24. The mass flow rate of steam required for humidification is

    (A) 49.6 lbmv/hr. (C) 109 lbmv/hr.

    (B) 89.6 lbmv/hr. (D) 509 lbmv/hr.

25. The enthalpy of the required steam to accomplish this heating and humidification process is

    (A) 1,200 Btu/lbmv. (C) 1,500 Btu/lbmv.

    (B) 1,300 Btu/lbmv. (D) 1,700 Btu/lbmv.

Questions 26 and 27 refer to the following situation.

A piece of mechanical equipment is to be supported at the top of a 5-inch nominal diameter steel pipe. The equipment has a mass of 2,500 kg. The base of the pipe will be anchored in a concrete pad while the top end will be unsupported. Use a factor of safety (S.F.) of 2.5 against buckling.

26. The moment of inertia of the column is

    (A) 235 cm^4. (C) 635 cm^4.

    (B) 435 cm^4. (D) 935 cm^4.

27. The maximum allowable height of the column is

(A) 5.15 m.
(B) 7.15 m.
(C) 9.15 m.
(D) 11.15 m.

Questions 28–30 pertain to the following situation.

A pump is used to move 10,000 kg/hr of water into a boiler. The water enters the pump at 30°C and 100 kPa and leaves at 30°C and 1 MPa. The water leaves the boiler at 600°C and 1 MPa. Both the pump and boiler are 80% efficient. The boiler is fired with natural gas, which has a heating value of 106 MJ/therm.

28. The work required to operate the pump is

(A) 1.13 kW.
(B) 2.13 kW.
(C) 3.13 kW.
(D) 4.13 kW.

29. The total heat transfer to the working fluid that occurs in the boiler is

(A) 9.92 kW.
(B) 9.92 MW.
(C) 19.2 kW.
(D) 19.2 MW.

30. The number of therms per hour required to fire the boiler is

(A) 121 therms/hr.
(B) 221 therms/hr.
(C) 321 therms/hr.
(D) 421 therms/hr.

Questions 31–33 pertain to the following situation.

An inherent problem with all temperature measuring sensors is that the sensor can only measure its own temperature. The reading of the sensor must be carefully evaluated to obtain reasonable results.

A large duct conveys a high-temperature stream of air at 4.6 m/s. The duct wall is at a temperature 100°C and a thermocouple with emissivity equal to 0.6 is placed in the middle of the duct. At a given instance, the thermocouple reads a temperature of 540°C. The thermocouple can be assumed to behave like a 0.6 cm cylinder in cross-flow and conduction up the thermocouple wire can be neglected. The following data applies to these situations:

$Re = 314$ (Reynolds number based on diameter of thermocouple)
$\nu = 88 \times 10^{-6}$ m²/s (kinematic viscosity of air in duct)
$k = 0.060$ W/m-K (thermal conductivity of air in duct)
$Pr = 0.716$ (Prandtl number of air in duct)

31. The convection heat transfer coefficient on the cylinder is

(A) 20 W/m^2-K.
(B) 40 W/m^2-K.
(C) 80 W/m^2-K.
(D) 90 W/m^2-K.

32. The actual gas temperature in this situation is

(A) 600°C.
(B) 700°C.
(C) 800°C.
(D) 900°C.

33. If a radiation shield could be installed between the thermocouple and the wall to reduce radiation losses by 90% and the thermocouple again reads 540°C, the temperature of the gas stream would be closest to

(A) 550°C.
(B) 630°C.
(C) 750°C.
(D) 830°C.

Questions 34–36 pertain to the following situation.

A horizontal high-pressure steam pipe is composed of 2-inch nominal Schedule 40 piping. The pipe passes through a large room whose wall and air temperatures are at 300 K. The steam is condensing at a pressure of 1 MPa as it passes through the pipe. Assume that the condensation heat transfer coefficient inside the pipe, h_1 to be 10,000 W/m^2-K, the free convection coefficient on the outside, h_2, to be 8.2 W/m^2-K, and the surface emissivity of the pipe to be 0.85.

34. The thermal resistance for a 1 meter length of pipe is

(A) 0.000424 K/W.
(B) 0.00724 K/W.
(C) 0.0124 K/W.
(D) 0.824 K/W.

35. The combined convection and radiation coefficient on the outer surface of the pipe is

(A) 8.2 W/m^2-K. (C) 18.9 W/m^2-K.

(B) 10.7 W/m^2-K. (D) 110 W/m^2-K.

36. The total heat loss from the 1 meter section of the pipe is closest to

(A) 150 W (C) 550 W

(B) 350 W (D) 950 W

Questions 37–39 pertain to the following situation.

The system shown represents a reservoir and associated piping and pumping equipment. The friction factor, *f*, is assumed to be 0.02. The entrance to the piping is noted as being a "sharp entrance." The remaining piping system has a loss coefficient, *C*, equal to 2.5.

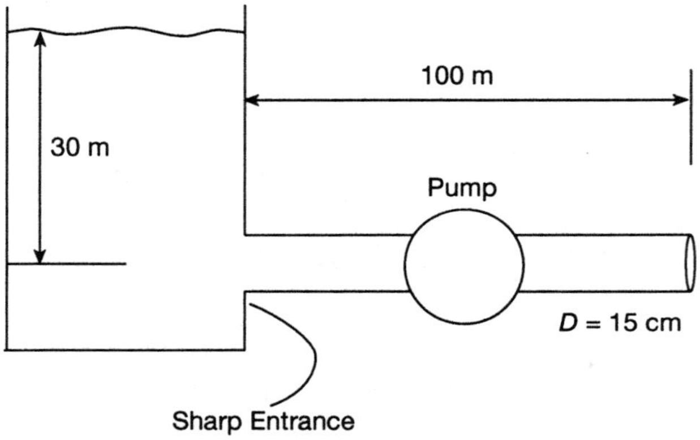

37. With the pump not operating, the volume flow rate through the pipe is

(A) 0.05 m^3/s. (C) 0.1 m^3/s.

(B) 0.08 m^3/s. (D) 0.2 m^3/s.

38. The minimum pumping power required to maintain a flow of 0.2 m^3/s is

(A) 53 kW. (C) 123 kW.

(B) 83 kW. (D) 162 kW.

39. The equivalent length of the minor losses in the piping system is

(A) 22.5 m.

(B) 42.5 m.

(C) 122.5 m.

(D) 1,020 m.

Questions 40–42 pertain to the following situation.

The cylinder pictured below rests on a horizontal surface. The coefficient of friction between the cylinder and surface is 0.5. It is attached to a block by a cord and a frictionless pulley. The cylinder has a radius of gyration of 400 mm and a mass of 70 kg. The block has a mass of 35 kg.

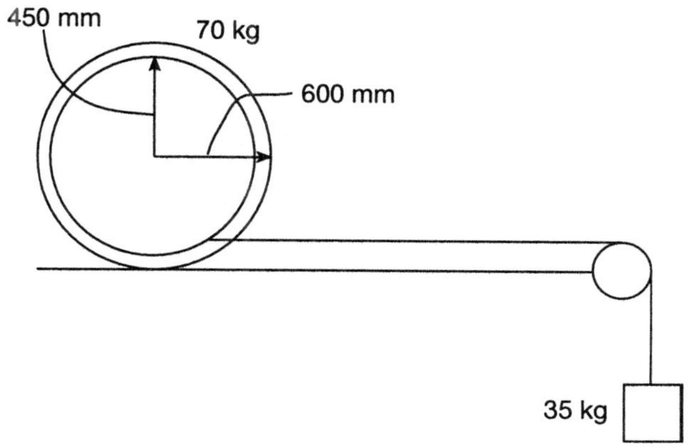

40. The mass moment of inertia of the cylinder about its center is

(A) 11.2 kg-m^2.

(B) 132 kg-m^2.

(C) 1,520 kg-m^2.

(D) 2,120 kg-m^2.

41. If the cylinder is assumed to roll without slipping, which of the following describes the relationship between the angular acceleration of the cylinder, α, and the acceleration of the block, a_1?

(A) $a_1 = 0.45 \times \alpha$

(B) $a_1 = 0.15 \times \alpha$

(C) $a_1 = 0.6 \times \alpha$

(D) $a_1 = \alpha$

42. The acceleration of the block, a_1, is closest to

(A) 0.5 m/s^2.

(B) 1.0 m/s^2.

(C) 1.5 m/s^2.

(D) 2.0 m/s^2.

Questions 43–45 pertain to the following situation.

The cylinder-piston shown below contains 0.1 m^3 of nitrogen at 300 kPa and 77°C. An electric heater located in the system passes a current of 1 amp through a potential of 50 volts for a period of 4 minutes. During this time the nitrogen expands polytropically, with n = 1.1, until the initial volume doubles.

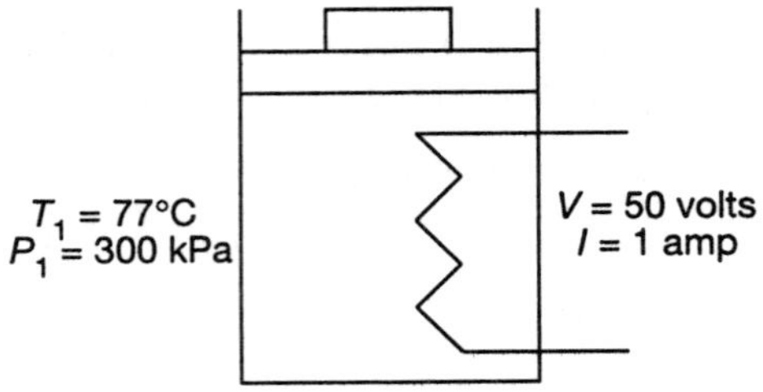

43. The boundary work done on or by the system is

(A) 10 kJ. (C) −10 kJ.

(B) 20 kJ. (D) −20 kJ.

44. The change in total internal energy of the nitrogen during the process is closest to

(A) −5 kJ. (C) −20 kJ.

(B) +5 kJ. (D) +20 kJ.

45. The heat transfer during this process is

(A) +1 kJ. (C) +3 kJ.

(B) −1 kJ. (D) −3 kJ.

Questions 46 and 47 pertain to the following situation.

An experiment is to be designed that requires measuring a flow rate of 2,000 liters per minute of water at 20°C through a schedule 40 5-inch pipe (*ID* = 128.1 mm). A vertical mercury manometer is available with a useful length of 1 m. The leads of the manometer are filled with water. The flow measurement is to be made with a horizontal venturi meter with a velocity coefficient C_v = 0.98.

46. The maximum pressure difference that this manometer can record is closest to

(A) 123 kPa.
(B) 200 kPa.
(C) 400 kPa.
(D) 800 kPa.

47. The required throat diameter (in centimeters) of the venturi for the maximum flow is most nearly

(A) 3.9.
(B) 5.2.
(C) 7.0.
(D) 8.9.

48. Two resistors are in series. Their measured values are as follows:

$$R_1 = 100.0\ \Omega \pm 0.1\ \Omega$$

$$R_2 = 50.0\ \Omega \pm 0.03\ \Omega$$

The estimated total error in the calculated result of the total resistance is closest to

(A) 0.170 Ω.
(B) 0.130 Ω.
(C) 0.104 Ω.
(D) 0.070 Ω.

49. A closed loop gain of –90 and stability of 97% is required from an inverting amplifier with a forward gain variance of 20%. What is the value of G that will satisfy the design requirements?

(A) –600.3
(B) +600.3
(C) 0.03
(D) –5.67

50. If a step input of height 10 occurs at $t = 0$, what is the transformed forcing function?

(A) $\frac{s}{10}$
(B) $\frac{10}{s}$
(C) $\frac{1}{10s}$
(D) $10s$

51. A mechanical system is acted upon by a constant force of 8 kilograms starting at $t = 0$. What is the response of the transfer function P(s)?

$$P(s) = 6/(s + 2)(s + 4)$$

(A) $R(t) = 6 - 12e^{-2t} + 6e^{-4t}$

(B) $R(t) = \dfrac{6}{(s + 8s + 8)}$

(C) $R(t) = 48 - 48e^{-2t} + 48e^{-4t}$

(D) $R(t) = 1 - e^{-2t} + e^{-4t}$

52. The endurance limit of a steel member is 120 MPa and the tensile strength is 400 MPa. The fatigue strength corresponding to a life of 80,000 cycles is closest to

(A) 120 MPa.

(B) 150 MPa.

(C) 170 MPa.

(D) 200 MPa.

53. The endurance limit of rotating-beam specimen, S'_e must be modified to obtain data for the endurance limit of an actual mechanical element, S_e, by multiplying S'_e by a series of correction factors. These factors take into account

(A) the quality of finish.

(B) temperature to which the material was exposed.

(C) stress-concentration effects.

(D) All of the above.

54. Fatigue occurs under which type of loading condition?

(A) Repeated load

(B) High load

(C) Concentrated load

(D) Static load

55. If a sample of test material reflects no strain after being subjected to loading, which of the following results can be deducted from the experiments?

(A) It has a high modulus of elasticity.

(B) It is elastic.

(C) It is plastic.

(D) It's ductile.

56. The diffusion coefficient of carbon in γ-iron at a temperature of 1,000°C is required. Test data has determined the proportionality constant, D_o, to be 25 mm²/s and the activation energy, Q, to be 34,500 cal/mole. The diffusion coefficient is closest to

(A) 7.20×10^{-7} mm²/s.
(B) 8.20×10^{-6} mm²/s.
(C) 9.20×10^{-5} mm²/s.
(D) 2.98×10^{-5} mm²/s.

57. It is desired to determine the size of a UNS G10500 cold-drawn steel bar to withstand a tensile preload of 35.6 kN and a fluctuating tensile load varying from 0 to 71.2 kN. Assume that the endurance limit of mechanical element to be 23.4% of the endurance limit of rotating-beam specimen. The diameter for an infinite life and a factor of safety of 2.0 is closest to

(A) 1.2 cm.
(B) 2.2 cm.
(C) 3.7 cm.
(D) 5.7 cm.

58. A cylinder head of a steam engine is held by 12 bolts. The diameter of the cylinder is 30 cm and the steam pressure is 1,034 kPa. What size bolts are required if the tensile strength of the bolts are 600 MPa and a factor of safety of 3.0 is required?

(A) 0.62 cm
(B) 1.6 cm
(C) 2.0 cm
(D) 2.5 cm

Questions 59 and 60 refer to the following situation.

A gear wheel is attached to a 2.5 cm diameter shaft which is to transmit 10.0 kw of power at a speed of 1,750 rpm. The gear wheel is 1.5 cm thick. The key is to be made of steel with an allowable shear stress of 200 MPa and a factor of safety of 3.0 is to be used.

59. The shear force acting on the key is closest to

(A) 2,180 N.
(B) 4,370 N.
(C) 6,220 N.
(D) 8,220 N.

60. The required width of the key assuming that shear stress controls is

(A) 2.18 mm.

(B) 4.36 mm.

(C) 8.72 mm.

(D) 9.69 mm.

TEST 1

ANSWER KEY

1. (A)
2. (C)
3. (D)
4. (A)
5. (B)
6. (C)
7. (C)
8. (B)
9. (C)
10. (B)
11. (B)
12. (C)
13. (B)
14. (B)
15. (C)
16. (D)
17. (C)
18. (C)
19. (A)
20. (B)
21. (C)
22. (C)
23. (D)
24. (A)
25. (B)
26. (C)
27. (B)
28. (C)
29. (B)
30. (D)
31. (D)
32. (B)
33. (A)
34. (A)
35. (C)
36. (C)
37. (C)
38. (D)
39. (A)
40. (A)
41. (B)
42. (D)
43. (B)
44. (A)
45. (C)
46. (A)
47. (B)
48. (C)
49. (A)
50. (B)
51. (A)
52. (C)
53. (D)
54. (A)
55. (B)
56. (D)
57. (C)
58. (A)
59. (B)
60. (B)

DETAILED EXPLANATIONS OF ANSWERS

TEST 1

1. **(A)**

In a spreadsheet, a macro is a grouping and recording of instructions that the user executes manually. (B) and (C) make no sense. (D) implies that macros exist in other applications, but not in a spreadsheet.

2. **(C)**

(A) would have been the correct response if $S = 1$ instead of $S = 0$ in line 2. (B) is incorrect since *n!* and not *n* is being evaluated. (D) gives the wrong sum to the series.

3. **(D)**

While (A) is true, the accumulative error of roundoff could make the answer incorrect to three significant figures. (B) Most computers will perform as stated in the problem with no correction for the roundoff error. (C) is incorrect since seldom is an engineering problem worked to eight significant figures.

4. **(A)**

$P_1 = 100$ kPa; $P_2 = 500$ kPa; $T_1 = 25°\text{C} = 298$ K. For a reversible polytropic process the work is determined from:

$$w_{comp} = \frac{nRT_1}{n-1}\left[1-\left(\frac{P_2}{P_1}\right)^{\frac{(n-1)}{n}}\right]$$

$$= \frac{(1.3)(0.287)(298)}{1.3-1}\left[1-\left(\frac{500}{100}\right)^{\frac{(1.3-1)}{1.3}}\right] = -167 \text{ kJ/kg}$$

(B) is incorrect since work output was required which is negative for a compressor. (C) is also incorrect. (D) would be the result if T_1 was not changed to an absolute temperature.

5. **(B)**

To find q, a First Law energy balance is required.

$$q - w_{comp} = h_2 - h_1 = c_P (T_2 - T_1)$$

$$c_P = 1.00 \text{ kJ/kg-K}$$

To find T_2, use the polytropic relations:

$$\frac{T_2}{T_1} = \left(\frac{P_2}{P_1}\right)^{\frac{n-1}{n}}$$

$$T_2 = T_1\left(\frac{P_2}{P_1}\right)^{\frac{n-1}{n}} = 298\left(\frac{500}{100}\right)^{\frac{1.3-1}{1.3}} = 432 \text{ K}$$

$$q = -167 \text{ kJ/kg} + (1.00 \text{ kJ/kg-K})[(432 - 298)\text{K}] = -33 \text{ kJ/kg}$$

(A) is incorrect due to the sign. (C) would be the result if the temperature T_1 was not changed to absolute. (D) is incorrect.

6. **(C)**

The intermediate pressure, P_x, between stages is required. The value of P_x that results in the minimum amount of work is given as follows:

$$P_x = (P_1 P_2)^{1/2} = (100 \times 500)^{1/2} = 224 \text{ kPa}$$

The minimum work occurs when the pressure ratio across each stage of the compressor is the same. When this condition is satisfied, $w_{comp1} = w_{comp2}$. The total work for both stages is merely twice the work for either stage.

$$w_{comp} = 2w_{comp1} = 2\frac{nRT_1}{n-1}\left[1 - \left(\frac{P_2}{P_1}\right)^{\frac{(n-1)}{n}}\right]$$

$$= \frac{2(1.3)(0.287)(298)}{1.3-1}\left[1 - \left(\frac{224}{100}\right)^{\frac{(1.3-1)}{1.3}}\right]$$

$$= -152 \text{ kJ/kg}$$

(A) is the work for a single stage. (B) has the wrong sign for a single stage. (D) has the wrong sign.

7. **(C)**

The shear diagram for the beam is required. To construct the shear diagram, the reactions R_1 and R_2 must be found. Assuming that the total of the distributed force, 525 kN is applied 3.5 m from R_1:

$$\sum M_{R_1} = 0 = (1.5\text{ m}) \times (350\text{ kN}) - (R_2) \times (3.0\text{ m}) + (525\text{ kN}) \times (3.5\text{ m});$$

$$R_2 = 788\text{ kN}$$

$$\sum Fy = 0 = R_1 + R_2 - 350\text{ kN} - 525\text{ kN } R_1;\ R_1 = 87\text{ kN}$$

With these forces, the shear diagram can be drawn and appears below. The maximum shear is 525 kN at the location of support R_2.

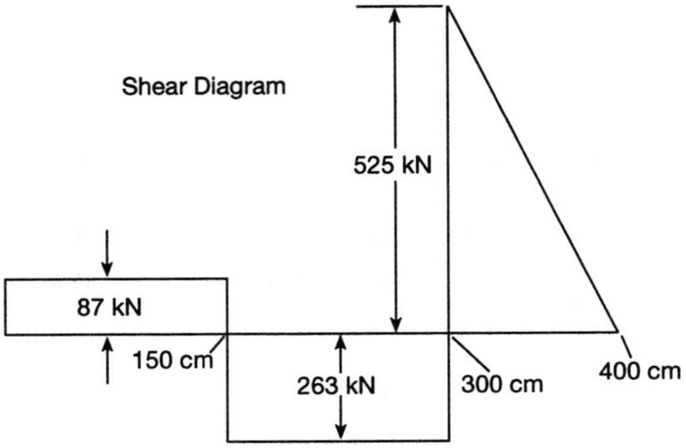

8. **(B)**

The stress, S, is found using the following equation:

$$S = \frac{My}{I}$$

where: M = moment at any beam location
y = distance from the neutral axis
I = moment of inertia

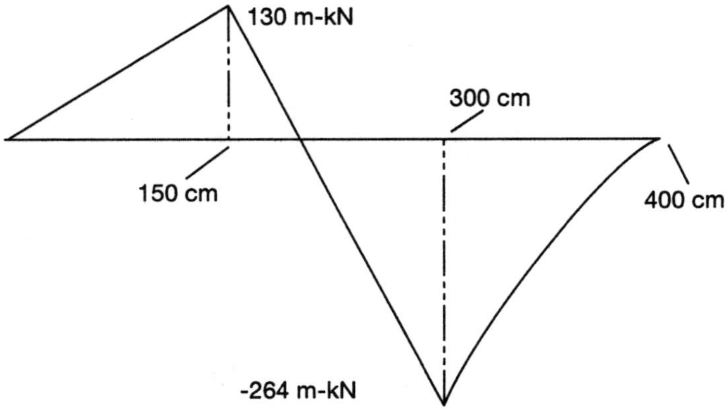

A moment diagram is constructed from the shear diagram. For uniform shear, the moment varies linearly from 0 at the left-hand side to 131 m-kN at 1.5 m. It decreases linearly to –264 m-kN at 3.0 m. It now varies parabolically to 0 at the right-hand side. The moment diagram is shown on the bottom of the previous page. The maximum moment is:

$$-264 \text{ m-kN at } R_2$$

At this point, the section of the beam above the neutral axis is in tension and below the neutral axis is in compression.

$$S = \frac{My}{I} = \frac{(264 \text{ m–kN}) \times (0.15 \text{ m})}{1.46 \times 10^{-4} \text{m}^4} = 271 \text{ MPa}$$

9. **(C)**

From the statement of the problem:

$$\sigma_x = 55 \text{ MPa}$$

$$\sigma_y = 0$$

$$\tau = 21 \text{ MPa}$$

$$\sigma_{1,2} = \frac{\sigma_x + \sigma_y}{2} \pm \sqrt{\left(\frac{\sigma_x - \sigma_y}{2}\right)^2 + \tau^2}$$

$$\sigma_{1,2} = \frac{55}{2} \pm \sqrt{\left(\frac{55}{2}\right)^2 + 21^2}$$

$$\sigma_{1,2} = 62.1 \text{ MPa}, -7.1 \text{ MPa}$$

10. **(B)**

The deflection under the loads will be found by superposition from known solutions. For the 350 kN load, the left diagram applies.

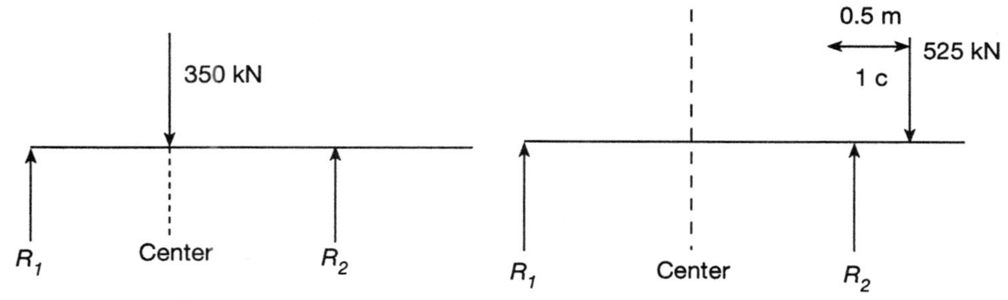

$$y_{center} = \frac{PL^3}{48\ \text{EI}}$$

$$y_{center} = \frac{(350\ \text{kN}) \times (3\ \text{m})^3}{(48) \times \left(172 \times 10^9\ \text{N/m}^2\right) \times \left(1.46 \times 10^{-4}\ \text{m}^4\right)}$$

$$y_{center} = 0.0078\ \text{m} = 0.78\ \text{cm down}$$

For the distribute load, the right diagram applies where the distributed load is replaced by an equivalent concentrated load of 525 kN.

$$y_{center} = \frac{PcL^2}{16\ \text{EI}}$$

$$y_{center} = \frac{(525\ \text{kN}) \times (0.5\ \text{m}) \times (3\ \text{m})^2}{(16) \times \left(172 \times 10^9\ \text{N/m}^2\right) \times \left(1.46 \times 10^{-4}\ \text{m}^4\right)}$$

$$y_{center} = 0.00588\ \text{m} = 0.588\ \text{cm up}$$

$$y = 0.78 - 0.588 = 0.192\ \text{cm} \downarrow \text{(down)}$$

11. **(B)**

To determine the force required to open the gate, the force of pressure, F_P, and the point of application of this force must be determined. F_P is determined as follows:

$F_P = \overline{P}A$ where $\overline{P}$ is the gage pressure at the centroid of the gate.

$\overline{P} = \gamma\overline{h}$ where $\overline{h}$ is the distance from surface to centroid. Thus:

$\overline{h} = [3.0 - (0.61/2)] = 2.70\ \text{m}$

$\overline{P} = \gamma\overline{h} = (9.8\ \text{kN/m}^3) \times (2.70\ \text{m}) = 26.4\ \text{kPa gage}$

$F_P = \overline{P}A = (26.4\ \text{kPa}) \times (\pi/4) \times [(0.61^2)\ \text{m}^2] = 7.72\ \text{kN}$

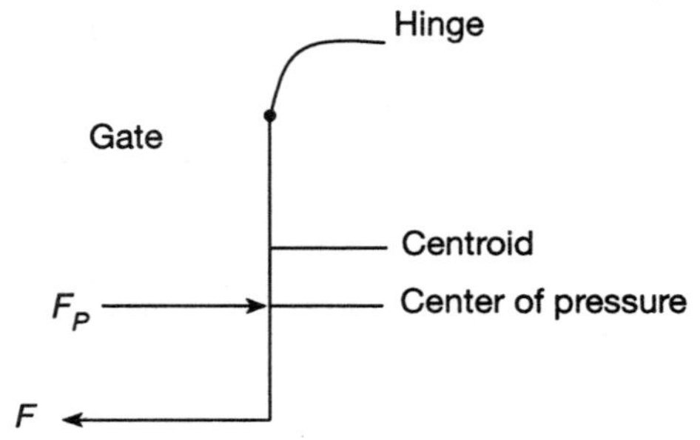

The application of this force is not at the centroid but at a point known as the center of pressure.

$$y_{CP} = \frac{I_{CG}}{y_{CG}A} + y_{CG}$$

$$I_{CG} = \frac{\pi r^4}{4} = \frac{(\pi)\times\left[(0.61/2)^4\right]}{4} = 0.00680 \text{ m}^4$$

$$y_{CP} - y_{CG} = \frac{0.00680 \text{ m}^4}{(2.70 \text{ m})\times(\pi/4)\times\left[(0.61)^2 \text{ m}^2\right]} = 0.00862 \text{ m}$$

The above represents the distance below the centroid that the pressure force acts. The moment arm for F_P becomes:

(0.61/2 m) + (0.0086 m) = 0.314 m

Taking moments about the hinge:

(0.61 m) × (F) = (7.72 kN) × (0.314 m)

F = 3.97 kN

(A) is the result if it is erroneously assumed that the pressure acts at the centroid. (C) is the pressure force and not the required force. (D) is incorrect.

12. **(C)**

With the pipe missing and the gate fully open, the Bernoulli equation can be used from the surface to the exit.

$$\frac{p_1}{\gamma} + z_1 + \frac{V_1^2}{2g} = \frac{p_2}{\gamma} + z_2 + \frac{V_2^2}{2g} \quad p_1 = p_2 = p_{atm}$$

$$V_1 = 0$$

$$z_2 = 0;\ z_1 = h = 3 \text{ m}$$

$$V_2 = \sqrt{2\times g\times h} = \sqrt{2\times 9.80 \text{ m/s}^2 \times 3 \text{ m}} = 7.67 \text{ m/s}$$

(A) and (B) are incorrect. (D) is the result if an incorrect value of g is used.

13. **(B)**
The energy equation for the system must be solved.

$$\frac{p_1}{\gamma} + z_1 + \frac{V_1^2}{2g} = \frac{p_2}{\gamma} + z_2 + \frac{V_2^2}{2g} + h_f + h_{f,fittings} + E_m$$

Since no information is given concerning the fittings, assume $h_{f,fittings} = 0$. However, h_f must be determined. Thus:

$$h_f = f\frac{L}{D}\frac{V^2}{2g}$$

The velocity in the pipe is found from the continuity equation, $\dot{Q} = VA$. A has been determined to be 0.292 m². With this information the velocity can be calculated.

$$V = \frac{\dot{Q}}{A} = \frac{\left(17 \text{ m}^3/\text{min}\right)}{\left(0.292 \text{ m}^2\right)} = 58.2 \text{ m/min} = 0.97 \text{ m/s}$$

The value of the friction factor, f, is found from the Moody diagram. First, the Re number must be found. $Re = VD/\mu$ where μ was given as 1.08×10^{-6} m² s.

$$Re = \frac{(0.97 \text{ m/s}) \times (0.61 \text{ m})}{1.08 \times 10^{-6} \text{m}^2/\text{s}} = 5.48 \times 10^5 \text{ (Turbulent flow)}$$

Next, the value of e/D is required:

$$\frac{e}{D} = \frac{0.00183 \text{ m}}{0.61 \text{ m}} = 0.003$$

With these two values the Moody diagram gives a value of $f = 0.026$.

$$h_f = f\frac{L}{D}\frac{V^2}{2g} = (0.026) \times \frac{30.5 \text{ m}}{0.61 \text{ m}} \times \frac{(0.97)^2 \text{m}^2/\text{s}^2}{2 \times 9.8 \text{ m/s}^2} = 0.0624 \text{ m}$$

$$\frac{p_1}{\gamma} + z_1 + \frac{V_1^2}{2g} = \frac{p_2}{\gamma} + z_2 + \frac{V_2^2}{2g} + h_f + h_{f,fittings} + E_m$$

$$p_1 = p_2$$

$$V_1 = V_2 = 0$$

$$z_1 = 3 \text{ m}$$

$z_2 = 1.2$ m

$3 \text{ m} = 1.2 \text{ m} + 0.0624 \text{ m} + E_m$

$E_m = 1.72$ m

$\dot{W} = \gamma \dot{Q} E_m = (9.8 \text{ kN/m}^2) \times (0.283 \text{ m}^3/s) \times (1.72 \text{ m}) = 4.77 \text{ kW}$

14. **(B)**

Since the track is frictionless, energy is conserved and the sum of potential and kinetic energy at 1 must equal the sum of potential and kinetic energy at 2.

$PE_1 + KE_1 = PE_2 + KE_2$

$KE_1 = {}^1/_2\, m\, v_1{}^2 = {}^1/_2\,(7.26 \text{ kg}) \times (3 \text{ m/s})^2 = 32.7 \text{ J};\ KE_2 = {}^1/_2\, mv_2{}^2$

$PE_1 = mgh_1 = (7.26 \text{ kg}) \times (9.8 \text{ m/s}^2) \times (3 \text{ m}) = 213 \text{ J}$

$PE_2 = mgh_2 = (7.26 \text{ kg}) \times (9.8 \text{ m/s}^2) \times (.3 \text{ m}) = 21.3 \text{ J}$

Substituting into the above equation:

$213 \text{ J} + 32.7 \text{ J} = 21.3 \text{ J} + {}^1/_2\, mv_2{}^2$

Solving for v_2 , $v_2 = 7.86$ m/s

Momentum is conserved for this central impact problem. Let m_1 be the mass of bowling ball and m_2 be the mass of the block.

$m_1 v_1 + m_2 v_2 = m_1 v_1{}' + m_2 v_2{}'$

where the $'$ stand for velocity after impact. Substituting:

$(7.26 \text{ kg}) \times (7.86 \text{ m/s}) + 0 = 0 + (14.5 \text{ kg}) \times (v_2{}')$

Solving, $v_2{}' = 3.94$ m/s.

(A) and (D) are incorrect. (C) is the speed of the bowling ball.

15. **(C)**

After the impact energy is again conserved. This time the conservation of energy must be applied to both the block and the spring.

$KE_{1B} + PE_{1B} + KE_{1S} + PE_{1S} = KE_{2B} + PE_{2B} + KE_{2S} + PE_{2S}$

$KE_{1B} = {}^1/_2\,(14.5 \text{ kg}) \times (3.94)^2 \text{ m}^2/\text{s}^2 = 113 \text{ J}$

$PE_{1B} = PE_{2B}$

and $KE_{2B} = 0$

$$PE_{S1} = {}^1/_2\, kx^2 = {}^1/_2\, (1{,}750 \text{ N/cm}) \times (5 \text{ cm})^2$$

$$= 21{,}900 \text{ N-cm} = 219 \text{ N-m} = 219 \text{ J}$$

$$113 \text{ J} + 219 \text{ J} = {}^1/_2\, (1.75 \times 10^5 \text{ N/m}) \times (x^2)$$

Solving, $x = 0.0616$ m = 6.2 cm.

16. **(D)**

If the spring is missing, the block will start moving in a circular motion at the end of the cable.

$$\sum F_n = T - W = \frac{mv^2}{r} = \frac{(14.5 \text{ kg}) \times (3.94 \text{ m/s})^2}{3 \text{ m}} = 75 \text{ N}$$

$$W = mg = 14.5 \text{ kg} \times 9.8 \text{ m/s}^2 = 142 \text{ N}$$

$$T = 142 + 75 = 217 \text{ N}$$

(A) is actually the mass (in kg). (B) is the weight of the block. This is the answer before motion. (C) is incorrect.

17. **(C)**

The heating process can be solved using the heat exchanger theory. The heat transfer between the cold liquid and the condensing steam is given by the following:

$$\dot{Q} = U_i A_i \Delta T_m = \dot{m}_w C_w \Delta T_w$$

$$= (0.9 \text{ kg/s}) \times (4.184 \text{ kJ/kg-K}) \times (49 - 4) \text{ K} = 169 \text{ kJ/s}$$

where:

$\dot{m}$ = mass flow rate of water

C_w = heat capacity of water

U_i = overall heat transfer coefficient based on inside diameter

A_i = inside area for heat transfer = $2\pi rl$ = πDl, where l is the length of pipe

ΔT_m = the log mean temperature difference, which is the average temperature difference for the heat exchanger

Since steam is condensing, the outside pipe temperature at both ends is 141°C. This makes:

$$\Delta T_i = (141 - 4) = 137°\text{C and } \Delta T_o = (141 - 49) = 92°\text{C}$$

$$\Delta T_m = \frac{\Delta T_i - \Delta T_o}{\ln\left(\dfrac{\Delta T_i}{\Delta T_o}\right)} = \frac{137 - 92}{\ln\left(\dfrac{137}{92}\right)} = 113 \text{ K}$$

Since U_i and A_i are unknown, find U_i and solve for A_i. This will require determination of the inside and outside convection heat transfer coefficients.

$$Nu = \frac{h_i D}{k_f} = 0.026 Re^{0.8} Pr^{1/3}\left(\frac{\mu_b}{\mu_w}\right)^{0.14} \qquad Re = \frac{4\dot{m}}{\pi D_i \mu}$$

Properties for the above are evaluated at the average water temperature of 27°C or 300 K.

$\rho = 997 \text{ kg/m}^3$

$Pr = 5.83$

$k = 0.613$ W/m-K

$\mu_w = 192 \times 10^{-6}$ N-s/m^2 (evaluated at $T_s = 141°$C)

$\mu_b = 855 \times 10^{-6}$ N-s/m^2

$$Re = \frac{(4)\times(0.9 \text{ kg/s})}{\pi(0.026 \text{ m})\times\left(855\times10^{-6}\text{N–s/m}^2\right)} = 5.15\times10^4$$

$$Nu = 0.026 Re^{\,0.8} Pr^{1/3}\left(\frac{\mu_b}{\mu_w}\right)^{0.14}$$

$$= (0.026) \times (5.15 \times 10^4)^{0.8} \times (5.83)^{1/3} \times (855/192)^{0.14} = 339$$

$$h_i = Nu \times k/D = 339 \times (0.613 \text{ W/m-K})/(0.0286)$$

$$= 7{,}300 \text{ W/m}^2\text{-K}$$

18. **(C)**

Since the steam is at 141°C and the water at 27°C, obtain the properties of the water at 84°C or 357 K for the following equation:

$$Nu_D = \frac{h_D D}{k} = 0.729\left[\frac{\rho_l g(\rho_l - \rho_v)\lambda D^3}{\mu_l k_l (T_{sat} - T_s)}\right]^{1/4}$$

$\mu = 343 \times 10^{-6}$ N-s/m^2

$\rho = 971$ kg/m^3

$k = 0.671$ W/m-K

$\lambda = 2.14 \times 10^6$ J/kg, heat of vaporization at 141°C

For this problem, $T_{sat} - T_s = \Delta T_m$ and $\rho_l(\rho_l - \rho_v) = \rho_l^2$

We can solve this equation for h_D as follows using the above listed changes:

$$h_D = 0.729\left[\frac{\rho^2 g\lambda k^3}{\mu_l D\Delta T_m}\right]^{1/4}$$

$$= 0.729\left[\frac{(971)^2 \times (9.8) \times (2.14 \times 10^6) \times (0.671)^3}{\left(343 \times 10^{-6}\right) \times (0.0286) \times (113)}\right]^{1/4} = 6{,}250 \text{ W/m}^2\text{-K}$$

19. **(A)**

To determine U_i, the thermal resistance must be determined. This is given by the following relation with the thermal conductivity of copper taken to be 80 W/m-K:

$$R_i = \frac{1}{h_i} + \frac{r_i \ln(r_o / r_i)}{k} + \frac{1}{(r_o / r_i)h_o}$$

$$= \frac{1}{7{,}270} + \frac{(0.0143)\ln(28.58/26.04)}{380} + \frac{1}{(28.58/26.04)6{,}250}$$

$R_i = 0.000283$ m^2-K/W

$U_i = 1/R_i = 3{,}530$ W/m^2-K

$$A_i = \frac{\dot{Q}}{U_i \Delta T_m} = \frac{169 \times 10^3 \text{ J/s}}{\left(3{,}530 \text{ W/m}^2\text{-K}\right) \times (113 \text{ K})} = 0.423 \text{ m}^2$$

$A_i = \pi DL$

$$L = \frac{A_i}{\pi D} = \frac{0.423 \text{ m}^2}{\pi(0.026 \text{ m})} = 5.18 \text{ m} \approx 5.2 \text{ m}$$

20. **(B)**

State 1, the inlet to the compressor is defined as a saturated vapor at –10°C. From the *P–v* diagram for HFC–134a, $h_1 = 392$ kJ/kg. The process through the compressor was described as being adiabatic. The most efficient adiabatic process is a reversible, adiabatic process. By definition a reversible, adiabatic process is isentropic. State 2, the outlet from the compressor is found by following a line of constant entropy from Point 1 to the 50°C line, giving $h_2 = 420$ kJ/kg.

$$w_{comp} = h_2 - h_1 = 420 - 392 = 28 \text{ kJ/kg}$$

21. **(C)**

The *COP* is defined as the ratio of heat removed by the evaporator divided by the work input by the compressor.

$$COP = \frac{q_{evap}}{w_{comp}}$$

To find q_{evap} state 4, the inlet state to the evaporator must be found. Since state 3, the outlet of the condenser is saturated liquid at 50°C, $h_3 = 273$ kJ/kg. Since no information is provided, assume the process through the throttling device to be of constant enthalpy. Hence, $h_3 = h_4 = 273$ kJ/kg.

$$q_{evap} = h_1 - h_4 = 392 - 273 = 119 \text{ kJ/kg}$$

$$COP = \frac{119}{28} = 4.25 \approx 4.3$$

22. **(C)**

To find the mass flow rate, the rate of heat removal in the evaporator must be first determined. One ton of refrigeration = 12,000 Btu/hr = 3.516 kW. Since $\dot{Q} = 2$ tons, $\dot{Q} = 7.03$ kW.

$$\dot{Q} = \dot{m} q_{evap}$$

$$\dot{m} = \frac{\dot{Q}}{q_{evap}} = \frac{(7.03 \text{ kJ / s})}{119 \text{ kJ / kg}} = 0.059 \text{ kg/s} = 3.54 \text{ kg/min} \approx 3.5 \text{ kg/min}$$

23. **(D)**

Since the process from 1–2 is a sensible heating process, $\omega_1 = \omega_2$. Point 1 is given and Point 2 can be established hence, their properties can be found from the standard atmosphere psychrometric chart.

State 1: 40°F and 80% relative humidity: $h_1 = 14.0$ Btu/lbma,

$\omega_1 = 0.004$ lbmv/lbma

State 2: 80°F and $\omega_2 = \omega_1 = 0.004$ lbmv/lbma; $h_2 = 23.3$ Btu/lbma

The heat required in the heating section is given by the following equation:

$$\dot{Q} = \dot{m}_a (h_2 - h_1)$$

where $\dot{m}$ is the mass flow rate of the air. This must be determined from the volume flow rate which is given in the problem statement.

$$\dot{m}_a = \frac{\dot{V}}{v_1} = \frac{(500 \text{ ft}^3/\text{min}) \times (60 \text{ min/hr})}{12.7 \text{ ft}^3/\text{lbma}} = 2{,}360 \text{ lbma/hr}$$

Where v_1, the specific volume of the moist air, was found in the psychrometric chart at point 1.

$$\dot{Q} = \dot{m}(h_2 - h_1) = (2{,}360 \text{ lbma/hr}) \times [(23.3 - 14.0) \text{ Btu/lbma}]$$

$$= 21{,}900 \text{ Btu/hr}$$

24. **(A)**

The mass flow rate of steam required is given by the following equation:

$$\dot{m}_w = \dot{m}_a(\omega_3 - \omega_2)$$

The value for ω_3 is found in the psychrometric chart and equals 0.025 lbmv/lbma.

$$\dot{m}_w = \dot{m}_a(\omega_3 - \omega_2) = (2{,}360 \text{ lbma/hr}) \times [(0.025 - 0.004)\text{lbmv/lbma}]$$

$$= 49.6 \text{ lbmv/hr}$$

25. **(B)**

The enthalpy of steam required can be found from a unique feature of the psychrometric chart. The process from 2–3 is an adiabatic, humidification process where adiabatic means that only energy from the steam is being added to the air. For this process:

$$h_w = \frac{\Delta h}{\Delta \omega}$$ where h_w is the required enthalpy of the steam.

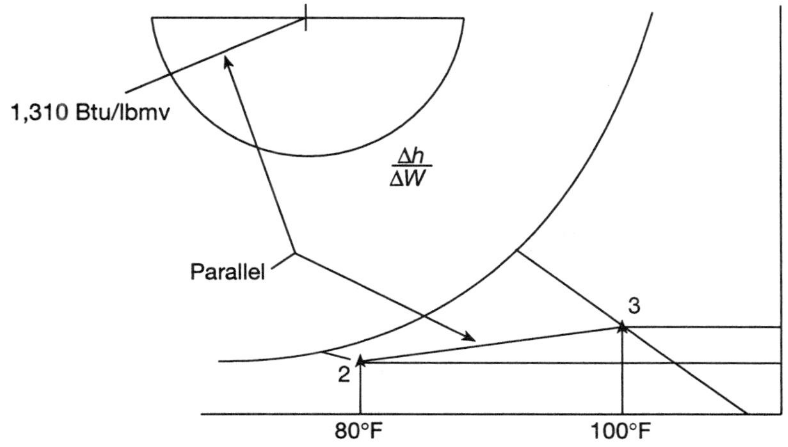

The value of h_w is determined as follows. First, Point 2 and Point 3 are connected. This establishes a slope on the psychrometric chart. Next, this slope is transferred to the protractor which appears in the upper left-hand corner of the chart. With the line passing through the crossed lines on the protractor, the line is extended to meet the outer edge of the protractor. The value of h_w is read as 1,310 Btu/lbmv, ≈ 1,300 Btu/lbmv.

26. **(C)**

Data for the 5-inch nominal steel pipe is as follows:

$OD = 141.33$ mm $\quad$ $ID = 128.1$ mm $\quad$ $E = 200 \times 10^9$ Pa

$$I = \frac{\pi\left(r_0^4 - r_i^4\right)}{4} = \frac{\pi\left[(14.13/2\text{ cm})^4 - (12.81/2\text{ cm})^4\right]}{4} = 635\text{ cm}^4$$

27. **(B)**

For a cantilever column:

$$P_{critical} = \frac{\pi^2 EI}{4L^2}$$

$$P_{allow} = \frac{P_{critical}}{S.F.} = \frac{\pi^2 EI}{(4)\times(2.5)\times\left(L^2\right)} = 24{,}500\text{ N}$$

where $P_{allow} = mg = 2{,}500\text{ kg} \times 9.8\text{ m/s}^2 = 24{,}500\text{ N}$

Solving for L:

$$L = 7.15\text{ m}$$

28. **(C)**

The required pump power is given by the following equation:

$$\dot{W} = \frac{\dot{m}v\Delta P}{\eta}$$

The value v is found from the steam tables to be 0.001003 m^3/kg. Substituting:

$$\dot{W} = \frac{\dot{m}v\Delta P}{\eta} = \frac{(10{,}000/3{,}600)\ \text{kg/s} \times \left(0.001003\ \text{m}^3/\text{kg}\right) \times (900\ \text{kPa})}{0.8} = 3.13\ \text{kW}$$

29. **(B)**

The total energy to the working fluid in passing through the boiler is given by:

$$\dot{Q} = \dot{m}(h_3 - h_2)$$

where h_2 is the enthalpy entering the boiler and h_3 is the enthalpy leaving the boiler. Since state 2 is in the compressed liquid state, $h_2 = h_f$ at 30°C and h_2 = 125.79 kJ/kg. Since state 3 is superheated, h_3 = 3,697.9 kJ/kg.

$$\dot{Q} = \dot{m}(h_3 - h_2)$$

$$= (10{,}000/3{,}600)\ \text{kg/s} \times [(3{,}697.9 - 125.79)\ \text{kJ/kg}]$$

$$= 9{,}923\ \text{kW} = 9.92\ \text{MW}$$

30. **(D)**

The fuel consumption, *FC*, is related to the heat input and the heating value of the fuel, *HV*, by the following relation:

$$FC = \frac{\dot{Q}}{HV \times \eta} = \frac{\left(9.92 \times 10^3\ \text{kJ/s}\right)}{(106{,}000\ \text{kJ/therm}) \times (0.8)}$$

$$= 0.117\ \text{therms/s} = 421\ \text{therms/hr}$$

31. **(D)**

To determine the gas temperature, a First Law energy balance is made on the thermocouple. The energy into the thermocouple by convection from the air stream equals the radiation loss of the thermocouple. Thus:

$$\frac{\sigma\left(T_{tc}^4 - T_W^4\right)}{\frac{1-\varepsilon_{tc}}{\varepsilon_{tc}A_{tc}} + \frac{1}{A_{tc}F_{tc-W}} + \frac{1-\varepsilon_W}{\varepsilon_W A_W}} = hA_{tc}\left(T_g - T_{tc}\right)$$

where $F_{tc\text{-}W} = 1$.

A_W is relatively large and thus makes the third term in the denominator equivalent to zero. Using algebraic manipulation:

$$\sigma A_{tc} \in_{tc} (T_{tc}^4 - T_W^4) = hA_{tc}(T_g - T_{tc})$$

To solve the above equation, h must be determined. For flow perpendicular to a constant-temperature cylinder:

$$Nu = cRe^n Pr^{1/3}$$

For this range of Re (314 given), $c = 0.683$ and $n = 0.466$

$$Nu = (0.683)(314)^{0.466}(0.716)^{1/3} = 8.89$$

$$h = \frac{Nu \times k}{d} = \frac{(8.89)\,(0.06 \text{ W/m-K})}{0.006 \text{ m}} = 88.9 \text{ W/m}^2\text{-K} \approx 90 \text{ W/m}^2\text{-K}$$

32. **(B)**
Solving the equation in Question 31 for T_g:

$$T_g = \frac{\sigma \in_{tc}}{h}\left(T_{tc}^4 - T_W^4\right) + T_{tc}$$

$$= \frac{(0.6)\left(5.67 \times 10^{-8} \text{ W / m}^2\text{-K}^4\right)}{\left(88.9 \text{ W / m}^2\text{-K}\right)}\left[\left(813^4 - 373^4\right)\text{K}^4\right] + 813 \text{ K}$$

$$= 973 \text{ K} = 700°\text{C}$$

33. **(A)**
If the radiation loss could be reduced by 90%, then the radiation loss would be 10% of the original. This is found by inserting a 0.1 term in the expression used in Question 32:

$$T_g = \frac{0.1\sigma \in_{tc}}{h}\left(T_{tc}^4 - T_W^4\right) + T_{tc}$$

$$= \frac{(0.1)(0.6)\left(5.67\times10^{-8}\ \text{W/m}^2\text{-K}^4\right)}{\left(88.9\ \text{W/m}^2\text{-K}\right)}\left[\left(813^4-373^4\right)\text{K}^4\right]+813\ \text{K}$$

$$= 829\text{K} = 556°\text{C}$$

34. **(A)**

A 2-inch nominal, schedule 40 steel pipe has the following SI dimensions:

$OD = 60.3\ \text{mm}$ $\qquad ID = 52.5\ \text{mm}$ $\qquad t = 3.91\ \text{mm}$

Since the steam is condensing, the steam temperature will be constant at 180°C or 453 K. Since the thermal resistance of a steel pipe is typically small, the temperature gradient through the pipe will be small. The thermal conductivity of the steel will correspond to 453 K.

$$k_s = 52\ \text{W/m-K}$$

The thermal resistance through the pipe is given by the following equation:

$$R_t = \frac{\ln(r_2/r_1)}{2\pi L k_s} = \frac{\ln(60.3/52.5)}{(2)(\pi)(1\ \text{m})(52\ \text{W/m-K})} = 0.000424\ \text{K/W}$$

35. **(C)**

The total thermal resistance from the steam inside to the air outside is as follows:

$$R_t = \frac{1}{2\pi r_1 L h_1} + \frac{\ln(r_2/r_1)}{2\pi L k_s} + \frac{1}{2\pi r_2 L h_2}$$

$$\frac{1}{2\pi r_1 L h_1} = \frac{1}{(2)(\pi)((0.0525/2)\ \text{m})\left(10{,}000\ \text{W/m}^2\text{-K}\right)} = 0.000606\ \text{K/W}$$

Note that these first two thermal resistances (from Questions 34 and 35) are very small which is typical for this type of problem. It is the third thermal resistance that controls the heat flow and will be investigated next. The outer surface heat transfer is composed of free convection and radiation. Since free convection is small for this problem, the radiation must be considered. That is, $h_2 = h_c + h_r$, where h_c has been given as 8.2 W/m^2-K. The equivalent radiation coefficient, h_r, is given by the following equation (with $T_s = T_{steam}$ since the thermal resistances are small):

$$h_r = \in \sigma (T_s + T_\infty)(T_s^2 + T_\infty^2)$$

$$= (0.85)(5.67 \times 10^{-8}\ \text{W/m}^2\text{-K}^4)(453 + 300)(453^2 + 300^2)$$

$$h_r = 10.7\ \text{W/m}^2\text{-K}$$

$$h_2 = h_c + h_r = 8.2 + 10.7 = 18.9\ \text{W/m}^2\text{-K}$$

36. **(C)**

The thermal resistance for the outer surface is given by:

$$\frac{1}{(2)(\pi)(r_2)(L)(h_2)} = \frac{1}{(2)(\pi)((0.603/2)\text{m})(1\ \text{m})(18.9\ \text{W/m}^2\text{-K})}$$

$$= 0.279\ \text{K/W}$$

Clearly this dominates the other terms. When combined $R_T = 0.280$ K/W

$$q = \frac{\Delta T}{R_T} = \frac{(453 - 300)\ \text{K}}{0.280\ \text{K/W}} = 546\ \text{W}$$

37. **(C)**

The volume flow rate can be found if the velocity in the pipe is known. Since there is friction involved in this problem, the energy equation will be used.

$$\frac{p_1}{\gamma} + z_1 + \frac{V_1^2}{2g} = \frac{p_2}{\gamma} + z_2 + \frac{V_2^2}{2g} + h_f + h_{f,fittings} + E_m$$

$$p_1 = p_2 = 0;\ V_1 = 0;\ \text{E}_\text{m} = 0$$

$$h_{f,fitting} = C\frac{V^2}{2g}$$

From the statement of the problem, $C = 2.5$. The "sharp entrance" must also be considered, resulting in a final value of $C = 3.0$. Also, the pipe friction is found from:

$$f\frac{L}{D}\frac{V^2}{2g} = (0.02)\frac{100\ \text{m}}{0.15\ \text{m}}\frac{V^2}{2g} = 13.3\frac{V^2}{2g}$$

Substituting into the energy equation:

$$0+0+30\text{ m}=0+\frac{V^2}{2g}+3\times\frac{V^2}{2g}+13.3\times\frac{V^2}{2g}$$

$$V=5.8\text{ m/s}$$

$$Q=VA=(5.8\text{ m/s})\times(\pi/4)(0.15)^2\text{ m}^2=0.103\text{ m}^3/\text{s}\approx 0.1\text{ m}^3/\text{s}$$

38. **(D)**

In this case, the velocity required for this value of the volume flow rate will be determined.

$$V=\frac{Q}{A}=\frac{0.2\text{ m}^2/\text{s}}{(\frac{\pi}{4})0.15^2\text{ m}^2}=11.3\text{ m}^3/\text{s}$$

While the increase in velocity may increase the friction factor f, we will assume that it remains the same in this problem. With an added term for the head increase from the pump, the energy equation becomes:

$$0+0+30+E_P=0+17.3\left(\frac{11.3^2}{2g}\right)$$

$$E_p=82.7\text{ m}$$

The pump power is given by:

$$\dot{W}=\frac{Q\gamma E_m}{\eta}$$

With $\eta = 1$ in this case, since the minimum power is required.

$$\dot{W}=\frac{Q\gamma E_m}{\eta}=(0.2\text{ m}^3/\text{s})\times(9{,}800\text{ N/m}^3)\times(82.7\text{ m})=162\text{ kW}$$

39. **(A)**

The equivalent length of the fittings is found by comparing the losses of the fittings to the losses of the pipe:

$$h_{f,fitting}=C\frac{V^2}{2g} \qquad h_f=f\frac{L}{D}\frac{V^2}{2g}$$

By inspection, $C = f\,(L/D)$; therefore,

$$L_{eq}=(C\times D)/f=(3)\times(0.15\text{ m})/(0.02)=22.5\text{ m}$$

40. **(A)**

The mass moment of inertia is given by the following:

$$I = mr_z^2$$

where r_z is the radius of gyration of the body about its mass center. From the problem statement, $r_z = 400$ mm $= 0.4$ m and the mass $= 70$ kg, hence:

$$I = (70 \text{ kg}) \times (0.4 \text{ m})^2 = 11.2 \text{ kg-m}^2$$

41. **(B)**

To analyze the motion of the cylinder and the block, a free-body diagram for both objects must be drawn. The cylinder can roll on the surface or slip. From the problem statement, the cylinder is assumed to be rolling. Point A, which represents the point of contact between the cylinder and the horizontal surface, becomes the instantaneous center of zero acceleration. This means that the acceleration of any point on the cylinder can be expressed by:

$$a = r \times a'$$

where r is the distance from the instantaneous center. The acceleration of Point B on the cylinder is the same as the acceleration of the cord. Since the distance r in this case is 0.15 m, the acceleration of Point B and of the cord is:

$$a_1 = 0.15 \times a'$$

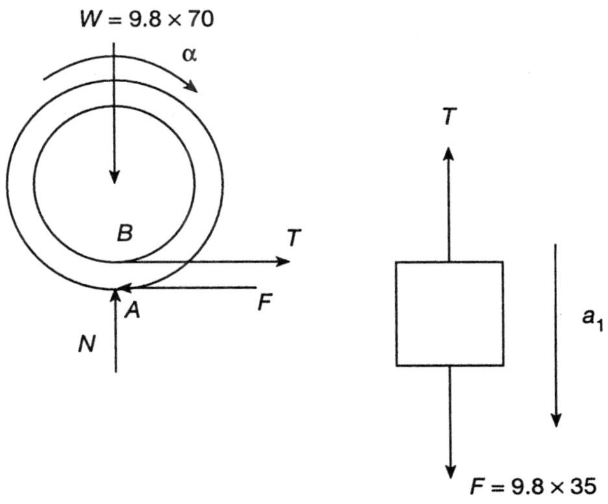

42. **(D)**

In this type of problem a motion is assumed, the problem is solved, and the assumption must then be verified. Here we will assume that rolling and not slipping occurs. This will be true provided the friction force, F, is equal to or less than the maximum allowable friction force. Let $\bar{a}$ be the acceleration of the mass center of the cylinder. Since the center of the cylinder is 0.6 m from Point A, $\bar{a} = 0.6\ \alpha$. From Question 41, $a_1 = 0.15\ \alpha$. With the accelerations defined, we can write $F = ma$ for the two bodies as follows:

$$(35 \times 9.8) - T = 35\ a_1$$

$$T - F = 70\bar{a}$$

If we substitute the expressions for angular acceleration in place of linear acceleration, we would have two equations and three unknowns. The third equation comes from summing moments about the center of the cylinder:

$$\sum \overline{M}_o = \bar{I}\alpha$$

$$(F \times 0.6) - (T \times 0.45) = 11.2\ \alpha$$

These equations are now solved simultaneously for α, which gives:

$$\alpha = 13.8 \text{ rad/s}^2$$

The solution must now be verified:

$$a_1 = 0.15 \qquad \alpha = (0.15) \times (13.8) = 2.07 \text{ m/s}^2$$

The first equation above can now be solved for T as follows:

$$T = (35 \times 9.8) - (35 \times 2.07) = 270 \text{ N}$$

The second equation can be solved for F as follows:

$$(270 \text{ N}) - F = (70) \times (0.6) \times (13.8)$$

$$F = -309 \text{ N}$$

As the result is negative, this means that the actual friction force in this problem is to the right rather than to the left. It remains to be established that rolling rather than slipping will occur.

$$N = mg = 70 \times 9.8 = 686 \text{ N}$$

$F_{max} = \mu$ N, where μ is the friction coefficient.

$$F_{max} = (0.5) \times (686) = 343 \text{ N}$$

Since the actual friction is less than the maximum friction, rolling will occur as assumed.

43. **(B)**

Since the process is described as being polytropic with $n = 1.1$, the following relations apply:

$$\frac{P_2}{P_1} = \left(\frac{V_1}{V_2}\right)^n ; \quad \frac{T_2}{T_1} = \left(\frac{P_2}{P_1}\right)^{\frac{n-1}{n}} ; \quad \frac{T_2}{T_1} = \left(\frac{V_1}{V_2}\right)^{n-1}$$

$$W_{1-2} = w_{1-2} = \frac{P_2 V_2 - P_1 V_1}{1-n}$$

The work can be determined from the second equation if P_2 were known. P_2 can be found from the first polytropic relationship since the volumes are known.

$$\frac{P_2}{P_1} = \left(\frac{V_1}{V_2}\right)^n$$

$$P_2 = (300 \text{ kPa}) \times \left(\frac{1}{2}\right)^{1.1} = 140 \text{ kPa}$$

Since the volume doubled:

$$\text{W} = \frac{P_2 V_2 - P_1 V_1}{1-n} = \frac{(140 \text{ kPa})(0.2 \text{ m}^3) - (300 \text{ kPa})(0.1 \text{ m}^3)}{1-1.1} = +20 \text{ kJ}$$

44. **(A)**

Using the cold-air assumption, the change in total internal energy is given by the following equation:

$\Delta U = m(u_2 - u_1) = mc_v(T_2 - T_1)$, where

$c_v = 0.743$ kJ/kg-K for nitrogen.

The mass m is found from the ideal gas law:

$$m = \frac{PV}{RT} = \frac{(300 \text{ kPa})(0.1 \text{ m}^3)}{(0.2968 \text{ kPa-m}^3/(\text{kg-K}))(350 \text{ K})} = 0.289 \text{ kg}$$

T_2 is found by solving the following equation:

$$\frac{P_1V_1}{T_1} = \frac{P_2V_2}{T_2}$$

$$T_2 = T_1\frac{P_2}{P_1}\frac{V_2}{V_1} = (350\text{ K})\frac{140}{300}(2) = 327\text{ K}$$

$$\Delta U = (0.289\text{ kg})(0.743\text{ kJ/kg-K})[(327 - 350)\text{k}] = -4.94\text{ kJ}$$

45. **(C)**

To find the total heat transfer Q, the First Law is used as follows:

$$Q = W_e + W_{1-2} + \Delta U$$

where W_e is the electrical work added to the system. W_e is negative since it is work added to the system.

$$W_e = -(1\text{ amp})(50\text{ volts})(240\text{ s}) = -12{,}000\text{ J} = -12\text{ kJ}$$

$$Q = -12\text{ kJ} + 20\text{ kJ} - 4.94\text{ kJ} = +3.06\text{ kJ}$$

46. **(A)**

When the legs of a manometer are water filled, the presence of the water will alter the directly observed reading. In the problem's diagram, suppose that Point 2 and Point 7 are attached to the venturi meter. The distance from 2-4 and 5-7 is 1 meter.

$$P_2 - (1\text{ m})\gamma_W + (1\text{ m})\,\gamma_{hg} = P_7$$

$$P_7 - P_2 = [(1)(13.6) - 1]\gamma_W$$

$$P_7 - P_2 = (12.6\text{ m})(9{,}800\text{ N/m}^3) = 123{,}000\text{ Pa} = 123\text{ kPa}$$

47. **(B)**

The equation for the volume flow rate as measured by a venturi meter is given by the following:

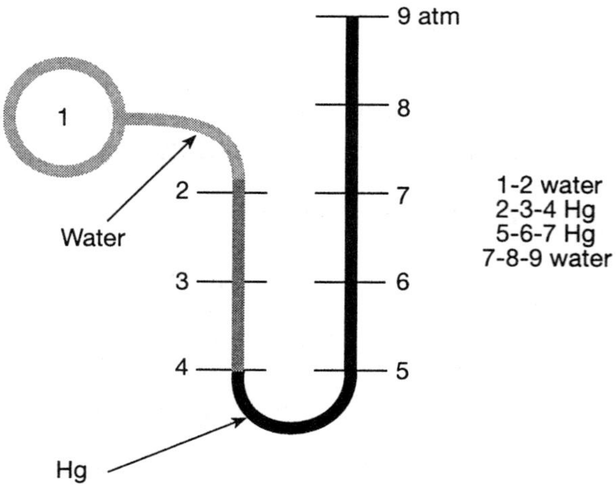

$$Q = \frac{c_v A_2}{\sqrt{1-(A_2/A_1)^2}}\sqrt{2g\left(\frac{P_1}{\gamma}-\frac{P_2}{\gamma}\right)}$$

$Q = 2{,}000$ liters/min $= 2/60$ m^3/s

$(P_1 - P_2)/\gamma = 12.6$ m from Question 46. Substituting the values:

$$\frac{2}{60} = \frac{(0.98)(A_2)}{\sqrt{1-\left(A_2/0.0129\ \text{m}^2\right)^2}}\sqrt{(2)\left(9.8\ \text{m/s}^2\right)(12.6\ \text{m})}$$

$A_2 = 0.00214$ m^2

$A_2 = (\pi/4)\, D_2^2$

$D_2 = 0.052$ m $= 5.2$ cm

48. **(C)**

If the uncertainty in the independent variables are all given with the same odds, then the uncertainty in the results is given by the following:

$$W_R = \left[\left(\frac{\partial R}{\partial x_1}W_1\right)^2 + \left(\frac{\partial R}{\partial x_2}W_2\right)^2 + \ldots + \left(\frac{\partial R}{\partial x_n}W_n\right)^2\right]^{1/2}$$

where: W_R = uncertainty in the results

$W_1, W_2, \ldots, W_n$ = uncertainty in the independent variables.

Since resistors in series $R_T = R_1 + R_2 = R$; $x_1 = R_1$; $x_2 = R_2$; $W_1 = 0.1\ \Omega$; $W_2 = 0.03\ \Omega$:

$$\frac{\partial R_T}{\partial R_1} = 1;\ \frac{\partial R_T}{\partial R_2} = 1$$

Substituting in the above equation:

$$W_R = \left[((1)\times 0.1)^2 + ((1)\times 0.03)^2\right]^{1/2} = 0.1044\ \Omega$$

49. **(A)**

This question tests steady-state and the transient performance analysis of a closed loop transfer function control system.

The variables of a closed loop system are: G_{loop} = closed loop gain;

V_{signal} = reference input signal; V_{output} = controlled output signal; G = dynamic transfer unit; H = feedback transfer unit; S = sensitivity; dG_{loop} = percent change of dynamic transfer unit; and dG = percent change of dynamic transfer unit.

The loop transfer function, G_{loop}, is the ratio of the output voltage to the signal voltage as follows:

$$G_{loop} = \frac{V_{output}}{V_{signal}} = \frac{G}{1-GH} = -90$$

Next, solve for the sensitivity, S, of the feedback system. The sensitivity of a system limits the maximum overshoot of the output variable which is necessary to protect the equipment and reduce the time of recovery of the system. Satisfactory performance can be obtained if the overshoot is set below 30%. The sensitivity is equal to the percent change in the loop transfer function divided by the percent change in the forward transfer function and is represented by the following equation:

$$S = \frac{dG_{loop} / G_{loop}}{dG / G} = \frac{G_{loop}}{1-GH}$$

The sensitivity requirements limit the value of dG_{loop}/G_{loop} to .03 (100% – 97% = 3% = 0.03) when dG/G is 0.2 (20%), therefore, substituting in the equation for S, results as follows:

$$0.03 = \frac{1(0.2)}{1-GH}$$

$$0.03(1 - GH) = 0.2$$

$$0.03 - 0.03GH = 0.2$$

$$GH = \frac{-0.2 + 0.03}{0.03} = \frac{-0.17}{0.03} = 5.67$$

The dynamic unit, G, is found by substituting the value of GH in the equation of the loop gain, G_{loop}:

$$G_{loop} = \frac{G}{1-GH} = -90$$

$$G = -90(1 - GH)$$

$$G = -90(1 - (-5.67)) = -90(6.67) = -600.3$$

(B) is incorrect because of the incorrect sign. It is very important to keep track of the sign convention during these calculations. Choice (C) is incorrect. The answer is the sensitivity requirement limits. Choice (D) is incorrect. This answer is equal to the sensitivity.

50. **(B)**

This question tests the understanding of signal inputs and how to perform Laplace transformations.

The variables of the equation are:

$F(t)$ = input signal function to the system

s = signal variable

Select the Laplace transform for a unit step signal, which is represented by the form $1/s$. Therefore, the input signal function is:

$$F(s) = \mathcal{L}[F(t)] = 10(1/s) = 10/s$$

Choice (A) is incorrect. It is a constant input of 1/10. Choice (C) is incorrect. It is a step input of 1/10th. Choice (D) is incorrect. It is a constant input of 10.

51. **(A)**

The question tests the understanding of effects of force on a mechanical control system, exponential time functions, and how to solve Laplace transformations.

The variables are:

$P(s)$ = transfer function of the system

$R(t)$ = transfer function

s = function variable

$\mathcal{L}$ = Laplace transform

$\mathcal{L}^{-1}$ = inverse Laplace transform

The transformed function is the ratio of the transformed response function to the transformed forcing function.

$$P(s) = \frac{\mathcal{L}[R(t)]}{\mathcal{L}[F(t)]} = \frac{R(s)}{F(s)}$$

$$R(t) = \mathcal{L}^{-1}[R(s)] = \mathcal{L}^{-1}[L[F(t)P(s)]]$$

$$R(t) = \mathcal{L}^{-1}[(8/s)(6/(s + 2)(s + 4)]$$

$$R(t) = \mathcal{L}[48/(s)(s + 2)(s + 4)]$$

Solving for the coefficients:

$$\frac{48}{(s)(s+2)(s+4)} = \frac{A}{s} + \frac{B}{(s+2)} + \frac{C}{(s+4)}$$

$$48 = A(s + 2)(s + 4) + B(s)(s + 4) + C(s)(s + 2)$$

$$48 = A(s^2 + 6s + 8) + B(s^2 + 4s) + C(s^2 + 2s)$$

$$48 = (A + B + C)s^2 + (6A + 4B + 2C)s + 8A$$

Therefore, equating the coefficients on both sides of the equation, we obtain the following equations:

$$s^2: \quad A + B + C = 0 \qquad (1)$$

$$s: \quad 6A + 4B + 2C = 0 \qquad (2)$$

$$48 = 8A \qquad (3)$$

Solving Equation (3):

$$A = \frac{48}{8} = 6$$

Substitute in Equation (1):

$$A + B + C = 0$$

$$6 + B + C = 0$$

$$B + C = -6$$

$$B = -6 - C \qquad (4)$$

Substitute in Equation (2):

$$6(6) + 4(-6 - C) + 2C = 0$$

$$36 + (-24 - 4C) + 2C = 0$$

$$36 - 24 - 4C + 2C = 0$$

$$12 - 2C = 0$$

$$C = 6$$

Substitute in Equation (4):

$$B = -6 - (6) = -12$$

Therefore, the solution for the time domain response function, $R(t)$, is:

$$R(t) = 6 - 12e^{-2t} + 6e^{-4t}$$

Choice (B) is incorrect. This is not a response for the inverse Laplace transforms. Choices (C) and (D) are not correct. These answers have the incorrect coefficients.

52. **(C)**

The fatigue strength, S'_f, is given by the following equation:

$$S'_f = 10^C N^b$$

where: N = number of cycles

$$C = \log \frac{(0.8S_{ut})^2}{S_e}$$

$$b = -\frac{1}{3}\log \frac{0.8S_{ut}}{S_e}$$

S_{ut} is the ultimate tensile strength and S'_e is the endurance limit of a rotating-beam specimen. With $0.8S_{ut} = 0.8 \times 400$ MPa = 320 MPa,

$$b = -\frac{1}{3}\log \frac{320}{120} = -0.142$$

$$C = \log \frac{(320)^2}{120} = 2.928$$

$$S'_f = 10^{2.928}\,(80{,}000)^{-0.142} = 171 \text{ MPa}$$

53. **(D)**

All the answer choices are factors which determine the endurance limit of an actual machine element.

54. **(A)**

The question tests understanding of stress analysis, fatigue, and loading of materials. Fatigue occurs when a material is placed under repeated loading cycles which exceed the material's yield point and deform.

Choice (B) is incorrect. Ultimate stress failure occurs under the highest load. Choices (C) and (D) are also incorrect. Normal failure occurs under concentrated and static loads.

55. **(B)**

The question tests understanding of dynamic loading, stress analysis, and material behavior. By definition, a material that returns to its original dimension when unloaded is elastic.

Choice (A) is incorrect. The modulus of elasticity is the level below the proportional limit of failure. Choice (C) is incorrect. No permanent damage or change occurs if the material is plastic. Choice (D) is incorrect. Ductile means that the material is plastic. It can be permanently changed and will not return to its original shape or condition.

56. **(D)**

The diffusion coefficient, D, is given by the following equation:

$$D = D_o e^{-\frac{Q}{RT}}$$

where R is the universal gas constant and T is the *absolute* temperature. Substituting into the above equation:

$$D = \left(25 \text{ mm}^2/\text{s}\right) e^{-\frac{34{,}500 \text{ cal/mole}}{(1.987 \text{ cal/mole-K})(1273\ K)}} = 2.98 \times 10^{-5} \text{mm}^2/\text{s}$$

57. **(C)**

Data for the given sample is as follows: yield strength, $S_y = 579$ MPa and tensile strength, $S_{ut} = 689$ MPa. Without additional information, the rotating beam endurance limit is assumed to be 50% of the tensile strength, hence, $S'_e = 0.5\ S_{ut} = 0.5 \times 689$ MPa $= 345$ MPa. From the statement of the problem, the endurance limit of the machine element is found from the following:

$$S_e = .234\ S_e{'} = 0.234 \times 345 \text{ MPa} = 80.8 \text{ MPa}.$$

The static stress is given by:

$$\sigma_s = \frac{F_s}{A} = \frac{35.6 \text{ kN}}{\pi d^2/4} = \frac{45.3 \text{ kN}}{d^2}$$

The stress range is given by:

$$\sigma_r = \frac{F_r}{A} = \frac{71.2 \text{ kN}}{\pi d^2/4} = \frac{90.6 \text{ kN}}{d^2}$$

$$\sigma_a = \frac{\sigma_r}{2} = \frac{45.3 \text{ kN}}{d^2}$$

Then:

$$\sigma_m = \sigma_s + \sigma_a = \frac{90.6 \text{ kN}}{d^2}$$

Therefore:

$$\frac{\sigma_a}{\sigma_m} = 0.5$$

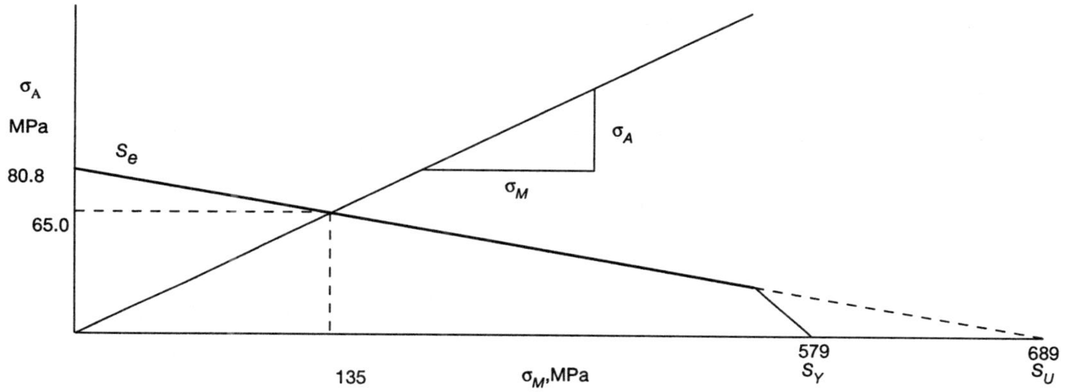

To relate the stresses and strengths, refer to the diagram which represents a modified Goodman diagram. Only the tensile side is needed. A line is drawn connecting the endurance strength of the mechanical element and the tensile strength. This line intersects a line connecting the yield strength line which is drawn between the ordinate and the abscissa. The intersection of the modified Goodman line with another line at a slope of $\sigma_a/\sigma_m = 0.5$ defines two values of strength. Since S_a is the strength corresponding to stress σ_a, for a factor of safety of 2.0 $\sigma_a \leq S_a/2.0$; therefore:

$$\frac{45.3 \text{ kN}}{d^2} \leq \frac{65}{2} \text{ MPa}$$

$$\frac{45{,}300 \text{ N}}{d^2} \leq 32.5 \times 10^6 \text{ N/m}^2$$

$$d \geq 0.0373 \text{ m} = 3.73 \text{ cm}$$

58. **(A)**

The force created by the pressure must be balanced by the bolts. $F_{head} = PA_{head}$ therefore:

$$F_{head} = P\frac{\left(\pi d^2\right)}{4} = \frac{\left(1.034 \times 10^6 \text{ N/m}^2\right)\left(\pi(0.3 \text{ m})^2\right)}{4} = 7.31 \times 10^4 \text{N}$$

The load per bolt will divide equally among the 12 bolts so that the force per bolt, F_{bolt}, is 6.09×10^3 N.

Assuming that the entire factor of safety is applied to the strength, $\sigma = S/n$, where σ is the bolt stress, n is the factor of safety, and S is the strength, 600 MPa in this case. The stress on the bolt is given by the bolt force divided by the area. Combining these two expressions:

$$\sigma = \frac{600 \text{ MPa}}{3} = \frac{6.09 \times 10^3 \text{ N}}{\frac{\pi}{4}d^2}$$

where d is the bolt diameter. Solving for d, $d = 0.62$ cm.

59. **(B)**

The power transmitted by a shaft is given by the following equation:

$$H = T \times \omega$$

where H is the power transmitted (W), T is the applied torque (N-m), and ω is the angular speed (rad/s).

ω can be found from the following relation:

$$\omega \text{ (rad/s)} = \frac{\pi \times \text{rpm}}{30} = \frac{\pi \times 1,750}{30} = 183 \text{ rad/s}$$

$$10.0 \text{ kW} = 10,000 \text{ W} = 10,000 \text{ j/s} = 10,000 \text{ N-m/s}$$

$T = F \times r$, where r is the distance from center of shaft to the application of force or 0.0125 m in this case. Combining, 10,000 N-m/s = $F \times$ (0.0125 m) $\times$ (182 rad/s); $F = 4,370$ N.

The correct choice is (B). (A) is the answer if the diameter of the shaft was used instead of the radius. (C) and (D) are incorrect.

60. **(B)**

The allowable shear stress on the key material is equal to 200 MPa/3 or 66.7 MPa. The shear stress is equal to the shear force acting on the key divided by the area. Hence:

$$66.7 \times 10^6 \text{ N/m}^2 = \frac{4{,}370 \text{ N}}{w \times 0.015 \text{ m}}$$

where 0.015 m is the length of the key and w is the required width of the key.

Solving, $w = 0.00436\, m = 4.36$ mm.

FE/EIT

FE: PM Mechanical Engineering Exam

Practice Test 2

FUNDAMENTALS OF ENGINEERING EXAMINATION

TEST 2

(Answer sheets appear in the back of this book.)

TIME: 4 Hours
60 Questions

DIRECTIONS: For each of the following questions and incomplete statements, choose the best answer from the four answer choices. You must answer all questions.

1. Cells A1...A5 of a spreadsheet contain the values of 1, 2, 3, 4, and 5 respectively, which represent the value of variable x. Cells B1...B5 are to contain the values of variable $y = x^2 - 1$. Which one of the following is the correct content for cell B3?

   (A) $x^2 - 1$
   (B) $y = x^2 - 1$
   (C) A1^2 – 1
   (D) A3 * A3 – 1

2. The flowchart for a computer program contains the following segment:

```
X = –1
Y = –2
Z = X * Y
IF Z < 0 Z = Z+1
RETURN
```

   What will be the value of Z returned by this program segment?

   (A) –1
   (B) 0
   (C) 2
   (D) 3

3. The flowchart for a computer program contains the following segment:

```
N = 7
S = 1; K = 0
FOR I = 3 TO N STEP 2
K = K + 1
S = S + 1/I * (–1)  ^ K
NEXT N
```

Which expression represents the resulting value of S?

(A) $1-\frac{1}{3}+\frac{1}{5}-\frac{1}{7}$

(B) $1+\frac{1}{3}+\frac{1}{5}+\frac{1}{7}$

(C) $1+\frac{1}{3^2}+\frac{1}{5^2}+\frac{1}{7^2}$

(D) $1-\frac{1}{3^2}+\frac{1}{5^2}-\frac{1}{7^2}$

Questions 4–6 pertain to the following situation:

At steady-state conditions and a total pressure of 101 kPa, an air conditioner takes in air at 24°C and 50% relative humidity and returns it to the room at a temperature of 4°C and 20% relative humidity. The volume flow rate of the incoming air is 500 l/s.

4. The mass flow rate of air of the incoming moist air stream is

(A) 2.30 kg/min.

(B) 7.30 kg/min.

(C) 53.4 kg/min.

(D) 35.1 kg/min.

5. The total cooling rate the air conditioner provides is closest to

(A) 1 ton.

(B) 7 tons.

(C) 10 tons.

(D) 100 tons.

6. The amount of water removed from the incoming moist air in a day of continuous operation is

(A) 8.57 kg_v.

(B) 45.3 kg_v.

(C) 425 kg_v.

(D) 8,470 kg_v.

Questions 7–9 pertain to the following situation:

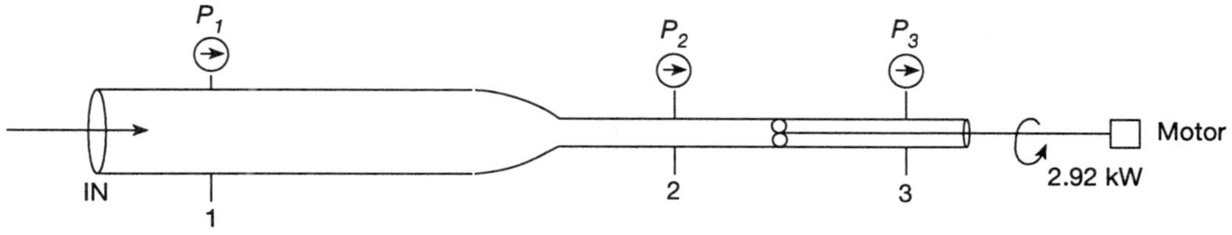

Water flows through the pipe system shown above. The following data applies:

$$D_1 = 0.3 \text{ m}$$

$$p_1 = 82.7 \text{ kPa gauge}$$

$$\gamma = 9{,}800 \text{ N/m}^3$$

$$D_2 = D_3 = 0.15 \text{ m}$$

$$V_1 = 3 \text{ m/s}$$

$$\dot{W} = \text{propeller input}$$

$$= 2.92 \text{ kW}$$

7. The velocity at point 2 is closest to

(A) 3 m/s.
(B) 6 m/s.
(C) 9 m/s.
(D) 12 m/s.

8. The pressure at point 2 is closest to

(A) 15.2 kPa.
(B) 30.4 kPa.
(C) 41.4 kPa.
(D) 165 kPa.

9. The pressure at point 3 is closest to

(A) 19.1 kPa.
(B) 28.9 kPa.
(C) 49.5 kPa.
(D) 69.3 kPa.

Questions 10–12 pertain to the following situation:

A centrifugal pump with a 7.5 cm diameter impeller delivers 2.27 m^3/min of 16°C water at a total head of 107 m when operating at 1,750 rpm. It is desired to deliver 3.78 m^3/min when operating at 3,500 rpm.

10. The diameter of a geometrically similar pump is approximately

    (A) 3.8 cm. (C) 10 cm.

    (B) 7.1 cm. (D) 15 cm.

11. The head developed by this new pump would be closest to

    (A) 57 m. (C) 210 m.

    (B) 110 m. (D) 380 m.

12. If the original pump had an operating efficiency of 60%, the power requirement of the new pump would be

    (A) 50 kW. (C) 391 kW.

    (B) 66.2 kW. (D) 522 kW.

Questions 13–15 refer to the following situation:

A 4-stroke, 12-cylinder diesel engine with a 16:1 compression ratio is to drive an AC generator rated at 500 kW at 1,200 rpm. Generator efficiency is 90% and the engine is to operate at 80% of its rated power. At its intermittent rated power, the air/fuel ratio is 20:1 (afr) and the brake specific fuel consumption (bsfc) is 0.243 kg/kW-hr. The intake pressure and temperature are 632 mm Hg and 35°C. The volumetric efficiency of the engine under these conditions is 85%. The stroke of the engine is 1.2 times the bore.

13. The air consumption for this engine is

    (A) 26.2 kg/min. (C) 56.2 kg/min.

    (B) 36.2 kg/min. (D) 76.2 kg/min.

14. The cylinder displacement in cm^3 is closest to

    (A) 4,700 cm^3. (C) 7,700 cm^3.

    (B) 6,700 cm^3. (D) 9,700 cm^3.

15. The clearance volume of each cylinder in cm^3 is

(A) 444 cm^3.
(B) 544 cm^3.
(C) 644 cm^3.
(D) 844 cm^3.

Questions 16 and 17 pertain to the following situation:

Two high-strength rods of different sizes are attached at *A* and *C* and support a load *W* at *B* as shown. The ultimate strength of the rods is 1.1×10^9 Pa and the factor of safety is to be 4. The rod *AB* has a cross-sectional area of 1.29 cm^2 and the rod *BC* has a cross-sectional area of 0.65 cm^2.

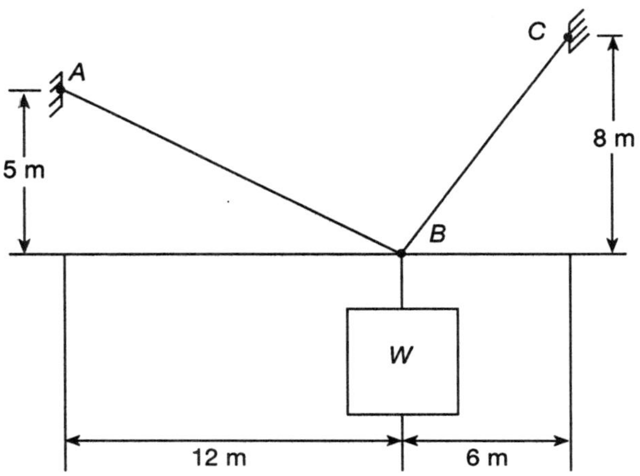

16. The load in member *AB* is what fraction of *W*?

(A) 13/21
(B) 20/21
(C) 7/21
(D) 25/21

17. The maximum load *W* that can safely be supported is

(A) 8,000 N.
(B) 18,700 N.
(C) 36,700 N.
(D) 57,300 N.

18. A bar of steel and a bar of aluminum have the dimensions shown. The magnitude of force P that will cause the total length of the two bars to decrease by 0.025 cm is

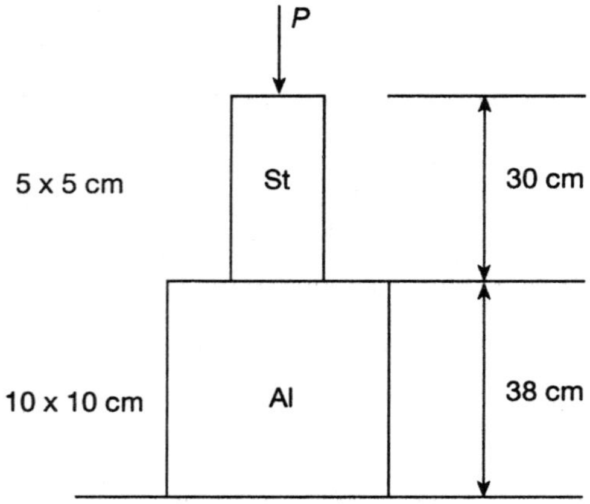

(A) 37,000 N.
(B) 57,000 N.
(C) 117,000 N.
(D) 217,000 N.

Questions 19 and 20 pertain to the following situation:

A copper sphere having a diameter of 3.0 cm is initially at a uniform temperature of 50°C. It is suddenly exposed to an airstream at 10°C with a resulting convective heat transfer coefficient of 15 W/m^2-°C.

19. The Biot number, Bi, for this system is

(A) 1.94×10^{-4}.
(B) 2.94×10^{-2}.
(C) 0.005.
(D) 3.84.

20. The time required for the center of the sphere to reach a temperature of 25°C is closest to

(A) 1 hour.
(B) 2 hours.
(C) 3 hours.
(D) 5 hours.

21. A metal plate is placed in sunlight. The incident radiant energy G is 788 W/m^2. The air and surroundings are at 10°C. The heat transfer coefficient by free convection from the upper surface of the plate is 17 W/m^2-K. The plate has an average emissivity of 0.9 at solar wavelengths and 0.1 at long wavelengths. Neglecting conduction losses on the lower surface, the equilibrium temperature of the plate is

(A) 20°C.
(B) 30°C.
(C) 40°C.
(D) 50°C.

Questions 22–24 refer to the following situation:

At the instant shown, the total acceleration of point B is 4 m/s^2, the tangential acceleration of point B is 1.5 m/s^2, and ω_{AB} is clockwise as indicated.

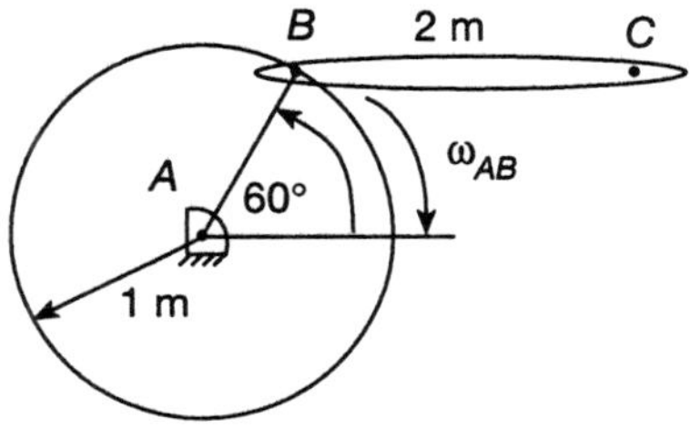

22. The magnitude of the angular velocity of the disk is

(A) 1 rad/s.
(B) 1.93 rad/s.
(C) 4 rad/s.
(D) 5 rad/s.

23. If rod BC has a mass of 15 kg, the radius of gyration of rod BC about point B is

(A) 0.65 m.
(B) 0.95 m.
(C) 1.15 m.
(D) 2.25 m.

24. The angular velocity of rod BC is closest to

(A) 0.483 rad/s clockwise.
(B) 0.483 rad/s counter-clockwise.
(C) 2.0 rad/s clockwise.
(D) 2.0 rad/s counter-clockwise.

Questions 25–27 pertain to the following situation:

Wet steam at the rate of 1.5 kg/s at 2 MPa is flowing through a pipe. A small portion of the steam is throttled in a calorimeter to a final pressure of 100 kPa and a final temperature of 150°C.

25. Which one of the following best describes the process through a calorimeter?

    (A) An adiabatic saturation process

    (B) An isentropic process

    (C) A throttling process

    (D) A constant volume process

26. The quality of the steam in the line is close to

    (A) 10%.

    (B) 40%.

    (C) 80%.

    (D) 99%.

27. The volume flow rate of the steam is

    (A) 0.78 m^3/min.

    (B) 8.86 m^3/min.

    (C) 98.6 m^3/min.

    (D) 442 m^3/min.

Questions 28–30 pertain to the following situation:

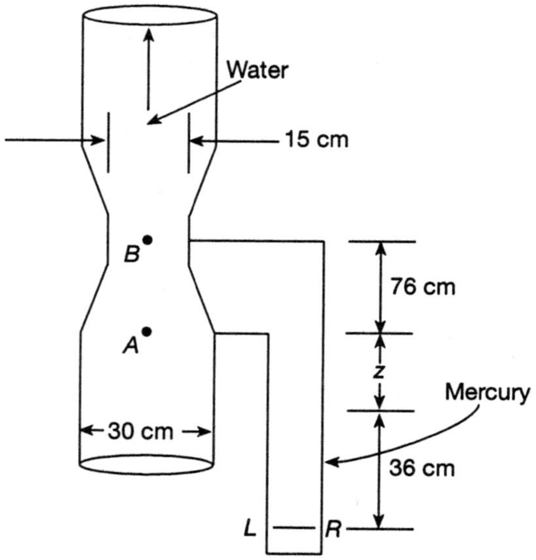

For the venturi meter shown on the previous page, the deflection of the mercury in the differential gage is 36 cm.

28. The pressure difference between point *A* and point *B* is approximately

    (A) 20 kPa. (C) 52 kPa.

    (B) 40 kPa. (D) 92 kPa.

29. The velocity at point *A* is approximately

    (A) 0.4 m/s. (C) 2.4 m/s.

    (B) 1.4 m/s. (D) 4.4 m/s.

30. The volume flow rate at point *B* is

    (A) 10.3 m^3/min. (C) 31.2 m^3/min.

    (B) 20.6 m^3/min. (D) 41.3 m^3/min.

Questions 31–33 pertain to the following situation:

The instrumentation to measure the volume flow rate of 27°C air at atmospheric pressure flowing through a rectangular duct 20 cm by 10 cm is to be investigated. The flow rate will be determined by multiplying the duct area by the flow velocity. A pitot tube connected to a water manometer will be used to measure the velocity of the air. The maximum manometer reading is limited to 15 cm. The measurements of length, width, and manometer head can only be read to an accuracy of ±0.1 cm.

31. The maximum air velocity that can be read is

    (A) 50 m/s. (C) 100 m/s.

    (B) 75 m/s. (D) 150 m/s.

32. The maximum volume flow rate is closest to

    (A) 0.5 m^3/s. (C) 1.0 m^3/s.

    (B) 0.75 m^3/s. (D) 2.0 m^3/s.

33. Assuming that the only errors in this experiment are due to the length measurements, the maximum error in reading the volume flow rate is closest to

(A) 1.2%.
(B) 3.2%.
(C) 7%.
(D) 10%.

Questions 34 and 35 pertain to the following situation:

A solid shaft is required to transmit 22.4 kW of power at 100 rpm. The shear modulus of the shaft is $G = 77.2 \times 10^9$ N/m^2 and the length of the shaft is 3 m.

34. The diameter of the shaft, so that the shearing stress does not exceed 4.14×10^4 kPa, is closest to

(A) 4.4 cm.
(B) 5.4 cm.
(C) 6.4 cm.
(D) 7.4 cm.

35. The diameter of the shaft so that the angle of twist is no more than 5.73° is closest to

(A) 4.4 cm.
(B) 5.4 cm.
(C) 6.4 cm.
(D) 7.4 cm.

36. A steel rod 254 cm long and 12.9 cm^2 in cross section is secured between two walls. The load on the rod is zero at 21°C. Assuming the walls are rigid, the stress developed in the rod when the temperature drops to –18°C is

(A) 41.3 MPa.
(B) 61.3 MPa.
(C) 81.3 MPa.
(D) 91.3 MPa.

Questions 37–39 pertain to the following situation:

An instrument weighing 11 N is set on three rubber mounts rated at 0.036 cm deflection per N each.

37. The effective spring constant for this system is

(A) 28 N/cm.
(B) 56 N/cm.
(C) 84 N/cm.
(D) 96 N/cm.

38. The natural frequency of the vibrating system is closest to

(A) 13.8 cps. (C) 83.8 cps.

(B) 33.8 cps. (D) 93.8 cps.

39. The maximum deflection for this system is closest to

(A) 0.0152 cm. (C) 4.71 cm.

(B) 0.131 cm. (D) 23.1 cm.

Questions 40 and 41 pertain to the following situation:

The plane wall pictured below is made of three materials of differing widths and thermal conductivity. T_1 is maintained at 150°C while T_4 is maintained at 10°C. The wall has the following values of thermal conductivities:

$$k_1 = 0.074 \text{ W/m-K}$$

$$k_2 = 0.69 \text{ W/m-K}$$

$$k_3 = 0.067 \text{ W/m-K}$$

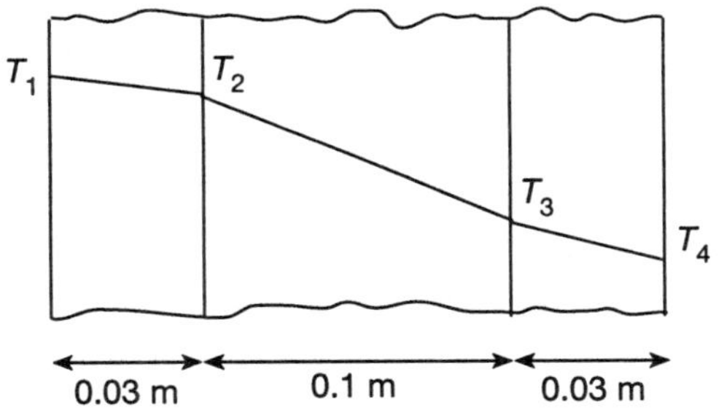

40. The heat flow through a 1 m^2 section of the wall is closest to

(A) 40 W. (C) 120 W.

(B) 80 W. (D) 140 W.

41. T_2, the temperature at the interface between the first and second sections, is closest to

(A) 120°C. (C) 93°C.

(B) 100°C. (D) 45°C.

42. Two large parallel plates are maintained at 727°C and 227°C respectively. Both plates have emissivities equal to 0.5. The radiation heat transfer between the two plates is

(A) 1,000 W/m^2.
(B) 2,000 W/m^2.
(C) 5,000 W/m^2.
(D) 17,700 W/m^2.

Questions 43 and 44 pertain to the following situation:

A rigid tank 0.1 m^3 in volume contains R-134a in a saturated vapor state at a temperature of 60°C. The R-134a is then cooled to a temperature of 30°C.

43. The mass of the R-134a in the tank is approximately

(A) 1 kg.
(B) 5 kg.
(C) 9 kg.
(D) 20 kg.

44. The quality, x, of the R-134a in the final state is closest to

(A) 41%.
(B) 61%.
(C) 81%.
(D) 100%.

45. In a cylinder-piston device, 0.1 kg of nitrogen at 100 kPa and 27°C is compressed isothermally to a final pressure of 400 kPa. The work and heat transfer for this process is closest to

(A) $Q_{1-2} = W_{1-2} = +12.3$ kJ
(B) $Q_{1-2} = W_{1-2} = -12.3$ kJ
(C) $Q_{1-2} = +12.3$ kJ; $W_{1-2} = -12.3$ kJ
(D) $Q_{1-2} = -12.3$ kJ; $W_{1-2} = +12.3$ kJ

Questions 46–48 pertain to the following situation:

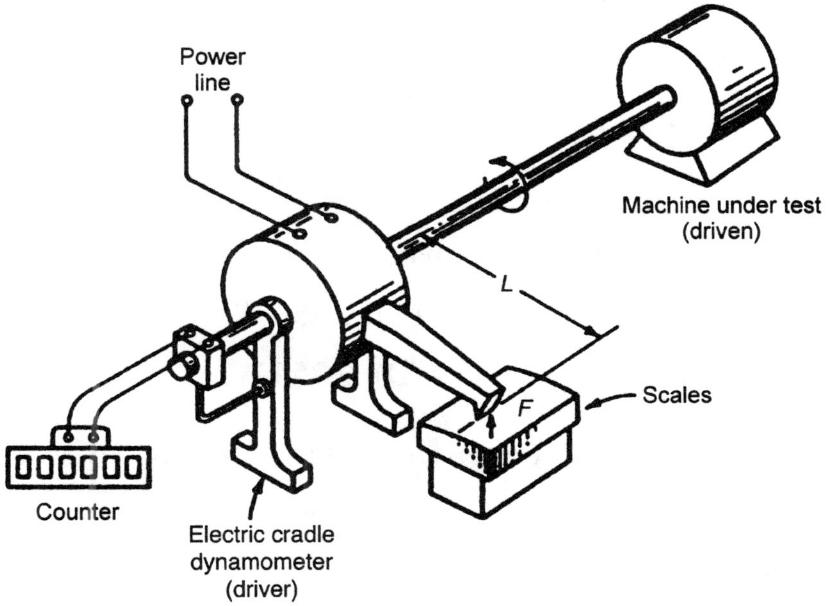

Consider an experiment for measuring by means of a dynamometer the average power transmitted by a rotating shaft. The formula for power can be written as:

$$\dot{W} = \frac{2\pi RFL}{t}$$

where:

R = revolutions of shaft during t

F = force at end of torque arm, N

L = length of torque arm, m

t = time, s

For a specific run the data are:

$R = 1{,}202 \pm 1.0$ revolution

$F = 45.01\text{ N} \pm 0.18\text{ N}$

$L = 39.7\text{ cm}$

$t = 60.0 \pm 0.50\text{ s}$

The length of the torque arm was measured with a ruler whose smallest division was 2.5 mm.

46. The error that should be assigned to the torque arm is closest to

    (A) 0.063 cm.
    (B) 0.125 cm.
    (C) 0.25 cm.
    (D) 0.50 cm.

47. The power transmitted by the shaft is

    (A) 1.25 kW.
    (B) 1.75 kW.
    (C) 2.25 kW.
    (D) 2.75 kW.

48. The combined error for this experiment is closest to

    (A) 0.8%.
    (B) 1.0%.
    (C) 1.6%.
    (D) 2.6%.

Questions 49–51 are based on the following situation:

One helical spring is nested inside another: the dimensions are as tabulated. Both springs have the same free length and carry a total maximum load of 2,500 N. The shear modulus of the spring material is 82.7 GPa.

	Outer Spring	**Inner Spring**
No. active coils	6	10
Wire diameter	1.3 cm	0.65 cm
Mean spring diameter	9.0 cm	5.8 cm

49. The maximum load carried by either spring is

    (A) 307 N.
    (B) 1,200 N.
    (C) 1,600 N.
    (D) 2,190 N.

50. The total deflection of each spring is

    (A) 0.45 cm.
    (B) 2.0 cm.
    (C) 3.2 cm.
    (D) 23.1 cm.

51. The maximum shear stress in the outer spring is

(A) 125 MPa.
(B) 34.3 MPa.
(C) 245 MPa.
(D) 425 MPa.

52. Rhodium has a FCC structure, an atomic weight of 102.9 g/mole, and an atomic radius of 1.345 Å. Its density is closest to

(A) 1.67 g/cm^3.
(B) 4.56 g/cm^3.
(C) 12.4 g/cm^3.
(D) 14.2 g/cm^3.

53. Suppose we have a 323 mm^2 bar that is 60% glass and 40% polyester by volume. The bar is subjected to a load of 44,500 N. Take the modulus of elasticity for glass, E_g, to be 70 GPa and the modulus of elasticity for the polyester, E_p, to be 2.76 GPa. The strain in the composite is closest to

(A) 1.19×10^{-3}.
(B) 2.19×10^{-3}.
(C) 3.19×10^{-3}.
(D) 1.00×10^{-4}.

54. What is the transfer function of a pole-zero diagram with a pole of $s = -3$ and a zero at $s = -9$?

(A) $T(s) = \dfrac{K(s-3)}{s-9}$
(B) $T(s) = \dfrac{K(s-9)}{s-3}$
(C) $T(s) = \dfrac{K(s+3)}{s+9}$
(D) $T(s) = \dfrac{K(s+9)}{s+3}$

55. Which answer best describes the function shown below?

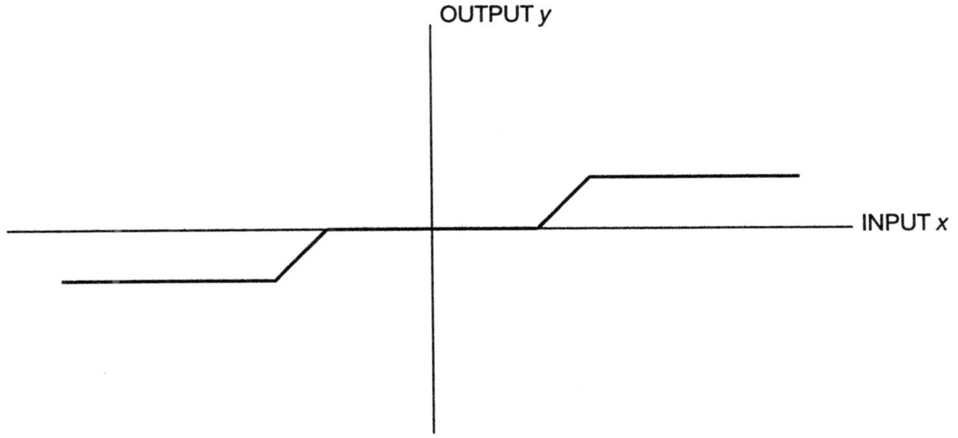

(A) There is both a dead and a saturated zone.

(B) It has an impulse zone.

(C) It has a dead zone.

(D) It has a ramp system.

56. A control system has a control system response ratio of

$$\frac{T(s)}{R(s)} = \frac{1}{s^2 - 0.3s + 1}$$

If the damping ratio is less than 0.707, how many peaks occur in the transient prior to reaching steady-state conditions?

(A) None, only a minimum

(B) None, it is a flat response

(C) One

(D) Two

Questions 57 and 58 are based on the following situation:

A power screw has 2.5 square threads per cm, double threads, and a major diameter, d, of 2.5 cm. Assume that $\mu = \mu_c = 0.08$, d_c, the mean collar diameter, equals 3.2 cm and the load per screw, F, equals 6,700 N.

57. The torque required to rotate the screw against the load is closest to

(A) 207 N-cm.

(B) 2,300 N-cm.

(C) 5,000 N-cm.

(D) 7,000 N-cm.

58. The overall efficiency of the power screw is

(A) 10%.

(B) 40%.

(C) 60%.

(D) 80%.

59. A spring composed of hard drawn wire has a wire diameter of 2.0 mm. The maximum allowable torsional stress for the wire is

(A) 590 MPa.

(B) 1,020 MPa.

(C) 1,310 MPa.

(D) 1,555 MPa.

60. Refer to the phase diagram below of Cu-Ag alloy.

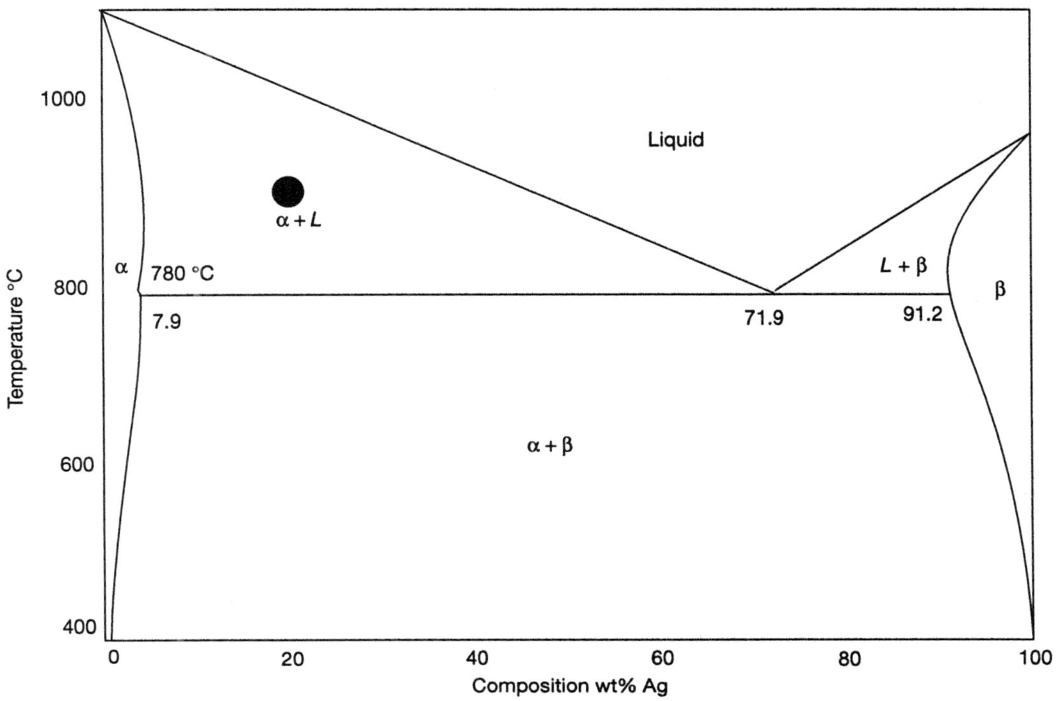

For the point shown on the figure, the weight percent of α and *L* (liquid) is closest to

(A) $\alpha = 20\%, L = 80\%$.

(B) $\alpha = 59\%, L = 41\%$.

(C) $\alpha = 80\%, L = 20\%$.

(D) $\alpha = 41\%, L = 59\%$.

TEST 2

ANSWER KEY

1. (D)
2. (C)
3. (A)
4. (D)
5. (B)
6. (C)
7. (D)
8. (A)
9. (B)
10. (B)
11. (D)
12. (C)
13. (C)
14. (D)
15. (C)
16. (A)
17. (B)
18. (D)
19. (A)
20. (C)
21. (D)
22. (B)
23. (C)
24. (B)
25. (C)
26. (D)
27. (B)
28. (C)
29. (C)
30. (A)
31. (A)
32. (C)
33. (A)
34. (C)
35. (B)
36. (D)
37. (C)
38. (A)
39. (B)
40. (D)
41. (C)
42. (D)
43. (C)
44. (A)
45. (B)
46. (B)
47. (C)
48. (B)
49. (D)
50. (C)
51. (C)
52. (C)
53. (C)
54. (D)
55. (A)
56. (C)
57. (B)
58. (B)
59. (A)
60. (B)

DETAILED EXPLANATIONS OF ANSWERS

TEST 2

1. **(D)**

Choices (A) and (B) are incorrect since the spreadsheet would interpret them as labels and not formulas. (C) would give the value corresponding to y_1 when the value corresponding to y_3 is required. A3 * A3 – 1 means the contents of cell B3 is the contents of cell A3 times the contents of cell A3 minus 1 which is the required result.

2. **(C)**

$Z = 2$ and as such the conditional statement $Z < 0$ is false which means that the statement $Z = Z + 1$ will not be executed. (A) would be the correct response if $Z = -2$ since the conditional statement would apply in this case. (D) would be the correct response if the conditional statement was "if $Z > 0$" in which case the 2 would be increased by 1.

3. **(A)**

This program segment gives a four-term series that has alternating plus and minus signs. (B) and (C) are incorrect since the plus and minus signs do not alternate. (D) is incorrect since the terms in the denominator are not to be squared.

4. **(D)**

$$\dot{Q} = 500 \text{ l/s} = 0.5 \text{ m}^3/\text{s}$$

$$\dot{m} = \frac{\dot{Q}}{v_a}$$

where v_a is the specific volume of the dry air and v_a can be found from a psychrometic chart. At $T = 24°\text{C}$ and $\Phi = 50\%$, $v_a = 0.854 \text{ m}^3/\text{kg}$. Thus:

$$\dot{m} = \frac{\dot{Q}}{v_a} = \frac{\left(0.5 \text{ m}^3/\text{s}\right) \times (60 \text{ s/min})}{0.854 \text{ m}^3/\text{kg}_\text{a}} = 35.1 \text{ kg}_\text{a}/\text{min}$$

5. **(B)**

The air conditioner is treated as a simple thermodynamic open system with kinetic and potential energy neglected. Since work is zero for this process, the energy equation becomes:

$$\dot{Q} = \dot{m}(h_2 - h_1)$$

where h_1 and h_2 are found in the psychrometric chart. T_1 = 24°C, Φ_1 = 50%, T_2 = 4°C, Φ_2 = 20%, h_1 = 48 kJ/kg$_a$, and h_2 = 6 kJ/kg$_a$

$$\dot{Q} = \dot{m}(h_2 - h_1) = \frac{(35.1 \text{ kg}_a/\text{min}) \times [(6.0 - 48) \text{ kJ/kg}]}{60 \text{ s/min}} = -24.6 \text{ kW}$$

The negative implies heat is being removed from the air. As 1 kW = 0.284 tons of cooling, then:

$$\dot{Q} = (24.6 \text{ kW}) \times (0.284 \text{ tons/kW}) = 6.99 \text{ tons}$$

6. **(C)**

The mass flow rate of the vapor is related to the mass flow rate of the air by:

$$\dot{m}_v = \dot{m}_a(W_2 - W_1)$$

where W_1 = 9.4 gm$_v$/kg$_a$ and W_2 = 1 gm$_v$/kg$_a$ (values from psychrometric chart).

$$\dot{m}_v = \dot{m}_a(W_2 - W_1)$$
$$= (-35.1 \text{ kg}_a/\text{min}) \times [(1 - 9.4) \text{ gm}_v/\text{kg}_a] -295\text{g}_v/\text{min}$$
$$= -0.295 \text{ kg}_v/\text{min}$$

Again the negative sign indicates the moisture is being removed from the air. To find the moisture removed per day, multiply the previous result by 60 × 24 which yields 425 kg/day.

7. **(D)**

The continuity equation is used to find V_2. For an incompressible fluid:

$$A_1 V_1 = A_2 V_2$$
$$V_2 = (A_1/A_2)\ V_1 = ((\pi/4)\ D_1^2/((\pi/4)\ D_2^2\ )) \times V_1$$
$$= (0.3)^2 / (0.15)^2 \times 3 = 12 \text{ m/s}$$

8. **(A)**

Since no information was provided concerning loss, assume the flow to be frictionless. We can write Bernoulli's equation from point 1 to point 2 as follows:

$$\frac{p_1}{\gamma} + z_1 + \frac{V_1^2}{2g} = \frac{p_2}{\gamma} + z_2 + \frac{V_2^2}{2g} = \frac{\left(82.7 \times 10^3 \text{ N/m}^2\right)}{\left(9{,}800 \text{ N/m}^3\right)} + 0 + \frac{(3 \text{ m/s})^2}{(2) \times \left(9.8 \text{ m/s}^2\right)}$$

$$= \frac{p_2}{\left(9{,}800 \text{ N/m}^3\right)} + 0 + \frac{(12 \text{ m/s})^2}{(2) \times \left(9.8 \text{ m/s}^2\right)}$$

$$p_2 = 15{,}200 \text{ Pa} = 15.2 \text{ kPa}$$

9. **(B)**

Assume that the propeller input can be treated as a pump. The equivalent head input of the pump will first be found and then the energy equation will be used to find the final pressure. The pump equation is as follows:

$$\dot{W} = \frac{Q\gamma E_m}{\eta}$$

Assume that η is 100%.

$$Q = V_2 A_2 = (12 \text{ m/s}) \times (\pi/4) \times (0.15)^2$$

$$Q = 0.212 \text{ m}^3\text{/s}$$

Substituting:

$$2.92 \times 10^3 \text{ J/s} = (9{,}800 \text{ N/m}^3) \times (E_m) \times (0.212 \text{ m}^3\text{/s})$$

$$E_m = 1.4 \text{ m}$$

The energy equation for this case becomes:

$$\frac{p_2}{\gamma} + E_m = \frac{p_3}{\gamma};\ \frac{\left(15{,}200 \text{ N/m}^2\right)}{9{,}800 \text{ N/m}^3} + 1.4\text{m} = \frac{p_3}{9{,}800 \text{ N/m3}};$$

$$p_3 = 28{,}900 \quad \text{Pa} = 28.9 \text{ kPa}$$

10. **(B)**

Fans and pumps follow laws based on dynamic similitude. For the case in point, the following equations apply:

$$\frac{Q_2}{Q_1} = \left(\frac{N_2}{N_1}\right)\left(\frac{D_2}{D_1}\right)^3$$

$$\frac{H_2}{H_1} = \left(\frac{N_2}{N_1}\right)^2\left(\frac{D_2}{D_1}\right)^2$$

$$\frac{\dot{W}_2}{\dot{W}_1} = \left(\frac{N_2}{N_1}\right)^3\left(\frac{D_2}{D_1}\right)^5$$

The first equation can be solved for D_2, the required new diameter.

$$D_2 = D_1\left[\left(\frac{Q_2}{Q_1}\right)\left(\frac{N_1}{N_2}\right)\right]^{1/3} = 7.5\text{ cm}\left[\left(\frac{3.78}{2.27}\right)\left(\frac{1{,}750}{3{,}500}\right)\right]^{1/3} = 7.06\text{ cm}$$

11. **(D)**

The second equation in Question 10 can be used to solve for the new head.

$$H_2 = 107 \times \left(\frac{3{,}500}{1{,}750}\right)^2 \times \left(\frac{7.06}{7.5}\right)^2 = 379\text{ m}$$

12. **(C)**

Before we can use the third equation from Question 10 to find the power requirement, the power requirement for the original pump is required. If the pumps are geometrically similar, then they will have the same efficiency.

$$\dot{W} = \frac{Q\gamma E_m}{\eta} = \frac{\left(9{,}800\text{ N/m}^2\right)\times(107\text{ m})\times\left(2.27/60\text{ m}^3\text{/s}\right)}{0.6}$$

$$= 6.62 \times 10^4\text{ W} = 66.2\text{ kW}$$

$$\dot{W}_2 = 66.2\text{ kW}\left(\frac{7.06}{7.5}\right)^5 \times \left(\frac{3{,}500}{1{,}750}\right)^3 = 391\text{ kW}$$

13. **(C)**

This problem requires several steps.

Generator input = 500 kW/0.9 = 556 kW

Engine intermittent power rating = 556 kW/0.8 = 694 kW

Fuel consumption (fc) = bsfc × power rating = (694 kW) × (0.243 kg/kW-hr) × (1/60 hr/min) = 2.81 kg/min (fuel)

Air consumption = afr × fc = 20 × 2.81 = 56.2 kg/min (air)

14. **(D)**

The volume flow rate of air into the engine is required. This can be found by dividing the mass flow rate of the air by its density. The density of the air is found from the ideal gas law as follows:

$$\rho = \frac{P}{RT}$$

where P = 632 mm Hg = 84 kPa and T = (35°C + 273) = 308 K and R = 0.287 kPa-m^3/kg-K

$$\rho = \frac{(84 \text{ kPa})}{(0.287 \text{ kPa-m}^3/\text{kg-K}) \times (308 \text{ K})} = 0.95 \text{ kg/m}^3$$

$$\text{Air volume} = \frac{\dot{m}}{\rho} = \frac{56.3 \text{ kg/min}}{0.95 \text{kg/min}^3} = 59.2 \text{ m}^3\text{/min}$$

$$\text{Engine displacement} = \frac{59.2}{0.85} = 69.6 \text{ m}^3\text{/min}$$

Displacement of cylinder in cm^3 = (69.6 m^3/min) × (10^6 cm^3/m^3) ×

$$\frac{1}{12} \times \frac{1}{(1{,}200 / 2)} = 9{,}666 \text{ cm}^3$$

Note: The factor of $^1/_{12}$ is due to the 12-cylinder engine. The 1,200 rpm is divided by 2 since this is a four-stroke engine (two revolutions are required to complete a cycle).

15. **(C)**

The ideal volume of the cylinder (not counting the clearance volume) would be (π/4) × bore^2 × stroke. The problem gives the stroke as 1.2 times the bore, thus:

$$(\pi/4) \times \text{bore}^2 \times \text{stroke} = (\pi/4) \times \text{bore}^2 \times 1.2 = 9{,}666$$

where the bore = 21.7 cm and the stroke = 26 cm. To find the clearance volume, V_{cl}, the following equation is used:

$$CR = \frac{V_{disp} + V_{cl}}{V_{cl}},\ \text{where } CR \text{ is the compression ratio.}$$

$$16 = \frac{9{,}666 + V_{cl}}{V_{cl}}$$

Solving:

$$V_{cl} = 644 \text{ cm}^3$$

16. **(A)**

Using the free-body diagram, the tensions in the rods, P_1 and P_2, can be determined.

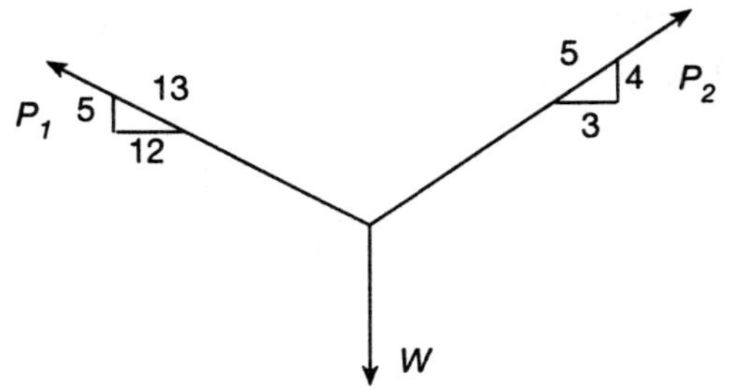

$$\Sigma F_H = 0$$

$$P_1\ (12/13) = P_2\ (3/5)$$

$$\Sigma F_V = 0$$

$$P_1\ (5/13) + P_2\ (4/5) = W$$

Substituting:

$$P_2\ (3/5)\ (13/12)\ (5/13) + P_2\ (4/5) = W$$

$$P_2 = (20/21)\ W$$

Therefore:

$$P_1 = (13/21)\ W$$

Answer choice (B) is P_2, while (C) and (D) are simply incorrect.

17. **(B)**

From the maximum allowable stress, the maximum weight W can be found. The maximum allowable load on CB is:

$$\sigma_2 = \frac{P_2}{A_2} = \frac{\frac{20}{21}W}{0.645 \times 10^{-4}\ \text{m}^2} = \frac{U.S.}{F.S.} = \frac{1.1 \times 10^9\ \text{Pa}}{4}$$

$$W = 18{,}700\ \text{N}$$

The maximum allowable load on *AB* is:

$$\sigma_1 = \frac{P_1}{A_1} = \frac{\frac{13}{21}W}{1.29\times10^{-4}\ \text{m}^2} = \frac{U.S.}{F.S.} = \frac{1.1\times10^9\ \text{Pa}}{4}$$

$$W = 57{,}300\ \text{N}$$

The smaller of the two loading scenarios, that is *CB*, governs the maximum loading allowed. (A) and (C) are incorrect. (D) is the maximum weight for *AB*.

18. **(D)**

The total deformation is the sum of the individual deformations and is given by the following equation:

$$\Delta u = \sum \frac{PL}{AE}$$

$$\Delta u = 0.00025\ \text{m}$$

The values of *E* for aluminum and steel are available in reference manuals:

$$E_{steel} = 200\times10^9\ \text{N/m}^2 \text{ and } E_{al} = 69\times10^9\ \text{N/m}^2$$

Substituting:

$$0.00025\ \text{m} = \frac{P(0.3\ \text{m})}{(0.05\ \text{m})^2\times\left(200\times10^9\ \text{N/m}^2\right)} + \frac{P(0.38\ \text{m})}{(0.1\ \text{m})^2\times\left(69\times10^9\ \text{N/m}^2\right)}$$

$$P = 217{,}000\ \text{N}$$

19. **(A)**

The Biot number is the parameter that is used to determine the appropriate solution method for transient heat transfer problems.

$$Bi = \frac{h(V\,/\,A)}{k}$$

where (*V*/*A*) is the characteristic length of the system. For a sphere, the characteristic length is (*r*/3). The following properties are for copper:

$$\rho = 8{,}954\ \text{kg/m}^3$$

$$c = 383\ \text{J/kg}$$

$$k = 386 \text{ W/m-K}$$

$$Bi = \frac{h(V/A)}{k} = \frac{hr}{3k} = \frac{\left(15 \text{ W/m}^2\text{-K}\right) \times (0.015 \text{ m})}{(3) \times \left(386 \text{W/m-K}\right)} = 1.94 \times 10^{-4}$$

Answer choices (B) and (C) are incorrect, while (D) is the characteristic length.

20. **(C)**

Since the "lumped capacity method" is appropriate, the following equation applies:

$$\frac{T - T_\infty}{T_i - T_\infty} = e^{-\frac{hA_s}{\rho c V} t}$$

where:

T_i = initial temperature of body

T_∞ = ambient temperature

T = body temperature at a given time

ρ = density of body, kg/m^3

c = heat capacity of body, J/kg-K

t = time

$$\frac{hA}{\rho c V} = \frac{h}{\rho c L} = \frac{15 \text{ W/m}^2\text{-K}}{\left(8{,}954 \text{ kg/m}^3\right)\left(383 \text{ J/kg-K}\right)\left(0.005 \text{ m}\right)} = 8.75 \times 10^{-4} s^{-1}$$

$$\frac{T - T_\infty}{T_i - T_\infty} = \frac{25 - 10}{50 - 10} = e^{-\frac{hA_s}{\rho c V} t} = e^{-8.75 \times 10^{-4} t}$$

$$\tau = 1{,}121 \quad \text{s} = 3.11 \text{ hours}$$

21. **(D)**

A first law energy balance is performed on the plate. Since the plate is at equilibrium, Kirchhoff's law applies, which states that $\alpha = \epsilon$.

$$q_{in} = \alpha_{solar} \, G = 0.9 \times 788 \text{ W/m}^2 = 709 \text{ W/m}^2$$

$$q_{out} = q_{conv} + q_{rad}$$

$$q_{conv} = h(T_p - T_\infty)$$

$$q_{rad} = \sigma \in_{long} (T_p^4 - T_\infty^4)$$

Since absolute temperatures are required for the radiation part, the convection temperature difference should likewise be expressed in terms of absolute temperatures.

$$709 \text{ W/m}^2 = 17 \text{ W/m}^2\text{-K}(T_p - 283 \text{ K}) + 5.67 \times 10^{-8} \text{ W/m}^2\text{-K}^4)$$
$$\times (0.1) \times [T_p^4 - (283 \text{ K})^4]$$

Trial and error is required to complete the solution. By substituting different values it is found that:

$$T = 323 \quad \text{K} = 50°\text{C}$$

22. **(B)**

Since the rod and disk are pinned at point B, the motion of point B of the rod and of the disk are the same. Let's consider the disk. The total acceleration can be broken into normal and tangential components as follows:

$$\bar{a}_B = (\bar{a}_B)_n + (\bar{a}_B)_t$$

$$\bar{a}_B = 4.0 \text{ m/s}^2 \text{ and } (\bar{a}_B)_t = 1.5 \text{ m/s}$$

Since acceleration is a vector quantity, the scalar components add as follows:

$$a^2 = a_n^2 + a_t^2$$

Solving for a_n or $a_n = \sqrt{a^2 - a_t^2} = \sqrt{4^2 - 1.5^2} = 3.71 \text{ m/s}^2$

Since point B on the disk represents a body rotating about a fixed point, $a_n = r\,\omega_{AB}^2$. Solving for ω_{AB}:

$$\omega_{AB} = \sqrt{\frac{a_n}{r}} = \sqrt{\frac{3.71 \text{ m/s}^2}{1 \text{ m}}} = 1.93 \text{ rad/s}$$

23. **(C)**

The rod can be considered a bar with a mass of 15 kg. The radius of gyration is related to the mass moment of inertia by the following:

$$r_z = \sqrt{\frac{I}{m}}$$

To find the mass moment of inertia about the centroid of the rod use:

$$\bar{I} = \frac{1}{12} mL^2 = \frac{1}{12} \times 15 \text{ kg} \times 2 \text{ m}^2 = 5 \text{ kg-m}^2$$

To find the moment of inertia about the end, the parallel axis theorem is used:

$$I = \bar{I} + md^2 = 5 + 15 \text{ kg} \times 1 \text{ m}^2 = 20 \text{ kg-m}^2$$

$$r_z = \sqrt{\frac{I}{m}} = \sqrt{\frac{20 \text{ kg-m}^3}{15 \text{ kg}}} = 1.15 \text{ m}$$

24. **(B)**

First, the velocity of point B can be found from ω_{AB} as follows:

$$V_B = r_{AB} \times \omega_{AB} = (1 \text{ m}) \times (1.93 \text{ rad/s}) = 1.93 \text{ m/s}.$$

To find ω_{BC}, the instantaneous center of rotation of bar BC is located. This is shown in the sketch below. The instantaneous center is the imaginary point about which an object is perceived as rotating. In terms of this instantaneous center, the velocity of point B is expressed as follows:

$$V_B = r_{B/IC}\ \omega_{BC}$$

$$V_B = 1.93 \text{ m/s} = 4\ \omega_{BC}$$

Solving for ω_{BC}:

$$\omega_{BC} = \frac{1.93 \text{ m/s}}{4 \text{ m}} = 0.483 \text{ rad/s counter-clockwise}$$

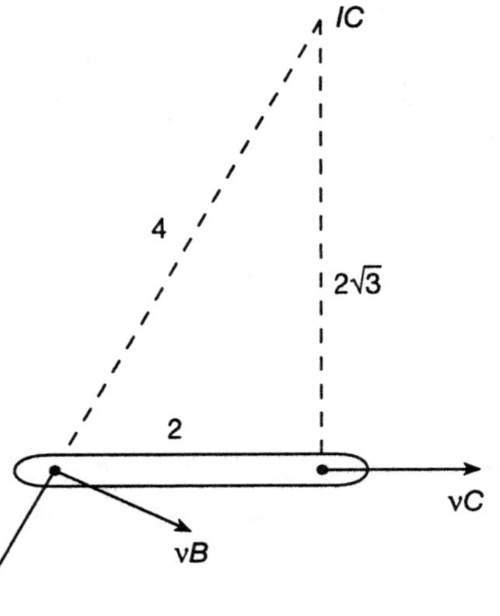

25. **(C)**

The calorimeter is based on the principle of a throttling process, which becomes isoenthalpic (constant enthalpy). (A) An adiabatic saturation process involves moist air and not steam. (B) The process, while being adiabatic, is highly non-isentropic. There is a great loss of entropy for this process. (D) This is an open system and as such the volumes will change.

26. **(D)**

For a throttling process, $h_1 = h_2$, where h_1 is the enthalpy of wet steam and h_2 is the enthalpy of the escaping steam. Provided the escaping steam is superheated, its enthalpy can be readily found from the steam tables. Hence, $h_2 = 2{,}776.4$ kJ/kg $= h_1$. At 2 MPa, this value of enthalpy is in the wet region, or:

$$h_f = 908.79 \text{ kJ/kg}$$

$$h_g = 2{,}799.5 \text{ kJ/kg}$$

$$h_{fg} = 1{,}890.7 \text{ kJ/kg}$$

$$h_l = h_f + xh_{fg}$$

where x is the quality of the steam.

Solving for x:

$$x = \frac{h_1 - h_f}{h_{fg}} = \frac{2{,}776.4 \text{ kJ/kg} - 908.79 \text{ kJ/kg}}{1{,}890.7 \text{ kJ/kg}}$$

$$= 0.988 \text{ or } 98.8\%$$

27. **(B)**

The volumetric flow rate, $\dot{Q}$, is given as follows:

$$\dot{Q} = \dot{m}v$$

where v is the specific volume of the steam in the pipe, $v = v_f + x(v_g - v_f)$. For a pressure of 2 MPa, this equation becomes:

$$v = v_f + x\,(v_g - v_f)$$

$$= (0.001177 \text{ m}^3\text{/kg}) + (0.988) \times (0.09963 \text{ m}^3\text{/kg}) - 0.001177 \text{ m}^3\text{/kg})$$

$$v = 0.0984 \text{ m}^3\text{/kg}$$

$$\dot{Q} = (1.5\text{kg/s}) \times (0.0984 \text{ m}^3\text{/kg}) = 0.148 \text{ m}^3\text{/s} = 8.86 \text{ m}^3\text{/min}$$

28. **(C)**

Since the problem states that no energy is lost between point *A* and point *B*, Bernoulli's equation applies.

$$\frac{p_A}{\gamma} + z_A + \frac{V_A^2}{2g} = \frac{p_B}{\gamma} + z_B + \frac{V_B^2}{2g}$$

To find $p_A - p_B$, the manometer that connects points *A* and *B* will be used. Moving down the leg of the manometer increases pressure while moving up the leg of the manometer decreases pressure as will be shown. The change in pressure will be expressed in meters of water.

$$\frac{P_A}{\gamma} + z + 0.36\text{ m} - (0.36)\times(13.6)\text{ m} - z - 0.76\text{ m} = \frac{P_B}{\gamma}$$

$$\frac{P_A - P_B}{\gamma} = 5.3\text{ m};\ P_A - P_B = (5.3m)\times\left(9{,}800\text{ N/m}^2\right) = 51.9\text{ kPa}$$

Note: The 13.6 term above for density is used for the section of the manometer containing mercury.

29. **(C)**

The continuity equation for an incompressible substance will be used to express V_A in terms of V_B. For an incompressible substance, density is assumed constant and the continuity equation becomes:

$$A_A V_A = A_B V_B$$

$$V_A = (A_B/A_A)V_B = (\pi/4)/(\pi/4) \times (D_B^2 / D_A^2)V_B = (1/4)V_B$$

Rearranging Bernoulli's equation:

$$\frac{P_A - P_B}{\gamma} + z_A + \frac{V_A^2}{2g} = z_B + \frac{V_B^2}{2g} = 5.3\text{ m} + 0 + \frac{V_A^2}{2g} = 0.76\text{ m} + \frac{(4V_A)^2}{2g}$$

$$V_A = 2.43\text{ m/s}$$

$$V_B = 9.74\text{ m/s}$$

30. **(A)**

The volume flow rate Q should not be confused with the mass flow rate $\dot{m}$. For an incompressible substance:

$$Q = AV = A_A V_A = (\pi/4)(D_A^2)V_A = (\pi/4)(0.3^2\text{ m}^2)(2.43\text{ m/s})$$

$$Q = 0.171\text{ m}^3\text{/s} = 10.3\text{ m}^3\text{/min}$$

31. **(A)**

The velocity of a fluid stream as read by a Pitot tube is given by the following equation:

$$V = \sqrt{(2/\rho)(p_o - p_s)}$$

where ρ is the density of the fluid being measured and $p_o - p_s$ is the difference between the static and stagnation pressure. This pressure difference is measured by the water manometer, which has a maximum head reading of 0.15 m of water. From the manometer equation:

$$\Delta p = p_o - p_s = \gamma_W \times h = 9{,}800 \text{ N/m}^3 \times 0.15 \text{ m} = 1{,}470 \text{ N/m}^2$$

Substituting:

$$V = \sqrt{(2/\rho)(p_o - p_s)} = \sqrt{\frac{2}{1.16 \text{ kg/m}^3}\left(1{,}470 \text{ N/m}^2\right)} = 50.3 \text{ m/s}^2$$

Note that the density of air was determined using the ideal gas law, atmospheric pressure, and a temperature of 27°C.

32. **(C)**

For a rectangular duct, the volume flow rate is given by the following:

$$Q = V \times L \times W = (50.5 \text{ m/s})\ (0.2 \text{ m})(0.1 \text{ m}) = 1.01 \text{ m}^3\text{/s}$$

33. **(A)**

In terms of the manometer head h, the volume flow rate can be written as follows:

$$Q = \sqrt{(2/\rho)(\gamma_w)(h)} \times L \times W$$
$$= \sqrt{(2/\rho)(\gamma_w)}\sqrt{(h)} \times L \times W = 129.9\sqrt{h} \times L \times W$$

The values of γ_w and ρ will be treated as constants in the following analysis. As was stated in the problem, only the errors in measurements will be considered.

$$h = 0.15 \text{ m} \pm 0.001 \text{ m}$$

$$L = 0.2 \text{ m} \pm 0.001 \text{ m}$$

$$W = 0.1 \text{ m} \pm 0.001 \text{ m}$$

The combined error for this situation is found by applying the following equation:

$$W_R = \left[\left(\frac{\partial R}{\partial x_1} W_1\right)^2 + \left(\frac{\partial R}{\partial x_2} W_2\right)^2 + \ldots + \left(\frac{\partial R}{\partial x_n} W_n\right)^2\right]^{1/2}$$

where:

W_R = uncertainty in the results

W_1, W_2, ..., W_n = uncertainty in the independent variables $x_1, x_2, \ldots, x_n$

$R = Q = 1.01 \text{ m}^3/\text{s}$

$x_1 = h \qquad W_1 = 0.001 \text{ m}$

$x_2 = L \qquad W_2 = 0.001 \text{ m}$

$x_3 = W \qquad W_3 = 0.001 \text{ m}$

$$\frac{\partial Q}{\partial h} = \frac{1}{2} \times h^{-1/2} \times 129.9 \times L \times W = \frac{Q}{2h}$$

$$\frac{\partial Q}{\partial L} = \frac{1}{2} \times h^{-1/2} \times 129.9 \times W = \frac{Q}{L}$$

$$\frac{\partial Q}{\partial W} = \frac{1}{2} \times h^{-1/2} \times 129.9 \times L = \frac{Q}{W}$$

$$W_R = \left[\begin{array}{l} \left(\dfrac{1.01 \text{ m}^3/\text{s}}{2 \times 0.15 \text{ m}} \times 0.001 \text{ m}\right)^2 + \left(\dfrac{1.01 \text{ m}^3/\text{s}}{0.2 \text{ m}} 0.001 \text{ m}\right)^2 + \\ + \left(\dfrac{1.01 \text{ m}^3/\text{s}}{0.1 \text{ m}} \times 0.001 \text{ m}\right)^2 \end{array}\right]^{1/2}$$

$$= 0.0118 \text{ m}^3/\text{s}$$

$$\sigma\% \text{ error} = \frac{0.0118}{1.01} \times 100 = 1.2\%$$

34. **(C)**

The torque being transmitted by the shaft must first be determined.

$$T = \frac{\dot{W}_b}{2\pi N} = \frac{(22.4 \times 10^3 \text{ N-m/s})}{(2)(\pi)(100 / 60 \text{ rev/s})} = 2{,}140 \text{ N-m}$$

$$\tau_{max} = \frac{Tc}{J} = \frac{(T)(d/2)}{(\pi)\left(\frac{d^4}{32}\right)} = \frac{16T}{\pi d^3}$$

Note that J = polar moment of inertia and c = outer radius.

Since it has been given that $\tau_{max} = 4.14 \times 10^4$ kPa, it can be substituted into the equation:

$$4.14 \times 10^7 \text{ N/m}^2 = \frac{(16)(2{,}140 \text{ N-m})}{\pi d^3}$$

$$d = 0.064 \text{ m} = 6.4 \text{ cm}$$

35. **(B)**

The formula for the angle of twist is as follows:

$$\Theta = \frac{TL}{JG}$$

To convert the degrees to radians:

$$(5.73°) \times (\text{rad} / 57.3°) = 0.10 \text{ rad}$$

Substituting:

$$0.10 \text{ rad} = \frac{(2{,}140 \text{ N-m}) \times (3 \text{ m})}{\left(\frac{\pi d^4}{32}\right) \times (77.2 \times 10^9 \text{ N/m}^2)}$$

$$d = 0.0539 \text{ m} = 5.39 \text{ cm}$$

36. **(D)**

Imagine that one end of the rod disconnects from the wall due to the temperature drop. This will cause it to shorten an amount e_T. The expansion due to the temperature change is expressed by:

$$e_T = \alpha L\ (\Delta T).$$

where:

α = linear thermal expansion coefficient (from reference tables)

L = length of the rod

ΔT = temperature change

To reattach the rod, a force P is needed to stretch the rod back into place.

$$e_T = \alpha L\ (\Delta T) = (11.7 \times 10^{-6}\ \text{K}^{-1}) \times (254\ \text{cm}) \times (39\ \text{K}) = 0.116\ \text{cm}$$

$$e = \frac{PL}{AE};\ \rho = \frac{P}{A}\ \text{combine these equations}$$

$$r = \frac{eE}{L} = \frac{(0.0116\ \text{m})\ (200 \times 10^9\ \text{N/m}^2)}{2.54\ \text{m}}$$

$$= 0.0913 \times 10^9\ \text{N/m}^2 = 91.3\ \text{MPa}$$

37. **(C)**

For a vibrating mass system the following equations apply:

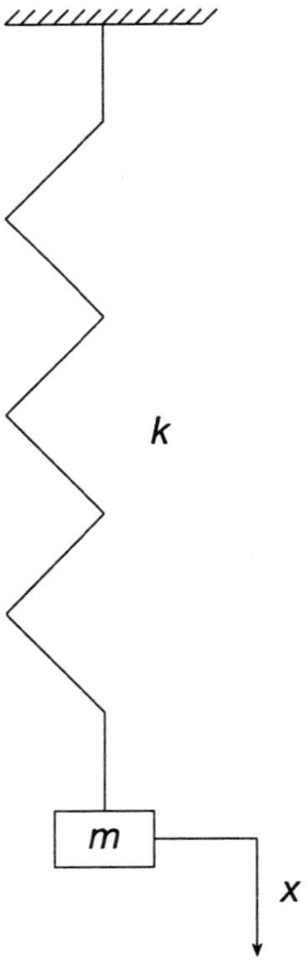

$$m\ddot{x} + kx = 0$$

$$\omega_n = \sqrt{\frac{k}{m}}\ \text{rad/s}$$

$$f_n = \frac{1}{2\pi}\sqrt{\frac{k}{m}}\ \text{cycles/s}$$

$$\tau = \text{period} = \frac{1}{f_n}\text{s}$$

Since the three rubber mounts behave as three springs in parallel, an effective spring constant must be found.

$$k = \text{N/cm} = 1/0.036 = 28\ \text{N/cm}$$

Springs in parallel are combined as follows:

$$k_e = k_1 + k_2 + k_3 = 3 \times 28 = 84\ \text{N/cm}$$

38. **(A)**
The natural frequency can be found from one of the equations in Problem 37.

Convert the weight to its mass:

$$m = W/g = 1.12 \text{ kg}$$

$$f_n = \frac{1}{2\pi}\sqrt{\frac{k}{m}} = \frac{1}{2\pi}\sqrt{\frac{(84 \text{ N/cm})(100 \text{ cm/m})}{1.12 \text{ kg}}} = 13.8 \text{ cycles/s}$$

39. **(B)**

$$x_{\max} = \frac{W}{K} = \frac{11 \text{ N}}{84 \text{ N/cm}} = 0.131 \text{ cm}$$

40. **(D)**
The heat transfer through the wall is given by the following equation:

$$q = \frac{\Delta T}{R_{total}}$$

$$R_{total} = \frac{L_1}{k_1 A} + \frac{L_2}{k_2 A} + \frac{L_3}{k_3 A}$$

$$= \frac{(0.03 \text{ m})}{(0.074 \text{ W/m-K})(1 \text{ m}^2)} + \frac{(0.1 \text{ m})}{(0.69 \text{ W/m-K})(1 \text{ m}^2)} + \frac{(0.03 \text{ m})}{(0.067 \text{ W/m-K})(1 \text{ m}^2)}$$

$$= 0.996 \text{ K/W}$$

$$q = \frac{\Delta T}{R_{total}} = \frac{(150 - 10) \text{ K}}{0.996 \text{ K/W}} = 140.5 \text{ W}$$

41. **(C)**
The heat flow q is the same through each section of the wall.

$$q = \frac{T_1 - T_2}{R_1}$$

$$R_1 = \frac{L_1}{k_1 A} = 0.405 \text{ K/W}$$

$$T_2 = T_1 - q \times R_1 = 150°\text{C} - 140\text{ W} \times (0.405\text{ C/W}) = 93.2°\text{C}$$

42. **(D)**

The radiation exchange between two, non-black bodies is given by the following equation:

$$q_{12} = \frac{\sigma\left(T_1^4 - T_2^4\right)}{\dfrac{1-\varepsilon_1}{\varepsilon_1 A_1} + \dfrac{1}{A_2 F_{12}} + \dfrac{1-\varepsilon_2}{\varepsilon_2 A_2}}$$

We can rewrite the equation on a unit area basis as follows:

$$\frac{q_{12}}{A} = \frac{\sigma\left(T_1^4 - T_2^4\right)}{\dfrac{1-\varepsilon_1}{\varepsilon_1} + \dfrac{1}{F_{12}} + \dfrac{1-\varepsilon_2}{\varepsilon_2}}$$

$$= \frac{5.67 \times 10^{-8}\text{ W/m}^2\text{-K}^4\left(1000\text{ K}^4 - 500\text{ K}^4\right)}{\dfrac{1-0.5}{0.5} + \dfrac{1}{1} + \dfrac{1-0.5}{0.5}} = 17{,}700\text{ W/m}^2$$

where $F_{12} = 1$ since we have infinite parallel plates.

(C) would be the result if the temperatures are not converted to absolute temperatures.

43. **(C)**

Since State 1 is defined as a saturated vapor, $P_1 = P_{sat} = 1.68$ MPa. The specific volume can be found from both tables or charts to be 0.0114 m^3/kg.

$$m = \frac{V}{v} = \frac{0.1\text{ m}^3}{0.0114\text{ m}^3\text{/kg}} = 8.77\text{ kg}$$

44. **(A)**

The rigid tank implies a constant volume process with $v_1 = v_2$. At 30°C and $v_2 = 0.0114$ m^3/kg, State 2 is in the 2-phase, liquid-vapor region. Thus:

$$v_f = 0.0008417\text{ m}^3\text{/kg}$$

$$v_g = 0.0265\text{ m}^3\text{/kg}$$

$$x = \frac{v_2 - v_f}{v_g - v_f} = \frac{0.0114 - 0.0008417}{0.0265 - 0.0008417} = 0.411 = 41.1\%$$

45. **(B)**

Assuming that the nitrogen behaves as an ideal gas, the work for an isothermal process is:

$$W_{1-2} = P_1 V_1 \ln \frac{V_2}{V_1}$$

For an isothermal process of an ideal gas:

$$\frac{V_2}{V_1} = \frac{P_1}{P_2} = \frac{100}{400} = 0.25$$

V_1 is found using the ideal gas law:

$$V_1 = \frac{mRT_1}{P_1} = \frac{(0.1 kg)(0.2968 \text{ kPa-m}^3/\text{kg-K})(300 \text{ K})}{100 \text{ kPa}} = 0.089 \text{ m}^3$$

$$W_{1-2} = P_1 V_1 \ln \frac{V_2}{V_1} = (100 \text{ kPa})(0.089 \text{ m}^3)\ln(0.25) = -12.3 \text{ kJ}$$

The First Law for a closed system states:

$$Q_{1-2} = W_{1-2} + \Delta U$$

Since the process is isothermal, $T_1 = T_2$ and $\Delta U = 0$ since $U = \mathrm{U}(T)$ for an ideal gas.

$$Q_{1-2} = W_{1-2} = -12.3 \text{ kJ}$$

46. **(B)**

When measurements are made with a linear scale, the convention is to assign an error to the measurement equal to $^1/_2$ the smallest scale division. In this case, the lowest scale division was 2.5 mm. Half of this is 1.25 mm or 0.125 cm.

47. **(C)**

The equation for the power transmitted is given in the problem. The units must be adjusted for the length in meters and the time in seconds.

$$\dot{W} = \frac{2\pi RFL}{t}$$

$$= \frac{(2)\times(\pi)\times(1{,}202 \text{ rev})\times(45.01 \text{ N})\times(0.397 \text{ m})}{60 \text{ s}}$$

$$= 2{,}249 \text{ W} = 2.25 \text{ kW}$$

48. **(B)**

The combined errors for this situation are found by applying the following equation:

$$W_R = \left[\left(\frac{\partial R}{\partial x_1}W_1\right)^2 + \left(\frac{\partial R}{\partial x_2}W_2\right)^2 + \ldots + \left(\frac{\partial R}{\partial x_n}W_n\right)^2\right]^{1/2}$$

where:

W_R = uncertainty in the results

W_1, W_2,…W_n = uncertainty in the independent variables x_1, x_2, …, x_n.

$R = \dot{W} = 2.25$ kW

$x_1 = F$ $\quad W_1 = 0.18$ N

$x_2 = R$ $\quad W_2 = 1.0$ rev

$x_3 = L$ $\quad W_3 = 0.125$ cm

$x_4 = t$ $\quad W_4 = 0.50$ s

$$\frac{\partial \dot{W}}{\partial F} = \frac{2\pi LR}{t} = \frac{(2\pi)(0.397 \text{ m})(1{,}202 \text{ rev})}{60 \text{ s}} = 50.0 \text{ W/N}$$

$$\frac{\partial \dot{W}}{\partial R} = \frac{2\pi FL}{t} = \frac{(2\pi)(45.0 \text{ N})(0.397 \text{ m})}{60 \text{ s}} = 1.87 \text{ W/rev}$$

$$\frac{\partial \dot{W}}{\partial L} = \frac{2\pi FR}{t} = \frac{(2\pi)(45.0 \text{ N})(1{,}202 \text{ rev})}{60 \text{ s}} = 5{,}664 \text{ W/m}$$

$$\frac{\partial \dot{W}}{\partial t} = -\frac{2\pi FLR}{t^2} = \frac{(2\pi)(45.0 \text{ N})(0.397 \text{ m})(1{,}202 \text{ rev})}{(60 \text{ s})^2} = 37.5 \text{ W/s}$$

$$W_R = [((50 \text{ W/N})(0.18 \text{ N}))^2 + ((1.87 \text{ W/rev})(1 \text{ rev}))^2 + ((5{,}664 \text{ W/m})(0.00125 \text{ m}))^2 + ((37.5 \text{ W/s})(0.5 \text{ s}))^2]^{1/2}$$

$$= 22.0 \text{ W}$$

$$\text{Percent of error} = \frac{22}{2{,}246} \times 100 = 0.978 = 1\%$$

(A) is the largest percent error of a single element. (C) is the summation of the percent error of the individual components. (D) is simply incorrect.

49. **(D)**

The deflection and force are related by the following:

$$F = kx$$

where F is the force on the spring, x is the deflection of the spring, and k is the spring constant. The spring constant is given by the following:

$$k = \frac{d^4 G}{8D^3 N}$$

where d is the wire diameter of the spring, D is the mean spring diameter, N is the number of coils, and G is the shear modulus of the spring material. The deflection of each spring can be found from:

$$x = \frac{F}{k} = \frac{8FD^3 N}{d^4 G}$$

Since the deflection of each spring is the same, we can equate them:

$$x = \frac{8F_i D_i^3 N_i}{G d_i^4} = \frac{8F_o D_o^3 N_o}{G d_o^4}$$

Solving this equation: $F_i = 0.14\ F_o$. Since the total applied force of 2,500 N is distributed between the two springs, $F_i + F_o = 2{,}500$ N. Solving these last two equations simultaneously gives $F_i = 2{,}190$ N and $F_o = 307$ N.

50. **(C)**

Either the inner or outer spring may be used to determine the answer. For the inner spring, the equation gives:

$$x = \frac{8F_iD_i^3N_i}{Gd_i^4} = \frac{8(307\text{ N})(5.8\text{ cm})^3 10}{\left(82.7\times10^9\text{ N/m}^2\right)(0.65\text{ cm})^4}\frac{10^4\text{cm}^3}{\text{m}^2} = 3.2\text{ cm}$$

The other answers represent the results if the data is improperly substituted into the equation. The answer is 23.1 if 2,193 N is used instead of 307 N.

51. **(C)**

The shear stress in the spring is found using the following series of equations:

$$\tau = \frac{K_S 8FD}{\pi d^3};\ K_s = \frac{(2C+1)}{2C};\ C = \frac{D}{d}$$

$$C = \frac{(9.0\text{ cm})}{(1.3\text{ cm})} = 6.92;\ K_s = \frac{(2\times6.92+1)}{2\times6.92} = 1.07$$

$$\tau = \frac{(1.07)\times(8)\times(2{,}193\text{ N})\times(9.0\text{ cm})}{\pi(1.3\text{ cm})^3}\frac{10^4\text{ cm}^2}{\text{m}^2}$$

$$= 2.45\times10^8\text{ N/m}^2 = 245\text{ MPa}$$

52. **(C)**

The density of an element can be found from the following relation:

$$\rho = \frac{NA}{V_cN_a}$$

where N is a property of cell type (4 for FCC), A is the atomic weight, V_c is the volume of the unit cell, and N_a is Avogadro's number. For a FCC unit cell,

$$V_c = 16r^3\sqrt{2}$$

where r is the cell radius. Substituting:

$$\rho = \frac{(4)(102.9\text{ g/mole})}{\left(16\sqrt{2}\right)\left(1.345\times10^{-10}\text{m}\right)^3\left(6.02\times10^{23}\right)}$$

$$= 1.24\times10^7\text{g/m}^3 = 12.4\text{ g/cm}^3$$

53. **(C)**

For a composite material the strain is given by the following equation:

$$\varepsilon = \frac{\sigma}{E_c} = \frac{P_c}{AE_c}$$

where the stress $\sigma = P_c/A$, P_c is the force applied to the composite material, A is the area over which the force is acting, and E_c is the elastic modulus for the composite material. E_c is given by the following relation: $E_c = E_gC_g + E_pC_p$, where the C's represent the volume fraction of the bar. Solving for E_c, E_c = (70 GPa)(0.6) + (2.76 GPa)(0.4) = 43.1 GPa.

$$\varepsilon = \frac{44{,}500 \text{ N}}{\left(323 \text{ mm}^2\right)\left(43.1\times 10^3 \text{ MPa}\right)} \frac{1}{\left(10^{-6}\text{m}^2/\text{mm}^2\right)} = 3.19\times 10^{-3}$$

54. **(D)**

This question tests knowledge of linear systems and polar plot analysis and representing the solution as a polar plot diagram.

The general form of the direct transfer function is $T(s) = KG(s)$. The variables of the problem are: $T(s)$ = input signal; K = gain constant; x_1 = zero; and x_2 = pole.

The frequency response of a closed-loop system can be derived from the polar plot of the direct transfer function. This method allows obtaining the response of the individual elements of the system composed of many elements, four or more, much more quickly.

In a polar plot diagram, poles are real values of x that will make the transfer function infinite, since they make the denominator of $T(s)$ equal to zero. Zeros are real values of x that make the transfer function equal zero, since they make the numerator equal zero.

Therefore, the general form of the direct transfer function is:

$$T(s) = \frac{K(s + x_1)}{(s + x_2)}$$

Since a pole makes the denominator equal zero:

$$s + x_1 = 0$$

$$-3 + x_1 = 0$$

$$x_1 = +3$$

$$s + x_1 = s + 3$$

Since a zero makes the numerator equal zero:

$$s + x_2 = 0$$

$$-9 + x_2 = 0$$

$$x_2 = +9$$

$$s + x_2 = s + 9$$

Therefore, the transfer function equals and is the only choice which has a zero at –3:

$$T(s) = \frac{K(s+9)}{(s+3)}$$

Answer choices (A), (B), and (C) are incorrect. None of these answers satisfy the definitions of poles and zeros.

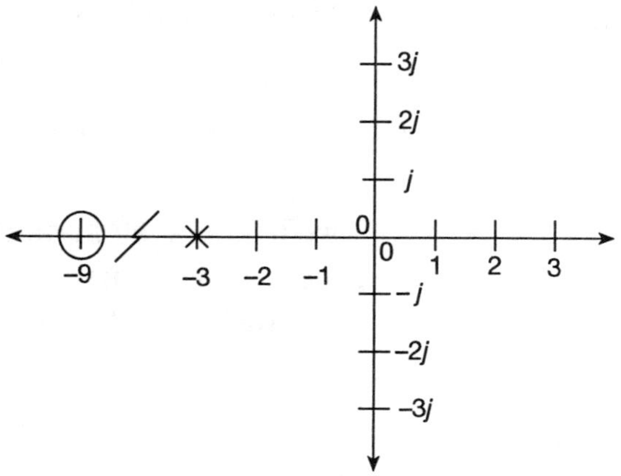

55. **(A)**

The question tests knowledge of input and output signals, amplitude, and analysis of plotted functions.

A dead zone occurs in the region where the amplitude is zero near the origin, and saturation occurs after a linear increase or decrease, respectively.

Answer choice (B) is incorrect. An impulse zone is a thin narrow major increase, such as a peak or single square wave pulsation, which is not shown here. Answer choice (C) is correct for the areas immediately near the origin, but does not define the entire signal. The figure is more complicated than just having a dead zone. Answer choice (D) is incorrect. A ramp is a single continuously increasing function and this figure is not simply a ramp.

56. **(C)**

The question tests knowledge of frequency (step-input) response analysis and viscous-damped control systems.

The variables of the equations are:

w_n = natural frequency

d = damping ratio

The control system response ratio has two complex poles but no zeros.

The roots of the denominator are:

$$s^2 - 0.3s + 1 = (s + (+\,0.15 + j0.99))\ (s + (+\,0.15 - j0.99))$$

Poles exist when the value of the roots make the denominator equal zero and, therefore, the function becomes infinite. Therefore, two poles exist at ($-0.15 - j0.99$ and $-0.15 + j0.99$).

The poles are equal to the complex roots of the equation $s^2 - 0.3s + 1$.

$$x = \frac{-b \pm \left(b^2 - 4ac\right)^{1/2}}{2a} = \frac{-(0.3) + \left((-0.3)^2 - 4(1)(1)\right)^{1/2}}{2(1)}$$

$$= -0.15 \pm \frac{(-3.91)^{1/2}}{2}$$

$$x = -0.15 \mp j0.99$$

Only one peak occurs for a damping ratio of 0.707.

Assume the control response equation is represented as:

$$\frac{T(s)}{R(s)} = \frac{1}{\left(s^2 + 2dw_n s + w_n^2\right)}$$

Therefore:

(1) $w_n^2 = 1$

(2) $2d = 0.3$

$$d = 0.15$$

The peak value is obtained by analyzing the root radical being real and positive.

$$R_{max} = \frac{1}{2d\left(1-d^2\right)^{1/2}} = \frac{1}{2(0.15)\left(1-(0.15)^2\right)^{1/2}} = 3.371$$

Answer choices (A) and (B) are incorrect. Answer choice (D) is incorrect. Only one peak occurs.

57. **(B)**

Since $N = 2.5$, the pitch, $p = 1/N = 1/2.5$.

d_m, the mean diameter, is given by the following:

$$d_m = d - \frac{p}{2} = 2.5 - \frac{1}{2 \times 2.5} = 2.3 \text{ cm}$$

The lead, 1, is given by 1 = *np,* where $n = 2$ since we have double threads. Hence, 1 = (2)(1/2.5) = 0.8 cm. The torque required to turn the screw against the load is:

$$T = \frac{Fd_m}{2}\left(\frac{l + \pi\mu d_m}{\pi d_m - \mu l}\right) + \frac{F\mu_c d_c}{2}$$

$$= \frac{(6{,}700 \text{ N})(2.3 \text{ cm})}{2}\left[\frac{0.8 \text{ cm} + \pi(0.08)(2.3 \text{ cm})}{\pi(2.3 \text{ cm}) - (0.08)(0.8 \text{ cm})}\right]$$

$$+ \frac{(6{,}700 \text{ N})(0.08)(3.2 \text{ cm})}{2}$$

$$= 2{,}340 \text{ N-cm}$$

58. **(B)**

The efficiency of a power screw is given by the following equation:

$$\eta = \frac{Fl}{2\pi T} = \frac{(6{,}700 \text{ N})(0.8 \text{ cm})}{(2\pi)(2{,}340 \text{ N-cm})} = 0.364 = 36\%$$

59. **(A)**

The minimum tensile strength of common spring steel may be determined from:

$$S_{ut} = \frac{A}{d^m}$$

where S_{ut} is the tensile strength in MPa, d is the wire diameter in mm, and A and m depend on the spring material. For hard-drawn wire, m = 0.201 and A = 1,510. Substituting:

$$S_{ut} = \frac{1,510}{(2.0)^{0.201}} = 1,310 \text{ MPa}$$

The strength of the wire is found from the following:

$$S_y = \sigma = 0.78$$

$$S_{ut} = (0.78)(1,310 \text{ MPa}) = 1,020 \text{ MPa}$$

The maximum allowable torsional stress, τ, is determined by applying the distortion energy theory which gives:

$$S_{sy} = 0.577$$

$$S_y = (0.577)(1,020) = 590 \text{ MPa}$$

60. **(B)**

Assuming that x represents the composition wt% Ag, then:

$$\text{wt\%}\alpha = \frac{x_L - x}{x_L - x_\alpha} = \frac{40-20}{40-6} = 0.588 = 59\%$$

The wt% L then must equal 100 – 59 = 41%.

FE/EIT

FE: PM Mechanical Engineering Exam

Answer Sheets

FE: PM MECHANICAL ENGINEERING

Test 1

ANSWER SHEET

1. (A) (B) (C) (D)
2. (A) (B) (C) (D)
3. (A) (B) (C) (D)
4. (A) (B) (C) (D)
5. (A) (B) (C) (D)
6. (A) (B) (C) (D)
7. (A) (B) (C) (D)
8. (A) (B) (C) (D)
9. (A) (B) (C) (D)
10. (A) (B) (C) (D)
11. (A) (B) (C) (D)
12. (A) (B) (C) (D)
13. (A) (B) (C) (D)
14. (A) (B) (C) (D)
15. (A) (B) (C) (D)
16. (A) (B) (C) (D)
17. (A) (B) (C) (D)
18. (A) (B) (C) (D)
19. (A) (B) (C) (D)
20. (A) (B) (C) (D)
21. (A) (B) (C) (D)
22. (A) (B) (C) (D)
23. (A) (B) (C) (D)
24. (A) (B) (C) (D)
25. (A) (B) (C) (D)
26. (A) (B) (C) (D)
27. (A) (B) (C) (D)
28. (A) (B) (C) (D)
29. (A) (B) (C) (D)
30. (A) (B) (C) (D)
31. (A) (B) (C) (D)
32. (A) (B) (C) (D)
33. (A) (B) (C) (D)
34. (A) (B) (C) (D)
35. (A) (B) (C) (D)
36. (A) (B) (C) (D)
37. (A) (B) (C) (D)
38. (A) (B) (C) (D)
39. (A) (B) (C) (D)
40. (A) (B) (C) (D)
41. (A) (B) (C) (D)
42. (A) (B) (C) (D)
43. (A) (B) (C) (D)
44. (A) (B) (C) (D)
45. (A) (B) (C) (D)
46. (A) (B) (C) (D)
47. (A) (B) (C) (D)
48. (A) (B) (C) (D)
49. (A) (B) (C) (D)
50. (A) (B) (C) (D)
51. (A) (B) (C) (D)
52. (A) (B) (C) (D)
53. (A) (B) (C) (D)
54. (A) (B) (C) (D)
55. (A) (B) (C) (D)
56. (A) (B) (C) (D)
57. (A) (B) (C) (D)
58. (A) (B) (C) (D)
59. (A) (B) (C) (D)
60. (A) (B) (C) (D)

FE: PM MECHANICAL ENGINEERING

Test 2

ANSWER SHEET

1. (A) (B) (C) (D)
2. (A) (B) (C) (D)
3. (A) (B) (C) (D)
4. (A) (B) (C) (D)
5. (A) (B) (C) (D)
6. (A) (B) (C) (D)
7. (A) (B) (C) (D)
8. (A) (B) (C) (D)
9. (A) (B) (C) (D)
10. (A) (B) (C) (D)
11. (A) (B) (C) (D)
12. (A) (B) (C) (D)
13. (A) (B) (C) (D)
14. (A) (B) (C) (D)
15. (A) (B) (C) (D)
16. (A) (B) (C) (D)
17. (A) (B) (C) (D)
18. (A) (B) (C) (D)
19. (A) (B) (C) (D)
20. (A) (B) (C) (D)
21. (A) (B) (C) (D)
22. (A) (B) (C) (D)
23. (A) (B) (C) (D)
24. (A) (B) (C) (D)
25. (A) (B) (C) (D)
26. (A) (B) (C) (D)
27. (A) (B) (C) (D)
28. (A) (B) (C) (D)
29. (A) (B) (C) (D)
30. (A) (B) (C) (D)
31. (A) (B) (C) (D)
32. (A) (B) (C) (D)
33. (A) (B) (C) (D)
34. (A) (B) (C) (D)
35. (A) (B) (C) (D)
36. (A) (B) (C) (D)
37. (A) (B) (C) (D)
38. (A) (B) (C) (D)
39. (A) (B) (C) (D)
40. (A) (B) (C) (D)
41. (A) (B) (C) (D)
42. (A) (B) (C) (D)
43. (A) (B) (C) (D)
44. (A) (B) (C) (D)
45. (A) (B) (C) (D)
46. (A) (B) (C) (D)
47. (A) (B) (C) (D)
48. (A) (B) (C) (D)
49. (A) (B) (C) (D)
50. (A) (B) (C) (D)
51. (A) (B) (C) (D)
52. (A) (B) (C) (D)
53. (A) (B) (C) (D)
54. (A) (B) (C) (D)
55. (A) (B) (C) (D)
56. (A) (B) (C) (D)
57. (A) (B) (C) (D)
58. (A) (B) (C) (D)
59. (A) (B) (C) (D)
60. (A) (B) (C) (D)

FE/EIT

FE: PM Mechanical Engineering Exam

Appendix

VARIABLES

a	=	acceleration
a_t	=	tangential acceleration
a_r	=	radial acceleration
d	=	distance
e	=	coefficient of restitution
f	=	frequency
F	=	force
g	=	gravity = 32.2 ft/sec^2 or 9.81 m/sec^2
h	=	height
I	=	mass inertia
k	=	spring constant, radius of gyration
KE	=	kinetic energy
m	=	mass
M	=	moment
PE	=	potential energy
r	=	radius
s	=	position
t	=	time
T	=	tension, torsion, period
v	=	velocity
w	=	weight
x	=	horizontal position
y	=	vertical position
α	=	angular acceleration
ω	=	angular velocity
θ	=	angle
μ	=	coefficient of friction

EQUATIONS

Kinematics

Linear Particle Motion

Constant velocity

$$s = s_o + vt$$

Constant acceleration

$$v = v_o + at$$

$$s = s_o + v_o t + \left(\frac{1}{2}\right)at^2$$

$$v^2 = v_o^2 + 2a(s - s_o)$$

Projectile Motion

$$x = x_o + v_x t$$

$$v_y = v_{yo} - gt$$

$$y = y_o + v_{yo} t - \left(\frac{1}{2}\right) g t^2$$

$$v_y^2 = v_{yo}^2 - 2g\,(y - y_o)$$

Rotational Motion

Constant rotational velocity

$$\theta = \theta_o + \omega t$$

Constant angular acceleration

$$\omega = \omega_o + \alpha t$$

$$\theta = \theta_o + \omega_o t + \left(\frac{1}{2}\right) \alpha t^2$$

$$\omega^2 = \omega_o^2 + 2\alpha\,(\theta - \theta_o)$$

Tangential velocity

$$v_t = r\omega$$

Tangential acceleration

$$a_t = r\alpha$$

Radial acceleration

$$a_r = r\omega^2 = \frac{v_t^2}{r}$$

Polar coordinates

$$a_r = \frac{d^2 r}{dt^2} - r\left(\frac{d\theta}{dt}\right)^2 = \frac{d^2 r}{dt^2} - r\omega^2$$

$$a_\theta = r\left(\frac{d^2\theta}{dt^2}\right) + 2\left(\frac{dr}{dt}\right)\left(\frac{d\theta}{dt}\right) = r\alpha + 2\left(\frac{dr}{dt}\right)\omega$$

$$v_r = \frac{dr}{dt}$$

$$v_\theta = r\left(\frac{d\theta}{dt}\right) = r\omega$$

Relative and Related Motion

Acceleration

$$a_A = a_B + a_{A/B}$$

Velocity

$$v_A = v_B + v_{A/B}$$

Position

$$x_A = x_B + x_{A/B}$$

Kinetics

$$w = mg$$

$$F = ma$$

$$F_c = ma_n = \frac{mv_t^2}{r}$$

$$F_f = \mu N$$

Kinetic Energy

$$KE = \left(\frac{1}{2}\right)mv^2$$

Work of a force = $\int Fds$

KE_1 + Work$_{1-2}$ = KE_2

Potential Energy

Spring $PE = \left(\frac{1}{2}\right)kx^2$

Weight $PE = wy$

$$KE_1 + PE_1 = KE_2 + PE_2$$

Power

Linear power $P = Fv$

Torsional or rotational power $P = T\omega$

Impulse-Momentum

$$mv_1 + \int Fdt = mv_2$$

Impact

$$m_A v_{A1} + m_B v_{B1} = m_A v_{A2} + m_B v_{B2}$$

$$e = \frac{v_{B2} - v_{A2}}{v_{A1} - v_{B1}}$$

Perfectly plastic impact ($e = 0$)

$$m_A v_{A1} + m_B v_{B1} = (m_A + m_B)v'$$

One mass is infinite

$$v_2 = ev_1$$

Inertia

Beam $$I_A = \left(\frac{1}{12}\right)ml^2 + m\left(\frac{1}{2}\right)^2 = \left(\frac{1}{3}\right)ml^2$$

Plate

$$I_A = \left(\frac{1}{12}\right)m\left(a^2 + b^2\right) + m\left[\left(\frac{a}{b}\right)^2 + \left(\frac{b}{2}\right)^2\right] = \left(\frac{1}{3}\right)m\left(a^2 + b^2\right)$$

Wheel $$I_A = mk^2 + mr^2$$

Two-Dimensional Rigid Body Motion

$$F_x = ma_x$$
$$F_y = ma_y$$
$$M_A = I_A\alpha = I_{cg}\,\alpha + m(a)d$$

Rolling Resistance

$$F_r = \frac{mga}{r}$$

Energy Methods for Rigid Body Motion

$$KE_1 + \text{Work}_{1-2} = KE_2$$

$$\text{Work} = \int Fds + \int Md\theta$$

Mechanical Vibration

Differential equation

$$\frac{md^2x}{dt^2} + kx = 0$$

Position

$$x = x_m \sin\left[\sqrt{\frac{k}{m}}\,t + \theta\right]$$

Velocity

$$v = \frac{dx}{dt} = x_m\sqrt{\frac{k}{m}}\cos\left[\sqrt{\frac{k}{m}}\,t + \theta\right]$$

Acceleration

$$a = \frac{d^2x}{dt^2} = -x_m\left(\frac{k}{m}\right)\sin\left[\sqrt{\frac{k}{m}}t + \theta\right]$$

Maximum values

$$x = x_m, v = x_m\sqrt{\frac{k}{m}}, a = -x_m\left(\frac{k}{m}\right)$$

Period

$$T = \frac{2\pi}{\left(\sqrt{\frac{k}{m}}\right)}$$

Frequency

$$f = \frac{1}{T} = \frac{\sqrt{\frac{k}{m}}}{2\pi}$$

Springs in parallel

$$k = k_1 + k_2$$

Springs in series

$$\frac{1}{k} = \frac{1}{k_1} + \frac{1}{k_2}$$

AREA UNDER NORMAL CURVE

$$\frac{1}{\sqrt{2\pi}}\int_{0}^{z} e^{-\frac{z^{2}}{2}}\,dz$$

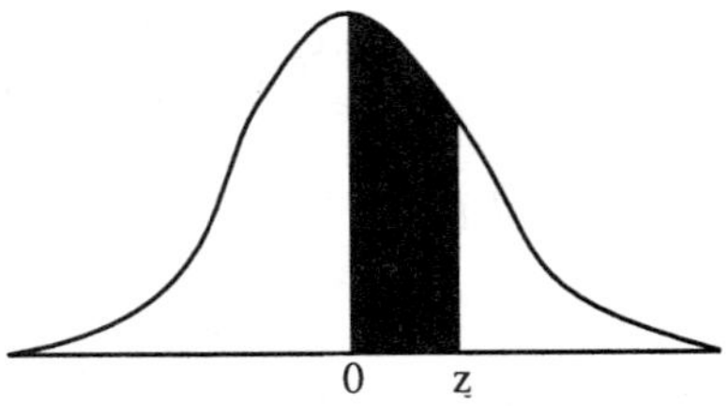

Z	0	1	2	3	4	5	6	7	8	9
0.0	.0000	.0040	.0080	.0120	.0160	.0199	0239	.0279	.0319	.0359
0.1	.0398	.0438	.0478	.0517	.0557	.0596	.0636	.0675	.0714	.0754
0.2	.0793	.0832	.0871	.0910	.0948	.0987	.1026	.1064	.1103	.1141
0.3	.1179	.1217	.1255	.1293	.1331	.1368	.1406	.1443	.1480	.1517
0.4	.1554	.1591	.1628	.1664	.1700	.1736	.1772	.1808	.1844	.1879
0.5	.1915	.1950	.1985	.2019	.2054	.2088	.2123	.2157	.2190	.2224
0.6	.2258	.2291	.2324	.2357	.2389	.2422	.2454	.2486	.2518	.2549
0.7	.2580	.2612	.2642	.2673	.2704	.2734	.2764	.2794	.2823	.2852
0.8	.2881	.2910	.2939	.2967	.2996	.3023	.3051	.3078	.3106	.3133
0.9	.3159	.3186	.3212	.3238	.3264	.3289	.3315	.3340	.3365	.3389
1.0	.3413	.3438	.3461	.3485	.3508	.3531	.3554	.3577	.3599	.3621
1.1	.3643	.3665	.3686	.3708	.3729	.3749	.3770	.3790	.3810	.3830
1.2	.3849	.3869	.3888	.3907	.3925	.3944	.3962	.3980	.3997	.4015
1.3	.4032	.4049	.4066	.4082	.4099	.4115	.4131	.4147	.4162	.4177
1.4	.4192	.4207	.4222	.4236	.4251	.4265	.4279	.4292	.4306	.4319
1.5	.4332	.4345	.4357	.4370	.4382	.4394	.4406	.4418	.4429	.4441
1.6	.4452	.4463	.4474	.4484	.4495	.4505	.4515	.4525	.4535	.4545
1.7	.4554	.4564	.4573	.4582	.4591	.4599	.4608	.4616	.4625	.4633
1.8	.4641	.4649	.4656	.4664	.4671	.4678	.4686	.4693	.4699	.4706
1.9	.4713	.4719	.4726	.4732	.4738	.4744	.4750	.4756	.4761	.4767
2.0	.4772	.4778	.4783	.4788	.4793	.4798	.4803	.4808	.4812	.4817
2.1	.4821	.4826	.4830	.4834	.4838	.4842	.4846	.4850	.4854	.4857
2.2	.4861	.4864	.4868	.4871	.4875	.4878	.4881	.4884	.4887	.4890
2.3	.4893	.4896	.4898	.4901	.4904	.4906	.4909	.4911	.4913	.4916
2.4	.4918	.4920	.4922	.4925	.4927	.4929	.4931	.4932	.4934	.4936
2.5	.4938	.4940	.4941	.4943	.4945	.4946	.4948	.4949	.4951	.4952
2.6	.4953	.4955	.4956	.4957	.4959	.4960	.4961	.4962	.4963	.4964
2.7	.4965	.4966	.4967	.4968	.4969	.4970	.4971	.4972	.4973	.4974
2.8	.4974	.4975	.4976	.4977	.4977	.4978	.4979	.4979	.4980	.4981
2.9	.4981	.4982	.4982	.4983	.4984	.4984	.4985	.4985	.4986	.4986
3.0	.4987	.4987	.4987	.4988	.4988	.4989	.4989	.4989	.4990	.4990
3.1	.4990	.4991	.4991	.4991	.4992	.4992	.4992	.4992	.4993	.4993
3.2	.4993	.4993	.4994	.4994	.4994	.4994	.4994	.4995	.4995	.4995
3.3	.4995	.4995	.4995	.4996	.4996	.4996	.4996	.4996	.4996	.4997
3.4	.4997	.4997	.4997	.4997	.4997	.4997	.4997	.4997	.4997	.4998
3.5	.4998	.4998	.4998	.4998	.4998	.4998	.4998	.4998	.4998	.4998
3.6	.4998	.4998	.4999	.4999	.4999	.4999	.4999	.4999	.4999	.4999
3.7	.4999	.4999	.4999	.4999	.4999	.4999	.4999	.4999	.4999	.4999
3.8	.4999	.4999	.4999	.4999	.4999	.4999	.4999	.4999	.4999	.4999
3.9	.5000	.5000	.5000	.5000	.5000	.5000	.5000	.5000	.5000	.5000

POWER SERIES FOR ELEMENTARY FUNCTIONS

$$\frac{1}{x} = 1 - (x-1) + (x-1)^2 - (x-1)^3 + (x-1)^4 - \ldots + (-1)^n (x-1)^n + \ldots, \qquad 0 < x < 2$$

$$\frac{1}{1+x} = 1 - x + x^2 - x^3 + x^4 - x^5 + \ldots + (-1)^n x^n + \ldots, \qquad -1 < x < 1$$

$$\ln x = (x-1) - \frac{(x-1)^2}{2} + \frac{(x-1)^3}{3} - \frac{(x-1)^4}{4} + \ldots + \frac{(-1)^{n-1}(x-1)^n}{n} + \ldots, \qquad 0 < x \leq 2$$

$$e^x = 1 + x + \frac{x^2}{2!} + \frac{x^3}{3!} + \frac{x^4}{4!} + \frac{x^5}{5!} + \ldots + \frac{x^n}{n!} + \ldots, \qquad -\infty < x < \infty$$

$$\sin x = x - \frac{x^3}{3!} + \frac{x^5}{5!} - \frac{x^7}{7!} + \frac{x^9}{9!} - \ldots + \frac{(-1)^n x^{2n+1}}{(2n+1)!} + \ldots, \qquad -\infty < x < \infty$$

$$\cos x = x - \frac{x^2}{2!} + \frac{x^4}{4!} - \frac{x^6}{6!} + \frac{x^8}{8!} - \ldots + \frac{(-1)^n x^{2n}}{(2n)!} + \ldots, \qquad -\infty < x < \infty$$

$$\arctan x = x - \frac{x^3}{3} + \frac{x^5}{5} - \frac{x^7}{7} + \frac{x^9}{9} - \ldots + \frac{(-1)^n x^{2n+1}}{2n+1} + \ldots, \qquad -1 \leq x \leq 1$$

$$\arctan x = x - \frac{x^3}{3} + \frac{x^5}{5} - \frac{x^7}{7} + \frac{x^9}{9} - \ldots + \frac{(-1)^n x^{2n+1}}{2n+1} + \ldots, \qquad -1 \leq x \leq 1$$

$$(1+x)^k = 1 + kx + \frac{k(k-1)x^2}{2!} + \frac{k(k-1)(k-2)x^3}{3!} + \frac{k(k-1)(k-2)(k-3)x^4}{4!} + \ldots, \qquad -1 < x < 1$$

$$(1+x)^{-k} = 1 - kx + \frac{k(k+1)x^2}{2!} - \frac{k(k+1)(k+2)x^3}{3!} + \frac{k(k+1)(k+2)(k+3)x^4}{4!} - \ldots, \qquad -1 < x < 1$$

TABLE OF MORE COMMON LAPLACE TRANSFORMS

$f(t) = L^{-1}\{F(s)\}$	$F(s) = L\{f(t)\}$
1	$\frac{1}{s}$
t	$\frac{1}{s^2}$
$\frac{t^{n-1}}{(n-1)!}; n = 1, 2, \ldots$	$\frac{1}{s^n}$
e^{at}	$\frac{1}{s-a}$
$t\, e^{at}$	$\frac{1}{(s-a)^2}$
$\frac{t^{n-1} e^{-at}}{(n-1)!}$	$\frac{1}{(s+a)^n}; n = 1, 2, \ldots$
$\frac{e^{-at} - e^{-bt}}{b-a}; a \neq b$	$\frac{1}{(s+a)(s+b)}$
$\frac{a\, e^{-at} - b\, e^{-bt}}{a-b}; a \neq b$	$\frac{s}{(s+a)(s+b)}$
sin *st*	$\frac{a}{s^2 + a^2}$
cos *at*	$\frac{s}{s^2 + a^2}$
sinh *at*	$\frac{a}{s^2 - a^2}$

$f(t) = L^{-1}\{F(s)\}$	$F(s) = L\{f(t)\}$
$\cosh at$	$\dfrac{s}{s^2 - a^2}$
$\dfrac{1}{a^2}(1 - \cos at)$	$\dfrac{1}{s(s^2 + a^2)}$
$\dfrac{1}{a^3}(at - \sin at)$	$\dfrac{1}{s(s^2 + a^2)}$
$\dfrac{t}{2a}\sin at$	$\dfrac{s}{(s^2 + a^2)^2}$
$\dfrac{1}{b}e^{-at}\sin bt$	$\dfrac{1}{(s+a)^2 + b^2}$
$e^{-at}\cos bt$	$\dfrac{s+a}{(s+a)^2 + b^2}$
$h_1(t - a)$ 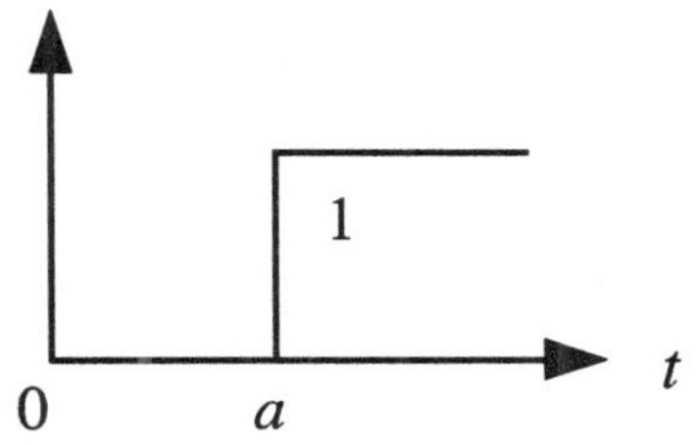	$\dfrac{1}{s}e^{-as}$
$h_1(t) - h_1(t - a)$ 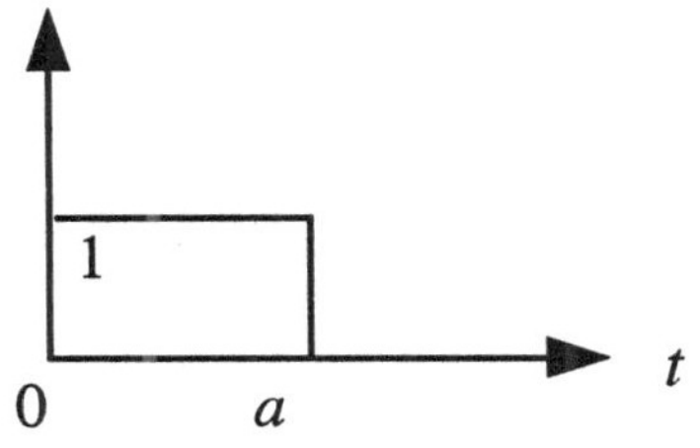	$\dfrac{1 - e^{-as}}{s}$
$\dfrac{1}{t}\sin kt$	$\arctan\dfrac{k}{s}$

REA's Test Prep Books Are The Best!

(a sample of the <u>hundreds of letters</u> REA receives each year)

" I am writing to congratulate you on preparing an exceptional study guide. In five years of teaching this course I have never encountered a more thorough, comprehensive, concise and realistic preparation for this examination. "

Teacher, Davie, FL

" I have found your publications, *The Best Test Preparation...*, to be exactly that. "

Teacher, Aptos, CA

" I used your *CLEP Introductory Sociology* book and rank it 99% – thank you! "

Student, Jerusalem, Israel

" Your GMAT book greatly helped me on the test. Thank you. "

Student, Oxford, OH

" I recently got the French SAT II Exam book from REA. I congratulate you on first-rate French practice tests."

Instructor, Los Angeles, CA

" Your AP English Literature and Composition book is most impressive."

Student, Montgomery, AL

" The REA LSAT Test Preparation guide is a winner! "

Instructor, Spartanburg, SC

(more on front page)